Structural Dynamics and Probabilistic Analyses for Engineers

Structural Dynamics and Probabilistic Analyses for Engineers

Giora Maymon

AMSTERDAM • BOSTON • HEIDELBERG • LONDON
NEW YORK • OXFORD • PARIS • SAN DIEGO
SAN FRANCISCO • SINGAPORE • SYDNEY • TOKYO

Butterworth-Heinemann is an imprint of Elsevier

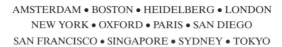

Butterworth-Heinemann is an imprint of Elsevier
30 Corporate Drive, Suite 400, Burlington, MA 01803, USA
Linacre House, Jordan Hill, Oxford OX2 8DP, UK

 Recognizing the importance of preserving what has been written, Elsevier
prints its books on acid-free paper whenever possible.

ANSYS® and all other ANSYS, Inc. product names are trademarks or registered
trademarks of ANSYS, Inc. or its subsidiaries in the United States or other countries.

Library of Congress Cataloging-in-Publication Data
Application submitted

British Library Cataloguing-in-Publication Data
A catalogue record for this book is available from the British Library.

ISBN: 978-0-7506-8765-2

For information on all Butterworth-Heinemann publications
visit our web site at *www.books.elsevier.com*

Printed in the United States of America

08 09 10 11 12   10 9 8 7 6 5 4 3 2 1

In memory of my father, who passed away before this book was published.

Acknowledgments

I wrote part of this book while on Sabbatical from RAFAEL Ltd (Israel) at the Center of Aerospace Research and Education (CARE) at the University of California at Irvine, under the direction of Professor S. N. Atluri. The financial support of RAFAEL Ltd and the supporting atmosphere at CARE are well appreciated. Another part of the manuscript was completed at home, with the tremendous support of my wife, who deserves all my gratitude. I am also greatly thankful to Professor Isaac Elishakoff, of Florida Atlantic University, for his encouragement and valuable help.

Giora Maymon
Haifa, Israel, 2008

Contents

CHAPTER

3 Dynamic Response of a Structure to Random Excitation — 70

CHAPTER

4 Contacts In Structural Systems — 146

CHAPTER

8 Some Important Computer Programs for Structural Analysis   300

CHAPTER

9 Conclusions—Do and Don't Do in Dynamic and Probabilistic Analyses   307

APPENDIX

Computer Files for the Demonstration Problems   313

* A CD-ROM with all these files is attached to this publication.

List of Figures

List of Tables

Foreword

I begin my observations with an obvious one: Most graduates of engineering schools take jobs in industry after graduation. At this stage, they are confronted with the task of designing structural elements of practical projects, with a view to a structure that will not fail in the course of the desired life of the element in question. To this end, they have to carry out both static and dynamic analyses of this element. The following question arises: How are these young engineers equipped for such tasks? The curriculum they are offered at the university comprises strength of materials, the theory of elasticity, and vibration analysis. Most undergraduate courses in structural dynamics are rather basic, and therefore engineers at the beginning of their career try to distance themselves as much as possible from the applications of the latter. Similarly, most omit the relatively new field of probabilistic structural analysis. Thus, the young engineers rarely benefit from the advantages of this modern analysis.

Failure of structural elements in real-life projects is always due to action of both static and dynamic loads. Moreover, the uncertainties involved in the design process call for a probabilistic approach. Thus, both dynamic and probabilistic structural analyses are a must in modern aerospace, mechanical, and civil engineering structures. Once the designer is no longer "afraid" to undertake such analyses, the structures designed by him/her will be more robust, less conservative, and of much higher quality.

The purpose of this book is to encourage the design engineer to use *both* these analyses. Once their basic and physical concepts are understood, their use can become routine.

The basic concepts of dynamic analysis are described in this book, and the reader is referred to relevant textbooks in which more information is available. The basic theory of vibrations; the responses of a structure to external practical deterministic loads and to random excitation forces; the concept of nondeterministic analysis of structures, and the probabilistic approaches to crack propagation are included. An extensive collection of examples either solved analytically or numerically is provided in a manner that enables the readers to apply the data to his/her own realistic cases. Issues like "factor of safety," combined (static and dynamic) failure criteria, and the problem of "reliability demonstration" of large projects are also included. Relevant commercially available computer codes, extensively used in industry are reviewed. A list of "do" and "do not do" advice is offered in the final chapter.

The presented material reflects Dr. Maymon's nearly 40 years of experience in the fields in question, which included coaching of generations of young engineers. Dr. Maymon is familiar with their "fears" of these types of analysis, and with the methodological approaches that enables these "fears" to be overcome.

It is believed that engineers who read this outstanding book will grasp the main issues involved in applying structural dynamics and probabilistic analyses. They will be able to follow the demonstrated examples, and "run" their own practical design problems, with the aid of the attached CD-ROM that contains the computer files for solving the examples. As a result, their designs will be improved, and their self-confidence enhanced, making them better professionals. Moreover, the material presented in the book can serve as a solid base for an excellent professional course for the interested industry establishments.

Dr. Maymon ought to be congratulated on this outstanding book that will benefit our engineers and, through them, all of us.

Dr. Isaac Elishakoff
J.M. Rubin Distinguished Professor
in Safety, Reliability and Security
Florida Atlantic University
Boca Raton, Florida
November 2007

Preface

Most engineers, especially the young ones, rely heavily on modern CAD programs, which make the lives of today's designers much easier than those of previous generations. These programs take care of drawings, tolerances, special views and cross-sections, lists of parts, check and approval procedures, and distribution and archiving of the design files. All these modern CAD programs are capable of performing a basic stress analysis, based on simple finite elements created for the relevant design. Most of them are also capable of computing the resonance frequencies of the designed parts.

In spite of the sophistication of these CAD programs, they cannot perform dynamic analysis of harmonic and random vibrations, responses to transient loading, and structural reliability analysis, based on probabilistic approaches to design. Performing such analyses is a major necessity in the design analysis of practical, real-world structures. Therefore, most of the young engineers avoid dynamic analyses, and almost all of them avoid probabilistic analysis. They usually think that this part of the design should be left for "special experts." Some also create "static equivalent models" for dynamic problems, in order to solve them with their available static tools, so a correct dynamic analysis is avoided. Most of them develop a "fear" of structural dynamics. Almost all never heard of probabilistic structural analysis.

This approach should be avoided if significant improvements in structural design and the safety of structural systems are desired. Most of the structural failures encountered in mechanical and aerospace real designs are the results of dynamic loading and dynamic behavior of the structural elements. Fatigue and crack propagation, response to transient blast loads, and other dynamic characteristics are typical reasons that cause failure in structures of mechanical, civil, and aerospace engineering systems.

The "fear" to do dynamic analyses is unjustified, and once young engineers understand the basic concepts and procedures of dynamic analysis, their participation in successful designs can be improved. The same argumentation holds for probabilistic analysis, mainly because it is a much newer field. Many engineers think that this field belongs to mathematicians and statisticians—which is really not the case. The reliability of a structural design, which is part of the total reliability of the product, depends mainly on the design process as a whole, including all kinds of analyses (analytical and numerical), product's model simulations, components tests, and systematic experiments. The mathematicians cannot achieve this reliability, and the role of the design engineers in this task is essential.

Many design engineers have access to one of the commercially available finite elements computer codes, like NASTRAN®, ANSYS®, ADINA™, ABAQUS®, and STAGS. They are using it, sometimes, to solve structural static problems during the design process. The moment these people understand that a structural dynamic problem can be easily solved with these same codes by using the same database file prepared for the static analysis, more and more dynamic analyses will be performed during the design, without applying "equivalent static solutions" that are, in most cases, erroneous. To do that, a better understanding of the physical meanings of structural dynamics concepts is essential. This is the main purpose of the present book.

The most important part of a computerized structural analysis is the preparation of the database file. In this file, the types of the elements are decided on, geometry of the structure and its dimensions are determined, material properties are introduced, and boundary conditions are set. This input data is required for the construction of the rigidity matrix. When material densities are included, the mass matrix can also be created. These two outcomes—the rigidity matrix and the mass matrix—are also the basic characteristics required for dynamic solutions, when the proper external dynamic forces are introduced to the solution phase. Therefore, the efforts invested in the correct preparation of the database for the static analysis include inherently all the information required for dynamic solutions, which then can be performed without much additional effort. This can and should be done when the designers will not be "afraid" of structural dynamics. The translators of finite elements codes that are included in today's probabilistic analysis commercially available programs, and the probabilistic analysis modules that are included in the present version of the large finite elements codes, enable the user to perform probabilistic analysis as well, and focus the attention to the failure criteria, which may be the most important parameters in a design analysis.

The professional and the intelligent analysis of the results of a numerical computation is a basic demand for a successful design. It is easier to interpret the results once the basic physics of the problem and the behavior of the structure are understood. The following chapters are an attempt to clarify the basic concepts of structural dynamics and probabilistic analysis. The presentation is accompanied by basic examples and extensive analysis of results, in an effort to decrease the "fear" that prevents young engineers from doing such analyses.

Although the examples presented are very simple, the procedures described are of a general nature. When an ANSYS file of a simple problem is described, solved, and understood, the solution of a more complex practical case can be done just by replacing relevant parts of that file with the new case data. Thus, the examples can be used as guidelines for many other cases that may be more practical and more realistic.

Most of the examples described in this book are based on the behavior of the cantilever beam. For many years, the author was fascinated by this basic structure. It is a structural element on which all the procedures required for a comprehensive structural analysis could be demonstrated, and many of the common mistakes engineers make in computing and testing structural elements can be demonstrated and explained. One reason why this structural element is "classical" for educational purposes is that when it vibrates, there are *high accelerations at the tip*, where almost *no stresses exist*, and the *maximum stresses are near the clamped edge*, where usually *no accelerations exist*. Usually, engineers should be concerned with high dynamic stresses (as the cause for failure), but it so happens that most of them are concerned with high acceleration responses that occur, for this structural element, in a location where almost no stresses exist. A generic cantilever beam used in most of the numerical examples is described at the beginning of Chapter 2. Nevertheless, to emphasize that the described methods are not unique to beams only, some examples of frames plates and shells are also included. The theory of simple beams is only briefly discussed in this book, as this is not the objective of this publication. There are dozens of textbooks describing beam theories, from the most simplified strength of material approach to much more rigorous and complex formulation, and the interested reader can complete his knowledge by referring to them.

In some simple cases demonstrated in the following pages, analytical solutions are available. Such solutions are rarely available in real practical designs. The structural analyst usually uses a numerical algorithm, such as a finite elements code. Those have been tremendously advanced during the last three decades. They are commercially available and accessible to most design establishments and individuals in many design disciplines such as aerospace, mechanical and civil engineering, and to engineering students, who can use the more limited, much cheaper educational versions. Solution of the presented examples using such codes can help designers in evaluating their own cases and writing their own solutions. Therefore, the input text files used for the simple examples are listed in the Appendix and included on the CD-ROM that accompanies this book. A few important basic computer codes used in practical structural design (certainly not all of them) are described in Chapter 8.

The use of the correct units for the physical properties of an analyzed structure is also an essential feature in structural analysis, and in any engineering discipline. The Standard International (SI) system is only one of the ways parameters can be expressed in an analysis, certainly not the most practical one for engineers. Any other units system can be used as long as the user is consistent. If loads are given in kgf (kilogram-force), and dimensions are given in cm (centimeters), stresses will be in kgf/cm^2. For such cases, Young modulus should be defined in kgf/cm^2. If Forces are given in lbs (pounds), and dimensions are in (inches), stress results are in lbs/in^2 (psi, pounds per square inch). Confusion between kgm (kilogram mass) and kgf (kilogram force) should be avoided. In some computer codes, the value of the gravitational acceleration g is required, and should be introduced with the proper units (e.g., $g = 9.8 \, m/sec^2 = 980 \, cm/sec^2 = 385.8 \, in/sec^2$). Introduction of the wrong value causes, naturally, erroneous results. In the examples described in this book, SI units are not used, as these are certainly not the units an engineer uses in his practice. Nevertheless, it should be emphasized again that the use of any set of units is appropriate, as long as there is a consistency along the entire solution process—analytical or numerical. Care should be taken when frequencies are expressed with the wrong units. Circular frequency, usually denoted f, is in cycles/seconds or Hertz (Hz), and angular frequency, usually denoted ω, is expressed in radians/seconds. Wrong use of frequency units changes the results by a factor of 2π and its powers! This question is discussed in more detail when the response of an elastic system to harmonic and random excitation is described.

It is also important, when long analytical expressions are evaluated, to check the units of the required quantity on the left-hand side of the equation against the units on the right-hand side of the obtained expression. Many algebraic manipulation mistakes can be detected and avoided by using this procedure.

In Chapter 1 of this book, some basics of the theory of deterministic vibrations are repeated to create a common knowledge base for the readers. Readers interested in more details about vibration theory should explore the suggested textbooks.

In Chapter 2, some basics of the theory of slender beams are quoted, as these beams form most of the numerical examples in this text. It should be noted that while beams are used in many of the numerical examples, there is no loss of generality for other cases, as the "beam" represents "database information," which can be replaced with any other structural model. Some examples of frame plates and shell structures have also been added to emphasize this point.

In Chapter 3, the responses of single-degree-of-freedom (SDOF), multiple-degrees-of-freedom (MDOF), and elastic (continuous) systems to random excitation are demonstrated. Again, references to appropriate textbooks are included.

In Chapter 4, dynamic contact problems are demonstrated and discussed. This kind of problem is encountered in practical design when one elastic subsystem is "packed" into another subsystem, and there is a possibility that during a routine service some parts knock on some other parts of the structure, thus creating excessive local stresses. The treatment of this sort of structural behavior is usually avoided in both academic courses and in practical design. This, in fact, leads engineers to very conservative (and there-fore inefficient) design solutions for cases where parts can knock against each other.

In Chapter 5, nondeterministic behavior of structures is discussed. Proba-bilistic analysis of structure is a fast-developing discipline in structural design, and any young engineer should be familiar with the basics of the theories and applications of this emerging field. According to our belief, this field will be an important part of the design requirements and specifications in the near future.

In Chapter 6, random behaviors of crack propagation under repeated loads are demonstrated and discussed. Crack propagation is highly influenced by the randomness of many material and design parameters, thus random behavior cannot be avoided.

In Chapter 7, some dynamic design criteria (including the classical and the new stochastic factors of safety) are described and discussed.

In Chapter 8, some practical computer programs used for analysis of structures (static and dynamic, deterministic and probabilistic) are reviewed, especially those used extensively in the industry and for the numerical examples described in this book.

Chapter 9 summarizes the book with some important "do" and "do not do" recommendations for someone who was afraid of structural dynamics and probabilistic analysis before reading this text and practicing its material. The author is sure that after reading and practicing, the reader will never again be afraid of these important design necessities, and his design methodology will be tremendously improved.

The Appendix contains a list of the computer files used in this book for the demonstration problems according to the book chapters, and file lists are included.

A CD-ROM is attached to this publication, in which the files mentioned and listed in the Appendix are included. The reader can copy these files to his computer. In many cases, the basic database part of the listed file can be easily replaced by one of the reader's files. If this is the case, care must be taken to modify the rest of the files, too.

As already mentioned, this publication is not intended for use as a textbook on the subject matter. There are many dozens of textbooks dealing with basic vibration theory, theory of beams, probabilistic analysis, and crack propagations. The interested reader is referenced to few of the existing texts. In addition, from the author's experience, the Internet can be used for an extensive search on many engineering problems, definitions, benchmark cases, and literature. This publication is directed mainly to the entry-level engineering practitioners, and to established engineering professionals who undergo a mid-career education cycle. As such, the text can be a solid base for professional and continuing education courses and for industry training workshops.

Giora Maymon
January 2008

Chapter 1 / Some Basics of the Theory of Vibrations

1.1 A Single Degree of Freedom System

A good physical understanding of the behavior of vibrating structures can be achieved by analyzing the behavior of a single degree of freedom (SDOF) oscillator. The SDOF system is covered by an extremely large number of textbooks (e.g., refs. [1–12]), and is the basis for every academic course in vibration analysis in aerospace, mechanical, and civil engineering schools. It will be discussed briefly in the first chapters to create a common baseline for the analysis of the behavior of the cantilever beam described in most of the examples, and for any other continuous structure.

The SDOF system can be excited either by a force (which is a function of time) acting on the mass, or by a forced movement of the support. The first type of excitation is usually called "force excitation" and the latter is called "base excitation." These two major types of excitations (loadings) are basic to the structural response analyses of both the SDOF and the continuous elastic systems.

The classical oscillator contains a point mass m (i.e., all the mass is concentrated in one point), which is connected to a rigid support through two elements: a linear massless spring with a stiffness k and a viscous damper c (which creates a force proportional to the velocity), or a structural damper h (which creates a force proportional to the displacement and in 90 degrees phase lag behind it). The system can be excited either by a force f acting on the mass or by a base movement x_s. The force-excited system is described in Figure 1.1. Note in the figure that two "elements" connect the mass to the

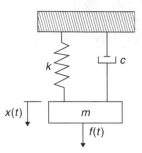

FIGURE 1.1 A force-excited SDOF.

support. This is only a schematic representation. The spring element represents the stiffness of the structure, while the internal viscous (or structural) damping of the spring is represented by the separate damper.

In the following evaluations, and in the rest of the book, only a viscous damper is assumed (although treatment of a structural damper is similar and can be found in many references; e.g., [10]). The basic equation of motion can be written by equilibrium of forces acting on the mass when it is moved in the x direction:

$$m\ddot{x} + c\dot{x} + kx = f(t) \tag{1.1}$$

The natural frequency of the undamped system is

$$\omega_0 = \sqrt{\frac{k}{m}} \tag{1.2}$$

The system is excited by a harmonic force of amplitude f_0 and frequency Ω:

$$f(t) = f_0 e^{i\Omega t} \tag{1.3}$$

Assume a solution in the form

$$x(t) = x_0 e^{i\Omega t} \tag{1.4}$$

which yield the following result for x_0:

$$x_0 = \frac{f_0}{\left(k - \Omega^2 m\right) + i(\Omega c)} = f_0 H(\Omega) \tag{1.5}$$

where

$$H(\Omega) = \frac{1}{\left(k - \Omega^2 m\right) + i(\Omega c)} \tag{1.6}$$

This is a complex expression (called also Transfer Function), which means that there is a phase difference between the excitation force and the displacement. Using classical solution methods one can obtain the absolute value of x_0,

$$|x_0| = \frac{f_0}{m\omega_0^2 \left[\left(1 - \left(\frac{\Omega}{\omega_0} \right)^2 \right)^2 + 4 \left(\frac{\Omega}{\omega_0} \right)^2 \zeta^2 \right]^{1/2}} = \frac{f_0}{m} H_m(\Omega) \qquad (1.7)$$

where

$$H_m(\Omega) = \frac{1}{\omega_0^2 \left[\left(1 - \left(\frac{\Omega}{\omega_0} \right)^2 \right)^2 + 4 \left(\frac{\Omega}{\omega_0} \right)^2 \zeta^2 \right]^{1/2}} \qquad (1.8)$$

ζ is the viscous damping coefficient given by

$$\zeta = \frac{c}{2\sqrt{km}} = \frac{c}{2\omega m} \qquad (1.9)$$

and is used extensively in engineering applications. The phase angle between the displacement and the force is

$$\theta = \arctan \left[2\zeta \left(\frac{\Omega}{\omega_0} \right) \bigg/ \left(1 - \left(\frac{\Omega}{\omega_0} \right)^2 \right) \right] \qquad (1.10)$$

Note that in Eq. (1.7) $f_0/m\omega_0^2 = f_0/k$ is the static deflection x_{static} of the SDOF under a static force of magnitude f_0, and therefore a dynamic load factor (DLF) can be defined by

$$\text{DLF} = \frac{|x_0|}{x_{static}} = \frac{1}{\left[\left(1 - \left(\frac{\Omega}{\omega_0} \right)^2 \right)^2 + 4 \left(\frac{\Omega}{\omega_0} \right)^2 \zeta^2 \right]^{1/2}} \qquad (1.11)$$

The DLF expresses the amplification of a static deflection because the load is dynamically applied. A plot of Eq. (1.11) is shown in Figure 1.2. These are the well-known "resonance curves." When the system is excited in resonance, $\Omega/\omega_0 = 1$, the maximum value of the DLF is

$$\text{DLF}_{max} = \frac{1}{2\zeta} \qquad (1.12)$$

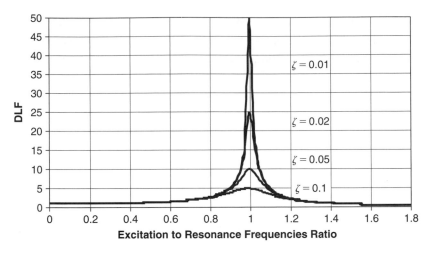

FIGURE 1.2 The DLF of a force-excited SDOF as a function of frequency ratio.

Examination of Eq. (1.10) shows that at very low excitation frequencies $(\Omega/\omega_0 \to 0)$, x is in phase with the excitation, and the DLF is close to 1. This means that the mass is "following" the excitation, with almost no dynamic effects. At resonance $(\Omega/\omega_0 = 1)$ the displacement is 90 degrees ahead of the excitation, and the amplification has a maximum. The value of this maximum is higher as the damping coefficient is lower. In high excitation frequencies $(\Omega/\omega_0 \to \infty)$ there is a 180-degree lag between the displacement and the excitation force, and the amplification tend to 0. This means that the dynamic force does not move the mass, so the SDOF is very rigid in these excitation frequencies. The length of the response vector and the size of the phase angle between the excitation force and the displacement are used extensively in many experimental methods for the determination of resonance frequencies and mode shapes of a structure (Ground Vibration Test—GVT; e.g., [13, 14]).

Many engineering applications involve base excitation rather than force excitation (see Figure 1.3). Such are the cases of structures subjected to earthquakes, vehicles moving on rough roads, and structural subsystems mounted on a main structure. It is convenient to express the excitation by a base input of acceleration $\ddot{x}_s(t)$.

It can be shown that $u = x(t) - x_s(t)$, the relative displacement between the mass and the support (which is the extension of the spring) obeys the following

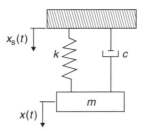

FIGURE 1.3   A base-excited SDOF.

differential equation:

$$m\ddot{u} + c\dot{u} + ku = -m\ddot{x}_s(t) \tag{1.13}$$

Thus, the differential equation for the relative displacement (between the mass and the base—the displacement that causes elastic forces in the spring) is similar to the differential equation of the basic mass-spring-damper excited by an external force, but with an "equivalent excitation force," which is equal to minus the mass multiplied by the base acceleration.

In most of the cases, the relative displacement between the mass and the support (and not the absolute displacement of the mass) is responsible for the stresses in the spring (or in the elastic structure) and therefore is of interest to the designer. Eq. (1.13) is similar to Eq. (1.1) of the force excitation, where an "equivalent force," equal to the mass multiplied by the base acceleration, in a direction opposite to the base excitation, is applied. Thus, when $\ddot{x}_s(t) = \ddot{x}_{s0}e^{i\Omega t}$ the term $(-m\ddot{x}_{s0})$ can replace f_0 in Eq. (1.1), whereas u, the relative displacement, replaces x, thus:

$$|u| = \frac{(-\ddot{x}_{s0})}{\omega_0^2 \left[\left(1 - \left(\dfrac{\Omega}{\omega_0}\right)^2\right)^2 + 4\left(\dfrac{\Omega}{\omega_0}\right)^2 \zeta^2 \right]^{1/2}} \tag{1.14}$$

When the support is moved by a harmonic displacement (rather than by harmonic acceleration) of amplitude x_{s0} and a frequency Ω,

$$x_s = x_{s0}e^{i\Omega t} \tag{1.15}$$

the acceleration is obtained by double time differentiation

$$\ddot{x}_s = -\Omega^2 x_{s0}e^{i\Omega t} \tag{1.16}$$

and, therefore,

$$\frac{|u|}{x_{s0}} = \frac{\left(\dfrac{\Omega}{\omega_0}\right)^2}{\left[\left(1 - \left(\dfrac{\Omega}{\omega_0}\right)^2\right)^2 + 4\left(\dfrac{\Omega}{\omega_0}\right)^2 \zeta^2\right]^{1/2}} \tag{1.17}$$

Eq. (1.17) is described in Figure 1.4. When the frequency of the base excitation is small ($\Omega/\omega_0 \to 0$), there is no relative displacement between the support and the mass, $u \to 0$, which means that the mass is moving almost as the support. When the support excitation frequency is high, ($\Omega/\omega_0 \to \infty$), u tends to x_{s0}, resulting in a relative displacement of $-x_{s0}$; i.e., the mass does not move relative to the "external world" and all the relative movement between the mass and the support is due to the base movement. This is the reason why shock absorbers are mounted on certain structural elements (like a car) and their frequency is designed to be smaller than the frequency of the expected base excitation (e.g., road surface roughness). The mass (the car body, in this case) does not move, while elastic stresses are created in the springs (the shock absorbers).

It is of major importance to understand and to master the solutions of SDOFs. In many practical engineering problems, systems can be represented

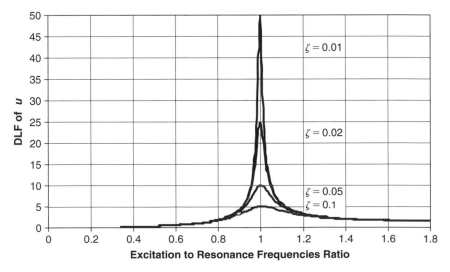

FIGURE 1.4   DLF of $u(t)$, the extension of the spring.

by equivalent SDOF, at least in the early stages of a design. Solutions of an equivalent SDOF system increase the designer's understanding of the behavior of the designed structure, and direct him in the right direction in the critical early stage of a project. Another advantage of the SDOF is that analytical solutions are usually possible, thus a quick preliminary design cycle can be obtained. Sometimes, a SDOF equivalent system is not possible, but in many cases equivalent multiple degrees of freedom (MDOF) models may be available for a simpler solution. Section 1.3 explains these systems.

1.2 Response of a SDOF to (Any) Transient Load

When the response of a SDOF system to a general (not necessarily harmonic) time dependent force is required, the most general solution can be obtained using the response $h(t)$ of such a system to a unit impulse. A unit impulse is defined mathematically as an infinitely large force acting during an infinitely small period, so that the total impulse is one unit. The solution for $h(t)$ is described in most of the available textbooks. For a SDOF system initially at rest, with a damping coefficient $\zeta < 1$ (which is usually the case in structural analyses) it is:

$$h(t) = \frac{1}{m\omega_0\sqrt{1-\zeta^2}} \cdot e^{-\zeta\omega_0 t} \cdot \sin\left(\omega_0\sqrt{1-\zeta^2} \cdot t\right) \qquad (1.18)$$

The response to a general force $f(t)$ can be represented by a series of repeated impulses of magnitude $f(\tau)\Delta\tau$, applied at $t = \tau$. In vibration textbooks, it is shown that the displacement response of a SDOF system can be written as

$$x(t) = \int_{\tau=0}^{\tau=t} f(\tau) \cdot h(t-\tau)\,d\tau \qquad (1.19)$$

This equation is known as the convolution or Duhamel's integral, and is used extensively in Fourier transforms. As $h(t-\tau)$ is identically zero for $t < \tau$, the Duhamel's integral can also be written as (e.g., [15])

$$x(t) = \int_{\tau=-\infty}^{\tau=t} f(\tau) \cdot h(t-\tau)\,d\tau \qquad (1.20)$$

In many cases, an analytical solution of Eq. (1.20) is possible. In other cases, a numerical integration scheme can be easily written to get the response of a SDOF system to an arbitrary transient excitation force. Both cases are demonstrated in the following pages.

In the first example, a constant step function force is applied to a SDOF system. This means that at time $t = 0$, the force $f_0 = $ constant is applied, with zero rise-time. Thus

$$f(\tau) = f_0 \quad t \geq 0 \tag{1.21}$$

Eq. (1.20) takes the form

$$x(t) = f_0 \frac{1}{m\omega_0\sqrt{1-\zeta^2}} \int_{\tau=0}^{\tau=t} \exp[-\zeta\omega_0(t-\tau)] \cdot \sin\left[\omega_0\sqrt{1-\zeta^2} \cdot (t-\tau)\right] d\tau \tag{1.22}$$

Performing the definite integration, the DLF can be found:

$$\text{DLF} = \frac{e^{-\zeta\omega_0 t}}{\omega_d} [-\zeta\omega_0 \sin(\omega_d t) - \omega_d \cos(\omega_d t)] + 1 \tag{1.23}$$

where $\omega_d = \omega_0\sqrt{1-\zeta^2}$ is the resonance frequency of the damped system. Note the difference between ω_d and ω_0, which is very small for small values of damping ζ. In Eq. (1.23), the rigidity $k = m\omega_0^2$ was used to express the static deflection of the system under a force f_0.

In Figure 1.5, the DLF for an undamped system ($\zeta = 0$) is shown. In Figure 1.6, results are shown for $\zeta = 0.05$. The solution is done with file **duhamel1.tkw** (see Appendix).

It is quite interesting to analyze the results shown in Figures 1.5 and 1.6. The applied force is a step function with zero rise-time. The response of the undamped system is between zero and twice the static deflection created by the same force, if applied statically. The maximum DLF equals two. It can be shown that in cases where the rise-time is finite, and not zero, the maximum value of the DLF is less than 2. Thus, a transient force that is not repetitive (i.e., has not a periodical component) causes a dynamic effect to a SDOF system whose maximum is twice the static effect. Therefore stresses are also twice of those obtained in the static applied force. When the system is damped,

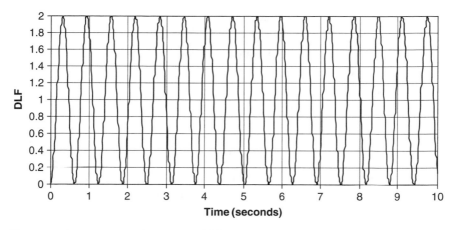

FIGURE 1.5 Response of undamped SDOF to a step force.

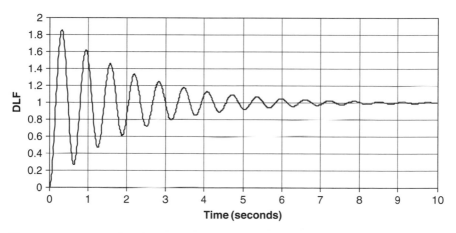

FIGURE 1.6 Response of damped SDOF to a step force.

there are decreasing fluctuations around the static value of 1, and after a long enough period, the system is settled with deflection and stresses that are equal to those created by applying the load statically.

In Section 1.4 it is shown that an elastic system can be described by an equivalent SDOF system, with generalized quantities (generalized mass, rigidity, damping, and forces). Thus, the same conclusions can be drawn for an elastic system. Consider the cantilever beam, very slowly and carefully loaded by a tip static force f_0. The tip deflection is w_{tip}, which can be calculated using beam theory. Suppose a weight of f_0 is suddenly (zero rise-time) put at the tip. The beam vibrates between a displacement of zero and twice the static displacement, when it is undamped. If damping exists, the amplitudes of the tip are damped when time passes and after long enough time the beam takes the static deflection under the force f_0.

Duhamel's integral can also be performed analytically for harmonic excitations. The results of such an analysis are more general than the one described earlier for harmonic excitations, as it expresses both the steady state (evaluated in Section 1.1) and the transient response. Such solutions are not demonstrated here, and are left to the interested reader [15].

The Duhamel's integral cannot always be evaluated explicitly. Numerical integration can always be performed to determine the response of a SDOF system (and therefore of any other system, using generalized parameters) to a general force. In Figure 1.7, a flowchart for such a computational procedure (cited from [16]) is shown for a trapezoidal integration (see file **duhamel2.tkw** in the Appendix). Any other algorithm can replace the integration block of this scheme. For practical complex structures, the response for any transient excitation (force, displacement, accelerations, etc.) can be numerically solved by one of the modules of a commercially available finite elements program, such as NASTRAN® or ANSYS®.

In some cases, it is easier to solve a SDOF system with a transient loading by "sewing" together two direct solutions, instead of using the Duhamel's integral method. Suppose a system is excited by a half sine force, with the amplitude of the half sine being F_0 and its duration being τ_0. In this case, the excitation is

$$
\begin{aligned}
F &= F_0 \sin(\Omega t) \quad 0 \leq t \leq \tau_0 \\
F &= 0 \qquad\qquad\quad t \geq \tau_0
\end{aligned}
\tag{1.24}
$$

In such a case, it is easier to solve for the system's behavior up to $t = \tau_0$ as if the excitation is fully harmonic, to find the displacement $x(\tau_0) = X_0$ and the velocity $\dot{x}(\tau_0) = V_0$, and then solve the homogenous equation for the zero excited SDOF using the homogenous solution with X_0 and V_0 as initial conditions. Then, the two solutions can be matched to form a single solution of displacement as a function of time.

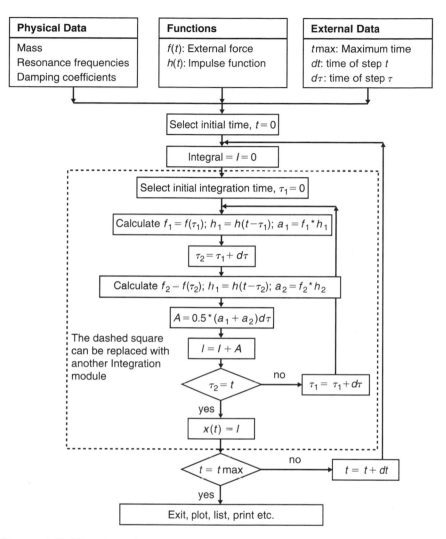

FIGURE 1.7   Flowchart for programming the Duhamel's integral (from [16]).

1.3 MULTIPLE-DEGREES-OF-FREEDOM (MDOF) SYSTEM

A multiple-degrees-of-freedom (MDOF) system contains several masses, interconnected by springs and dampers and excited by several external forces and/or base excitations. The number of degrees of freedom of such a system is determined by the number of masses n. Each mass m_i moves with a displacement x_i.

The analysis of many continuous systems (which, in fact, have an infinite number of degrees of freedom, as their mass distribution is continuous) can be analyzed, in many practical cases, by an equivalent system that has a finite number of degrees of freedom. It is possible to describe a continuous system by an approximation of a finite number of discrete masses. The main problem in the construction of such an equivalent system is, usually, the construction of the appropriate elastic elements (springs) that should approximate the elastic behavior of each mass relative to the other. In fact, whenever a numerical analysis of a structure is done (say, by a finite elements code), the continuous elastic system is approximated by a set of discrete elements, each with its mass and elasticity. The interested reader should refer to textbooks on the basics of the finite elements methods and other discretization methods (see, e.g., [17, 18]).

Therefore, it is essential to learn the behavior of MDOF systems and the methods used in order to analyze them. Most of the definitions and the methods of solution of MDOF systems are applied, in this book and in many others, to the solution of continuous elastic systems.

It can be shown (and it will be demonstrated later in this chapter) that the following differential equation can be written for a MDOF system in a matrix form:

$$[m]\{\ddot{x}(t)\} + [c]\{\dot{x}(t)\} + [k]\{x(t)\} = \{f(t)\} \tag{1.25}$$

$[m]$, $[c]$, and $[k]$ are mass, damping, and stiffness matrices, respectively. $\{.\}$ are vectors of the accelerations, velocities, displacements, and external forces. The main problem, of course, is to build these matrices properly.

The free undamped vibration of the system is governed by

$$[m]\{\ddot{x}(t)\} + [k]\{x(t)\} = 0 \tag{1.26}$$

Assume a solution

$$\{x(t)\} = \{x\}e^{i\omega t} \tag{1.27}$$

and substitute it into Eq. (1.26). A homogenous equation is obtained:

$$\left([k] - \omega^2[m]\right)\{x\}e^{i\omega t} = \{0\} \tag{1.28}$$

The nontrivial solution of Eq. (1.28) exists only when their determinant is zero,

$$\det \left| [k] - \omega^2 [m] \right| = 0 \tag{1.29}$$

This relation produces a polynomial equation of the order n for the n values of the eigenvalues ω^2, i.e., $\omega_1^2, \ldots, \omega_n^2$. These are the n modal angular resonance frequencies of the system. Substituting any of these frequencies into Eq. (1.28) yields a corresponding set of relative (but not absolute) values for $\{x\}$. This is the nth modal shape, also called normal mode. The n normal modes are described by a matrix $[\phi]$ in which the nth column corresponds to the nth mode shape of the respective frequency ω_n. The nth raw corresponds to the nth degree of freedom of the system. The matrix $[\phi]$ is in the form

$$[\phi] = \begin{bmatrix} \phi_{1,1} & \phi_{2,1} & \cdots & \phi_{n,1} \\ \phi_{1,2} & \phi_{2,2} & \cdots & \phi_{n,2} \\ \vdots & \vdots & \vdots & \vdots \\ \phi_{1,n} & \phi_{2,n} & \cdots & \phi_{n,n} \end{bmatrix} \tag{1.30}$$

The first index is the mode number, and the second index indicates the system coordinate (degree of freedom).

The resonance frequencies and the mode shapes are characteristic of the system and not of the loading. They depend only on the masses (or mass distribution, for a continuous system), rigidities, and boundary conditions. Therefore, they are attractive for applications of structural dynamics analyses. The mode shapes (sometimes called normal modes) possess an important property known as orthogonality. This means that

$$[\phi]^T [m] [\phi] = [M] \tag{1.31}$$

where the superscript T represents a transposed matrix and $[M]$ is a diagonal matrix with elements known as the generalized masses. In a similar way

$$[\phi]^T [k] [\phi] = [K] \tag{1.32}$$

where $[K]$ is a diagonal generalized stiffness matrix. From Eq. (1.29)

$$[K] = \left[\omega_n^2 M \right] \tag{1.33}$$

For a system with a damping proportional to the mass and/or the stiffness

$$[\phi]^T [c] [\phi] = [C] \tag{1.34}$$

where the elements of the diagonal matrix $[C]$ are called the generalized damping. The assumption of a proportional damping that yields a diagonal generalized damping matrix is not necessarily correct. Nevertheless, this assumption greatly simplifies the calculations of structural response and therefore is used extensively in practical engineering applications. When modal damping coefficients ζ_n are known (from experiments or from accumulated practical knowledge), it can be shown that

$$[C] = [2\zeta_n \omega_n M] \tag{1.35}$$

It should be noted that the generalized quantities M, K, and C are not unique. They depend on the value of ϕ, which is a relative set of displacements. Selection of ϕ determines the generalized quantities. In some cases, a normal mode is selected so that its maximum value is 1. In other cases (especially in the large finite elements computer codes), the normal modes are selected in such a way that all the generalized masses are equal to 1, so $[M]$ is a diagonal unit matrix. It is not important what selection is done, as long as the calculation is consistent.

A very useful transformation that is used extensively in structural analysis is

$$\{x\} = [\phi] \{\eta\} \tag{1.36}$$

i.e., the displacements vector $\{x\}$ is expressed as a linear combination of the normal modes $[\phi]$ and a generalized coordinate vector $\{\eta\}$. This transformation, sometimes called normal modes superposition, is used extensively in structural dynamics analysis, and therefore will be demonstrated here for a system with two DOFs.

The modal matrix for such a system is

$$[\phi] = \begin{bmatrix} \phi_{1,1} & \phi_{2,1} \\ \phi_{1,2} & \phi_{2,2} \end{bmatrix} \tag{1.37}$$

Thus, the transformation is

$$\begin{Bmatrix} x_1 \\ x_2 \end{Bmatrix} = \begin{bmatrix} \phi_{1,1} & \phi_{2,1} \\ \phi_{1,2} & \phi_{2,2} \end{bmatrix} \begin{Bmatrix} \eta_1 \\ \eta_2 \end{Bmatrix} \tag{1.38}$$

or explicitly

$$x_1 = \phi_{1,1}\,\eta_1 + \phi_{2,1}\,\eta_2$$
$$x_2 = \phi_{1,2}\,\eta_1 + \phi_{2,2}\,\eta_2 \tag{1.39}$$

Thus, the displacements are expressed as a linear combination of the generalized coordinates, weighted by the modal shapes.

Substituting the transformation, Eq. (1.36) into Eq. (1.25) and multiplying each term by ϕ^T yields

$$[\phi]^T\,[m]\,[\phi]\,\{\ddot{\eta}\} + [\phi]^T\,[c]\,[\phi]\,\{\dot{\eta}\} + [\phi]^T\,[k]\,[\phi]\,\{\eta\} = [\phi]^T\,\{f\} \tag{1.40}$$

or

$$[M]\,\{\ddot{\eta}\} + [C]\,\{\dot{\eta}\} + [K]\,\{\eta\} = [\phi]^T\,\{f\} = \{F\} \tag{1.41}$$

Eq. (1.41) can also be written as

$$[M]\,\{\ddot{\eta}\} + [2\zeta\omega M]\,\{\dot{\eta}\} + \left[\omega^2 M\right]\,\{\eta\} = \{F\} \tag{1.42}$$

where each square matrix [.] on the left side of Eq. (1.42) is diagonal.

The quantity on the right-hand side of Eq. (1.42) is called the generalized forces matrix $\{F\}$. The generalized forces represent the work done by the external forces when the masses of the system move in a modal displacement. When the external forces and the normal modes are known, the generalized forces can be calculated easily.

Eq. (1.42) is a set of n uncoupled equations of motion, where in each of these n equations only one η_i exists. Each equation can be solved separately for the generalized coordinate η_i. It can be seen that when a MDOF system is expressed by its set of generalized masses, generalized stiffnesses, generalized dampings, and generalized forces, a set of n equations for n separate equivalent SDOFs is obtained. Each equation can be solved using all the known techniques and algorithms for a SDOF system. When all the generalized coordinates η_i ($i = 1, 2, \ldots, n$) are solved as a function of time, the deflection of the MDOF system can be obtained using Eq. (1.36) for the displacement vector $\{x\}$ as a function of location and time.

The equations of motion for a base-excited MDOF present the same characteristics. It is more difficult to generalize them because there are too many possibilities. At the end of a process of writing force equilibrium for any of the masses, a set of n equations is obtained. It is important to note which masses (degrees of freedom) are connected (through springs and dampers) to the base ("the external world"), and which masses are connected to each other. It is possible, though, to get a set of uncoupled differential equations where the left-hand side (l.h.s.) is similar to the l.h.s. of Eq. (1.42), and the right-hand side (r.h.s.) contains terms that are a function of the rigidities and damping that exist between those masses that are directly connected to the base, together with the normal modes of the system. This will be demonstrated in the examples.

In Figure 1.8, a system with two DOFs is illustrated. The system has two masses, three springs, and three dampers. The system is excited either by two external forces acting on the masses, or by a base movement $x_s(t)$. Of course, a combination of both external forces and base movement is also possible, but will not be demonstrated here.

For the case of the two external excitation forces, the forces acting on the masses are shown in Figure 1.9.

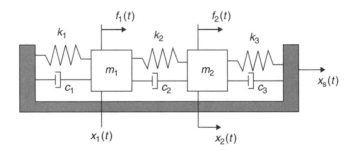

FIGURE 1.8 Two DOFs system.

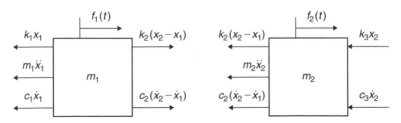

FIGURE 1.9 Forces equilibrium, two external excitation forces.

Formulating the two force equilibrium equations, it can be easily shown that the equations of motion can be written by

$$
\begin{bmatrix} m_1 & 0 \\ 0 & m_2 \end{bmatrix} \begin{Bmatrix} \ddot{x}_1 \\ \ddot{x}_2 \end{Bmatrix} + \begin{bmatrix} c_1 + c_2 & -c_2 \\ -c_2 & c_2 + c_3 \end{bmatrix} \begin{Bmatrix} \dot{x}_1 \\ \dot{x}_2 \end{Bmatrix}
$$
$$
\begin{bmatrix} k_1 + k_2 & -k_2 \\ -k_2 & k_2 + k_3 \end{bmatrix} \begin{Bmatrix} x_1 \\ x_2 \end{Bmatrix} = \begin{Bmatrix} f_1(t) \\ f_2(t) \end{Bmatrix}
$$
(1.43)

When the eigenvalues and eigenvectors of the undamped, unexcited system (the homogenous equation) are solved, two resonance frequencies and two normal modes (forming a 2×2 $[\phi]$ matrix) are obtained. The two uncoupled differential equations that are a special case of Eq. (1.25) can be solved using the methods used for one SDOF system.

For the case where there are no external forces, but the base ("the external world") is moved with a known time function $x_s(t)$, the force equilibrium is shown in Figure 1.10.

The two force equilibrium equations are

$$
\begin{bmatrix} m_1 & 0 \\ 0 & m_2 \end{bmatrix} \begin{Bmatrix} \ddot{x}_1 \\ \ddot{x}_2 \end{Bmatrix} + \begin{bmatrix} c_1 + c_2 & -c_2 \\ -c_2 & c_2 + c_3 \end{bmatrix} \begin{Bmatrix} \dot{x}_1 \\ \dot{x}_2 \end{Bmatrix}
$$
$$
\begin{bmatrix} k_1 + k_2 & -k_2 \\ -k_2 & k_2 + k_3 \end{bmatrix} \begin{Bmatrix} x_1 \\ x_2 \end{Bmatrix} = \begin{Bmatrix} k_1 x_s(t) + c_1 \dot{x}_s(t) \\ k_3 x_s(t) + c_3 \dot{x}_s(t) \end{Bmatrix}
$$
(1.44)

The r.h.s of the equation is a known function of time. Note that only the rigidities and damping between the masses and the support are included in the r.h.s. of the equation, and not those that are between the two masses.

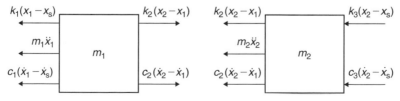

FIGURE 1.10   Forces equilibrium, base excitation.

With some algebraic manipulations, Eq. (1.44) takes the form

$$[M]\{\ddot{\eta}\} + [2\zeta\omega M]\{\dot{\eta}\} + \left[\omega^2 M\right]\{\eta\} = [\phi]^T \left\{ \begin{array}{l} k_1 x_{\rm s}(t) + c_1 \dot{x}_{\rm s}(t) \\ k_3 x_{\rm s}(t) + c_3 \dot{x}_{\rm s}(t) \end{array} \right\} \qquad (1.45)$$

On the l.h.s. of the equation are values that are typical to the system: generalized masses $[M]$ (obtained using the mass matrix and the modes matrix), resonance frequencies ω, and modal damping coefficients ζ. On the r.h.s. of these equations are some known parameters of the problem (k_1, k_3, c_1, and c_3), the transpose modal matrix $[\phi]^T$, and the known base movement. The equations should be solved for $\{\eta\}$, and then the transformation (Eq. (1.36)) can be used for time dependent deflection.

A continuous system is the limit of a MDOF system, when the number of DOFs tends to infinity. Most of the practical continuous (elastic) systems can be solved using a finite number of DOFs, as shown in the next chapters.

1.4 INFINITE-DEGREES-OF-FREEDOM (CONTINUOUS) SYSTEM

For continuous (elastic) systems, the equations of motions are obtained in details in Chapter 2 (Section 2.2). Just in order to complete the treatment of the basic theory of vibration, it should be noted here that continuous systems are treated similar to the way the MDOF systems were evaluated. The displacement perpendicular to the structure is expressed by the famous modal description:

$$w(x,t) = \sum_{n=1}^{\infty} \phi_n(\text{location}) \cdot \eta_n(\text{time}) \qquad (1.46)$$

Thus, the deflection is expressed by an infinite summation of general coordinates $\eta_n(t)$, weighted by an infinite number of modal shapes $\phi_n(x)$, where x is the location on the structure. This is because the continuous system is made of an infinite number of small masses. One can define a set of infinite number of generalized masses. These are obtained for one-, two-, and three-dimensional structures using

$$M_n = \int_{length} m_L(x) \cdot \phi_n^2(x) \cdot dx, \quad m_L(x) \text{ is mass per unit length}$$

$$M_n = \int_{area} m_A(x) \cdot \phi_n^2(x,y) \cdot dx \cdot dy, \quad m_A(x,y) \text{ is mass per unit area}$$

$$M_n = \int_{volume} m_V(x) \cdot \phi_n^2(x,y,z) \cdot dx \cdot dy \cdot dz, \quad m_V(x,y,z) \text{ is mass per unit volume}$$

$$(1.47)$$

In Chapter 2, it is shown that the continuous system can also be analyzed by solving the infinite set of uncoupled equations for $\eta_n(t)$:

$$[M]\{\ddot{\eta}\} + [2\zeta\omega M]\{\dot{\eta}\} + \left[\omega^2 M\right]\{\eta\} = \{F\} \tag{1.48}$$

where $[M]$ and $[F]$ are the generalized mass and generalized forces, respectively, and ω are the resonance frequencies. The actual deformation is then calculated using Eq. (1.46) with the known mode shapes.

Practically, a finite number of modes are used instead of the infinite number of modes that exist, unless an analytical solution is possible—a rare case in practical applications. Justification for this is also described in Chapter 2, in which the response of a continuous structure (a beam) to deterministic excitations is described and discussed.

1.5 MOUNTED MASS

One of the main purposes of a practical structure is to support some kinds of payloads. An aircraft wing produces lift, but also supports the engines and is used as a fuel tank. The floor of a building supports equipment mounted on it, as well as furniture and people. Car front structure supports the engine, a ship mast supports radar equipment, and many more examples can be listed. Sometimes, the supported payload also has elastic properties, but in many cases, it can be treated as a mass attached in some way to the main elastic structure. In these cases, the stiffness distribution of the main elastic structure is not changed significantly by adding the mass, but the mass distribution is not the original one. In most of the cases, the "attached mass" (the equipment) is connected to the structure through an elastic interconnection, such as shock mounts or another interconnecting structure, whose purpose is to support the payload and sometimes to suppress vibrations that are transferred from the main structure to the payload. The structural system is thus comprised of the main elastic structure, the elastic interface, and the attached mass. Because of the additional mass and elastic interface, the resonance frequencies of the original structure are changed, as well as the mode shapes. The change is significant when the mounted mass and elastic interface are not negligible compared to the prime structure. In some cases, when the mounted mass is small, the two subsystems—the original structure and the mounted mass—can be treated separately, but this is not the general case.

A mass mounted through an elastic structure (which can be represented by a spring and a damper) has a separate (uncoupled) resonance frequency that can be calculated by

$$\omega_{\mathrm{ms}} = \sqrt{\frac{k_{\mathrm{s}}}{m_{\mathrm{m}}}} \qquad (1.49)$$

where ω_{ms} is the angular frequency of the supported mass m_{m}, connected to the main structure through a spring with rigidity k_{s}.

This frequency can be observed in the solutions only when the main structure that supports the mass is not taken into account. Generally, when an analysis of the supported mass and the main supporting structure is performed, the (coupled) resonance frequencies of the coupled system will not exhibit the frequency of the uncoupled mounted mass.

Response of a continuous structure with a supported mass to harmonic excitations is demonstrated in Chapter 2. Response of such a structure to random excitation is shown in Chapter 3.

Chapter 2 / Dynamic Response of Beams and Other Structures to Deterministic Excitation

2.1 A GENERIC EXAMPLE OF A CANTILEVER BEAM

In many chapters of this book, the numerical examples treat a cantilever beam whose data is described in this section.

The data for the cantilever beam demonstrated in these examples is:

1. Steel cantilever beam of length 600 mm, width 80 mm and thickness 5 mm

2. Young modulus is $E = 2100000 \, \text{kg/cm}^2$

3. Specific weight is $\gamma = 7.8 \, \text{g/cm}^3$; therefore, specific density is $\rho = 7.959 \times 10^{-6} \, \text{kgf sec}^2/\text{cm}^4$

4. Weight of beam is $60*8*0.5*7.8 = 1872 \, \text{gramf} = 1.872 \, \text{kgf}$

As was mentioned in the Preface, the units of the parameters used in a solution are of major importance. Therefore, for convenience, Table 2.1 shows some of the typical parameters using three sets of units.

2.2 SOME BASICS OF THE SLENDER BEAM THEORY

It is not the purpose of this book to evaluate and to describe the theory of beams. Many dozens of textbooks (e.g., refs. [19–25]) describe the theory of beam. The interested reader should conduct a search using the subjects "strength of materials," "beam theory," and "theory of elasticity."

TABLE 2.1 Data for the cantilever beam.

Quantity	Cm, Kgf[A], sec	US Units[B]	SI Units[C]
L Length	60 cm	23.622 inches	0.6 m
b Width	8 cm	3.1406 inches	0.08 m
h Thickness	0.5 cm	0.19685 inches	0.005 m
E Young Modulus	$2100000\,\text{Kgf/cm}^2$	$29840\,\text{ksi}^{[D]}$	$20.58 \times 10^4\,\text{MPa}^{[E]}$
ρ Mass Density	$7.959 \times 10^{-6}\,\text{kgf sec}^2/\text{cm}^4$	$7.297 \times 10^{-4}\,\text{Lbs sec}^2/\text{in}^4$	$7800\,\text{Nsec}^2/\text{m}^4$
W Total Weight	1.872 Kgf	4.123 Lbs	$18.3456\,\text{N}^{[F]}$
q Weight/Length	0.0312 Kgf/cm	0.1746 Lbs/in	30.576 N/m

[A]Kgf is Kilogram Force.
[B]Units in Inches, Lbs, Seconds.
[C]Units in Meters (m), Newton (N), Seconds.
[D]ksi is 1000 psi (1000 Lbs per Square Inch).
[E]MPa is MegaPascal $= 1000000$ Pascal $= 1000000$ Newton/m^2.
[F]N is Newton $= \text{m kgm/sec}^2$, where kgm is kilogram mass.

The applications and methods described here are based on the slender beam theory. This is a beam whose cross section dimensions are small relative to the length of the beam, and the deflections are small relative to the cross section dimensions. Slender beam theory also assumes that a cross section plane, perpendicular to the neutral axis of the beam, remains a plane that is perpendicular to the deflected neutral axis after loads are applied to the beam system.

Nevertheless, the examples and applications described in this book can be performed on any linear elastic system, while using any other elastic element or any other elastic theory. The cantilever beam is used because many practical problems can be described and solved using this type of structural element, and the use of a simple elastic model provides the reader with a better way of understanding the important features of the problem and the described application.

In fact, the basic formulas describing the behavior of the slender beam can be derived from the basic rules of the theory of elasticity, by neglecting terms that are derived for phenomena that do not contribute to the general understanding of the beam's behavior. One should be careful when doing these neglections. In the large collection of papers that deal with the slender beam it is possible to find a very old paper (from the 1960s) in the *Journal of*

Irreproducible Results in which too many assumption are made, and the author is left with the beam's supports only.

The external loading acting on a beam can be concentrated forces, external moments, and distributed loads along the span. Local effects of concentrated forces and moments are not handled by the slender beam theory. These external loads create, along the length of the beam, bending moments and shear forces.

A beam has a span of given length (say L) and a cross section whose dimensions are small compared to the length. The cross section can be of any shape. It is possible, for a given cross section, to calculate its area A, its area's moment of inertia I, and the location of the neutral axis. It is also possible to calculate the maximum distance that exists in the cross section between the neutral axis (where the cross-section bending moment is zero) and the cross section contour. The beam usually has supports that are required in order to constrain its movements and to eliminate rigid body movement. In the vibration analysis of beams, sometimes there are no constrains at all. This is, for instance, the case of a slender missile in its free flight phase.

When a slender beam is loaded by external forces and external moments, one can describe (using the techniques provided in many strength of materials textbooks; e.g., [19–24]) the bending moments and the shear forces along the beam. The bending stresses along the beam are described by the following well-known expression:

$$\sigma_{\text{bending}}(x) = \pm \frac{M(x)}{I} \cdot y \tag{2.1}$$

where M is the bending moment at location x, I is the cross section area moment of inertia, and y is the distance from the cross section neutral axis in the direction perpendicular to the beam axis. The maximal and minimal bending stresses are at the top and the bottom of the cross section. If y_{max} is the maximal distance from the neutral axis, the maximal bending stress at a location x is given by

$$\sigma_{\text{bending, max}}(x) = \pm \frac{M(x)}{I} \cdot y_{\text{max}} \tag{2.2}$$

Note that bending stresses are not the only stresses in a cross section. Shear stresses also exist, and if the beam is also loaded in its axial direction, additional tension or compression stresses must be added when solving a practical case.

Another important expression in the theory of slender beams is the relation between the deflection line and the bending moments. For the slender beam deflected by small displacements, this relation is

$$E(x) \cdot I(x) \cdot y''(x) = -M(x) \qquad (2.3)$$

where $y''(x)$ is the second derivative (with respect to x), which approximates the curvature of the beam. The minus sign is due to definitions of the coordinate system. In fact, some of the textbooks omit this minus sign. For a beam of uniform material and uniform cross section, this equation becomes

$$E \cdot I \cdot y''(x) = -M(x)$$
$$y''(x) = -\frac{M(x)}{EI} \qquad (2.4)$$

Assume a cantilever beam, clamped at $x = 0$ and free at $x = L$, loaded by a force P at the free end. The bending moments are given by

$$M(x) = P(L - x) \qquad (2.5)$$

Inserting $M(x)$ into Eq. (2.4), integrating twice with respect to x, and calculating the constants of integration so that at $x = 0$, both $y'(0)$ and $y(0)$ are equal to zero, one obtains the well-known deflection of the cantilever beam loaded by a tip force [26]:

$$y = \frac{P}{6EI} \left(3Lx^2 - x^3\right)$$
$$y_{\text{tip}} = y(L) = \frac{PL^3}{3EI} \qquad (2.6)$$

Suppose the same beam is loaded by a continuous load w (force per unit length). The same procedure yields [26]

$$y = \frac{w}{24EI} \left(6L^2x^2 - 4Lx^3 + x^4\right)$$
$$y_{\text{tip}} = y(L) = \frac{wL^4}{8EI} \qquad (2.7)$$

More about deflection of beams, shear deformation, torsion, bending moment distribution, etc. can be found in many textbooks. Solutions for many cases are given in [26].

2.3 MODAL ANALYSIS OF A SLENDER CANTILEVER BEAM

There are many practical methods to obtain the resonance frequencies (eigenvalues) and the modes (eigenvectors) of a cantilever beam, or of a general structure. When the beam is uniform (that is, when the cross section along the beam does not vary), there is an analytical solution for both the frequencies and the mode shapes. These solutions can be found in the basic textbooks of vibration (e.g., refs. [1–12]). The solution is shown here without proof, as a reference to values computed using numerical procedures.

The basic differential equation of a uniform slender beam where the deflections are a function of both location and time is

$$EI\frac{\partial^4 w(x,t)}{\partial x^4} = -\rho A\frac{\partial^2 w(x,t)}{\partial t^2} \tag{2.8}$$

This is a result of a state of equilibrium between the elastic and the inertia forces. ρ is the uniform beam mass density, A is the cross section area, thus ρA is the mass per unit length of the beam, assumed uniform. It can be shown that the deflection $w(x,t)$ can be separated into two parts, one dependent on location only, and the other on time:

$$w(x,t) = \phi(x) \cdot \eta(t) \tag{2.9a}$$

Then two differential equations are obtained for $\phi(x)$ and $\eta(t)$:

$$\frac{\partial^4 \phi(x)}{\partial x^4} - k_n^4 \phi(x) = 0$$

$$\frac{\partial^2 \eta(t)}{\partial t^2} + \omega_n^2 \eta(t) = 0 \tag{2.9b}$$

k_n is associated with the resonance frequency ω_n:

$$k_n^4 = \frac{\omega_n^2 \cdot \rho \cdot A}{EI} \tag{2.10}$$

The result for a cantilever beam is obtained by solving Eq. (2.9a), taking into account zero deflection and zero slope at the clamped edge, and zero bending moment and zero shear force at the free end is (e.g., [10])

$$\phi_n\left(\frac{x}{L}\right) = \frac{1}{2}\left\{\begin{array}{c} \left[\cos\left(k_n L \cdot \frac{x}{L}\right) - \cosh\left(k_n L \cdot \frac{x}{L}\right)\right] \\ + \left[\frac{-\cos(k_n L) - \cosh(k_n L)}{\sin(k_n L) + \sinh(k_n L)}\right] \cdot \left[\sin\left(k_n L \cdot \frac{x}{L}\right) - \sinh\left(k_n L \cdot \frac{x}{L}\right)\right] \end{array}\right\} \tag{2.11}$$

The constant $\frac{1}{2}$ is introduced to the equation so that the absolute value of the tip deflection is normalized to 1 or -1. The eigenvalues are obtained by solving the equation

$$\tan(k_nL) = \tanh(k_nL) \tag{2.12}$$

The natural frequencies of the beam are then given by

$$\omega_n = 2\pi f_n = (k_nL)^2 \cdot \sqrt{\frac{EIg}{WL^3}} \tag{2.13}$$

where ω_n is the angular frequency (radians/sec), f_n is the circular frequency (cycles/sec $=$ Hz), W is the total weight of the uniform beam, g is the gravitation constant, E is the Young's modulus, I is the cross section moment of inertia, and L is the length of the beam.

Values for six normalized eigenvalues (k_nL) are given in Table 2.2. With these values, the first three resonance frequencies of the cantilever beam described in Table 2.1 are calculated and summarized in Table 2.3.

In Figure 2.1, the first three mode shapes of the cantilever beam are shown. It can be seen that the maximum absolute value of all the modes is 1. All modes have zero displacement and zero slope at the clamped edge (as derived from the boundary conditions). The second mode has one additional nodal point (a location where the modal deflection is zero). The third mode has two additional nodes, etc.

It is interesting to compare the normalized static deflection of the cantilever beam subjected to a force at the free tip, and the one subjected to a uniform

TABLE 2.2 Normalized eigenvalues for a uniform cantilever beam.

n	1	2	3	4	5	6
k_nL	1.875104	4.694091	7.854757	10.99554	14.13717	17.27876

TABLE 2.3 Three resonance frequencies of the generic numerical example.

Mode	1	2	3
Frequency (Hz)	11.5245	72.2228	202.2260

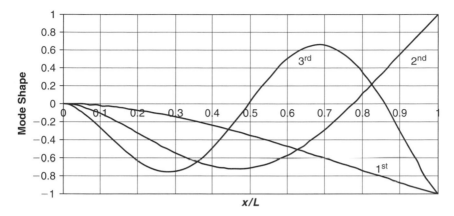

FIGURE 2.1 First three modes of a cantilever beam.

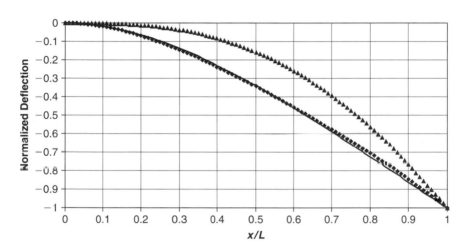

FIGURE 2.2 First vibration mode (full line), continuous normalized static load deflection (diamonds), and discrete normalized tip force deflection (triangles).

load along its length, to the normalized first mode deflection. In Figure 2.2, the normalized first mode is shown in full line, the normalized static deflection due to distributed uniform load is described by the diamonds, and the static deflection due to tip force by triangles. It can be seen that the normalized first mode (obtained from an expression that contains both trigonometric and hyper-trigonometric expressions, Eq. (2.11)) and the static deflection due to distributed load (obtained from a polynomial expression) are very similar. It must not be a surprise, as the inertia loading of a vibrating beam is characterized by a continuous load. In many applications, it can be much easier to approximate the deflection function of a vibrating beam by a "static equivalent" deflection. Algebraic manipulations of the expressions are much easier then. Care must be taken when stresses are to be computed, as these are an outcome of double differentiation (with respect to x) of the deflection line. This differentiation may sometimes introduce significant numerical errors.

Usually, it is not possible to obtain an analytical solution to the resonance frequencies and mode shapes of a practical, real structure. The analytical solution obtained above is for a beam with a uniform cross section, and a uniform material. When the cross section is not uniform, or when the material properties are changed along the beam, one has to apply numerical solutions, obtained in the industry today by one of the large finite element computer codes, like NASTRAN®, ANSYS®, or ADINA™. In the finite element codes, the continuous structure is replaced by discrete elements, with specific properties, and the resonance frequencies and modal shapes are obtained by a numerical procedure.

A short ANSYS input file (**beam1.txt**) (see Appendix) for the computation of the resonance frequencies for the cantilever beam is shown. This file refers to a case where the beam was replaced by 10 two-dimensional beam elements. The computation was also done for a beam of 2, 5, and 20 elements, and the calculated frequencies are shown in Table 2.4.

It is interesting to note that the first resonance frequency is (almost) not influenced by the number of elements. It is also clear why the computation of the third resonance with only two elements failed to produce a correct frequency. The third mode has three nodal points (including the clamp), and two elements cannot describe such a displacement. It is very encouraging,

TABLE **2.4** Resonance frequencies for the cantilever beam.

	2 Elements	5 Elements	10 Elements	20 Elements
1st Mode (Hz)	11.53	11.525	11.525	11.525
2nd Mode (Hz)	72.83	72.253	72.219	72.217
3rd Mode (Hz)	246.28	202.91	202.24	202.19

from an engineering point of view, to see that the first frequency is adequately obtained (within a very reasonably small error) even when only two elements are considered. As will be shown later, many practical dynamic response problems require the use of only the first resonance frequency. The smaller the number of elements in the model, the faster the computation, and the smaller the memory required of the computer for processing the computation.

For some examples shown in Chapter 3, the frequency of a beam that is clamped in one end and simply supported on the other hand is also required. These were calculated using the same ANSYS input file included in the Appendix (**beam1.txt**), with the addition of a support in the free end (see the file listing). Results for three resonances are shown in Table 2.5.

TABLE 2.5 Resonance frequencies for the clamped-supported beam.

	10 Elements
1st Mode (Hz)	50.536
2nd Mode (Hz)	163.78
3rd Mode (Hz)	341.86

2.4 STRESS MODES OF A SLENDER CANTILEVER BEAM

The main purpose of a structural analysis in a design process is to predict the stresses in the structural design. Most of the design criteria in engineering applications are related to stresses, or to a structural behavior that is an outcome of the stress fields in the structure. Stresses in an elastic system are a direct outcome of the relative displacement, and are obtained by using the material constitutive relations and the compatibility equations. The means by which the deflection was obtained, either statically or dynamically, are irrelevant. The effect of the dynamic load factor (DLF), described in Chapter 1, or the amplification factor, is introduced into the system when the deflections are calculated. The stress behavior of a dynamic response is better understood when "stress modes" are used.

The concept of stress modes was first introduced in [27], and better described in [28]. It was further explored in [16]. Some important highlights are described here, and stress modes for a cantilever beam are computed in this chapter.

For a specific type of linear structure, stresses can always be expressed by some differential operator in the spatial coordinates:

$$\sigma_i = K L_i[w(x,t)] \tag{2.14}$$

K is a constant that depends on the material properties and the structure's geometry, L_i is a differential operator, $w(x,t)$ is the structural deflection, which is a function of the location x and the time t. The index i indicates which stress is being calculated. For instance, $\sigma_i = \sigma_x$ may be the bending stress in the x direction, $\sigma_i = \sigma_{xy}$ may be the shear stress in the $x - y$ plane, $\sigma_i = \sigma_{sc}$ may be the (concentrated) stress at the edge of a hole in a plate. For a loaded deflected beam, the bending stress in the tensed side is

$$\sigma_x = -\frac{Eh}{2}\frac{\partial^2 w}{\partial x^2} \tag{2.15}$$

therefore

$$K = -\frac{Eh}{2}; \quad L_x = \frac{\partial^2}{\partial x^2} \tag{2.16}$$

The deflection can be described as a linear function of the normal modes and the generalized coordinates:

$$w(x,t) = \sum_{j=1}^{\infty} \phi_j(x) \cdot \eta_j(t) \tag{2.17}$$

thus

$$L_i[w(x,t)] = \sum_{j=1}^{\infty} L_i[\phi_j(x)] \cdot \eta_j(t) \tag{2.18}$$

According to Eq. (2.14)

$$\sigma_i = K \sum_{j=1}^{\infty} L_i[\phi_j(x)] \cdot \eta_j(t) \tag{2.19}$$

This expression can be rewritten as

$$\sigma_i = \sum_{j=1}^{\infty} \Psi_j^{(i)}(x) \cdot \eta_j(t) \tag{2.20}$$

where

$$\psi_j^{(i)} = K L_i \left[\phi_j (x) \right] \tag{2.21}$$

Eq. (2.20) resembles Eq. (2.17) and therefore the expressions $\Psi_j^{(i)}$ are called "stress modes." By the use of the stress modes, a certain stress (i) in the structure can be expressed by these modes and the generalized coordinates η_j, using Eq. (2.20).

The normal modes $\phi_j(x)$ are characteristics of the structure. It seems that using Eq. (2.21) one can calculate the stress modes $\Psi_j^{(i)}$, which are also characteristics of the structure.

Once the stress modes are calculated, a complete mapping of the stress field in any structural element due to dynamic loading can be evaluated. Nevertheless, it is usually impossible to calculate the stress modes using Eq. (2.21) for three major reasons:

1. The operator L_i is not always known in a closed form, although it is possible to express it for simple cases, as was done for the beam in Eq. (2.15).

2. The mode shapes of the structure are not always available in closed form expressions. In many cases, the modes are approximated by assumed expressions that satisfy the boundary conditions and give good approximation for the resonance frequencies (i.e., by using a Raleigh-Ritz method). Such was the approximation of the first mode of a cantilever beam by a polynom (see Eq. (2.7)). The double differentiation usually required to build an expression for the stresses may introduce significant errors in the computation.

3. In cases where the mode shapes are obtained by a numerical (say finite element) computer code, a double numerical differentiation is required. Such a process may usually introduce large numerical errors.

Examination of Eqs. (2.18) and (2.20) clearly shows that the stress mode is the stress distribution in the structure when it is deformed to its normal mode. As most of the commercially available structural computer code can calculate the stresses when all the structure's nodes are displaced to a prescribed deflection, it seems that this is the best way to practically obtain the stress modes of the structure. It is interesting that the stress mode, which is used for the solution of a dynamic problem, is calculated using a static problem. When a finite element solution for resonance frequencies and resonance modes is performed, the stress modes can be identified by searching the stresses in the eigenvector solution, in addition to the normal modes obtained by the modal solution. These stress modes depend on the normalization factor of the modal nodes. Thus, the normalized computed modes and the stresses created by these modes should be used consistently.

Experience has shown that a designer usually has a good intuitive feel for static loads and is able to identify weak points in his static design by looking at a static analysis. This intuitive feel is less reliable when dynamic problems are concerned. The meaning of the normal (displacement) modes may be understood, but visualization of a physical interpretation of a weighted combination of the normal modes is more difficult. This difficulty is enhanced when the design issue is the response of the designed structure to random vibration. By inspecting the stress modes, a better understanding of the dynamic stress distribution is obtained.

When one calculates the stress modes, special care must be taken to determine the dimensions of the quantities involved. In Eq. (2.17), the normal mode is assumed dimensionless. The dimensions of the deflection w are introduced through the generalized coordinate η. Thus, using the procedure described for the computation of the stress mode, the structure was really deflected by $A_0 \cdot \phi_j$, where A_0 is a constant that takes care of the length unit, and its value is 1 (cm, mm, inch, etc.). Practically, the numerical values obtained are the same. There is no need to be concerned if all the computations are done with consistent units. This precaution is required only in the cases where mixed dimensions exist (i.e., length in mm and in inches).

In Figure 2.3, three stress modes of the cantilever beam are shown. Note that:

1. All stress modes have a maximum at the clamp, the values of which are $512 \, \text{kgf/cm}^2$, $-3213 \, \text{kgf/cm}^2$, and $8997 \, \text{kgf/cm}^2$ for the first, second, and third mode, respectively. This does not mean the stress at the clamp will be the sum of the three values, because the real stress is obtained by a weighted combination of the three values, according to the participation of each mode in the response.

2. The stress modes were calculated assuming a modal displacement of 1 cm at the free tip of the beam. This is the reason why the third mode

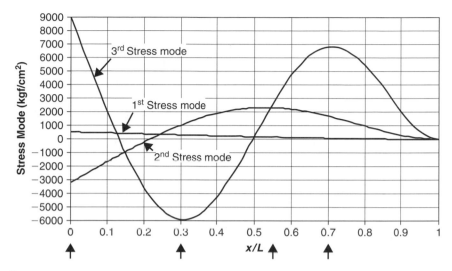

FIGURE 2.3 First three stress modes of the cantilever beam (arrows point to location of local maximum).

presents the highest values of stresses—the curvature of the beam is larger for this prescribed "deflection." The signs of the stresses at the clamped edge are an outcome of this selection. It will be shown later that usually, the first displacement mode dominates the response, and stresses due to the first mode are those that practically influence the final stress results.

3. Looking at the stress modes, one can estimate that the locations that will have higher chances for higher stresses are those marked with arrows at the bottom of Figure 2.3, located where the stress mode has maximal values. Of course, these are not necessarily the dangerous locations, as the final stresses are obtained by a weighted summation of the stress modes. Anyhow, a straight conclusion is that the bending stress at the free end will always be zero!

The stresses at the finite elements nodes of the cantilever beam can also be computed using the postprocessor of the finite element solution. In Figure 2.4, these stresses are depicted for the case of 10 elements along the beam. A very good agreement can be seen in the stresses at the clamped edge. Due to the discreteness of the elements, some differences can be seen at points along the beam, especially in the third mode. When this mode is important to the

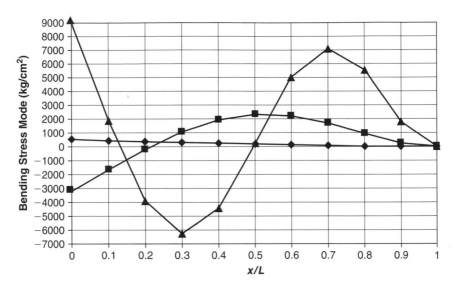

FIGURE 2.4 First three stress modes of the cantilever beam (finite element computation).

problem, the beam should be divided into more elements, so that the curvature can be expressed better.

Once the resonance frequencies and the mode shapes of the cantilever beam (in fact, all beams) are known, the generalized masses can be computed using

$$M_i = \int_0^L m(x) \cdot \phi_i^2(x) \cdot dx \tag{2.22}$$

$m(x)$ is the mass per unit length, and can be a function of x—a nonuniform beam—as long as the correct modal shape $\phi_i(x)$ is calculated.

Integration of Eq. (2.22) with the mode shapes given in Eq. (2.11) shows that the generalized mass for the cantilever beam is

$$M_i = \frac{1}{4} m \cdot L = \frac{1}{4} \cdot \frac{W}{g} \tag{2.23}$$

where W is the weight of the whole beam, and g is the gravitation constant. This result does not depend on the mode number.

In most of the large commercial finite element computer codes (e.g., ANSYS, NASTRAN, etc.) the modes are not normalized to 1. These programs normalize the mode shape in such a way that the generalized mass for each mode is equal to 1. Thus, when one obtains a numerical solution for a modal analysis, one can assume that the generalized masses in this solution are all equal to 1. The most important thing for the analyst is to be consistent. When modes are computed by a finite element code (say ANSYS), the solution of the response problem (response to external excitation forces) should use the non-normalized modal shapes obtained by the program, and generalized masses equal (all) to 1.

2.5 RESPONSE OF A SLENDER BEAM TO HARMONIC EXCITATION

The differential equation of a one-dimensional structural element (like any beam) is derived by using Lagrange equation approach (see, e.g., [10] and [16]).

First, the lateral deflection of the beam w (which is a function of the location x and the time t) is represented by the modal superposition. Any deflection is a superposition of a modal function $\phi_i(x)$ (i is the mode number), and the generalized coordinate $\eta_i(t)$. A separation between the variables x and t is performed. It is assumed that $\phi_i(x)$ is a known function, found from a modal analysis of the structure either by analytical or by a numerical computation analysis. The mode superposition states, in fact, that the total lateral displacement of the structure is a weighted sum of the modal shapes. The weighting functions are the general coordinates $\eta_i(t)$. Once the general coordinates are known, the displacements at any location x at any time t can be calculated by

$$w(x,t) = \sum_{i=1}^{\infty} \phi_i(x) \cdot \eta_i(t) \tag{2.24}$$

In fact, there is no need to do the infinite sum. A finite sum can be written, which will approximate the lateral deflection. Later this approximation will be justified.

$$w(x,t) \cong \sum_{i=1}^{N} \phi_i(x) \cdot \eta_i(t) \tag{2.25}$$

In order to find a differential equation for the general coordinate $\eta_i(t)$, the Lagrange equation is applied:

$$\frac{d}{dt}\left(\frac{\partial T}{\partial \dot{\eta}_r}\right) - \frac{\partial T}{\partial \eta_r} + \frac{\partial U}{\partial \eta_r} + \frac{\partial D}{\partial \eta_r} = N_r \tag{2.26}$$

in which

$$T = \frac{1}{2}\int m\dot{w}^2 dx \tag{2.27}$$

is the kinetic energy of the structure

$$U = \frac{1}{2}\int EIw''^2 dx \tag{2.28}$$

is the potential elastic energy of the structure

$$D = \frac{1}{2}\int c\dot{w}^2 dx \tag{2.29}$$

is the dissipation energy of the structure, due to internal damping, and N_r is the work done by the external loads.

The evaluation of the equations of motion is not presented here. It can be found in one of the many textbooks on the subject (i.e., [10] and [16]). During the evaluation, the following integrals appear:

$$\int m\phi_r(x)\phi_i(x)dx = 0 \quad \text{for} \quad i \neq r$$
$$\int m\phi_r(x)\phi_i(x)dx = M_r \quad \text{for} \quad i = r \tag{2.30}$$

These two equations are the result of the orthogonality of the mode shapes, which are the eigenvectors of the calculated system. M_r is the generalized mass of the r-th mode.

When the external loading is a continuous load per unit length (which may be also a function of time) $p(x,t)$, the external modal work ("generalized force") is given by

$$N_i = \int p(x,t)\phi_i(x)dx \tag{2.31}$$

When the external loading is of n discrete forces (that can be time dependent) $f_n(t)$ at $x = x_n$, the external modal work ("generalized force") is given by

$$N_i = \sum_1^n f_n(t) \cdot \phi_i(x = x_n) \tag{2.32}$$

The result of the mathematical evaluation [10] is the following set of N differential equations:

$$M_r\ddot{\eta}_r + 2\varsigma_r\omega_r M_r\dot{\eta}_r + \omega_r^2 M_r\eta_r = N_r(t) \quad r = 1, 2, \ldots N \qquad (2.33)$$

Two important facts should be noted:

(1) The N equations are decoupled. It means that each of them can be solved **separately** for $\eta_r(t)$.

(2) Each equation is identical to the single degree of freedom equation (SDOF) of motion, with generalized mass instead of a SDOF mass and generalized force instead of SDOF excitation force.

Thus, all the techniques used to solve a SDOF equation of motion can be used for the solution of a multiple degrees of freedom system, once the modes of the structure (resonance frequencies ω_r, eigenvectors ϕ_r, and damping factors) are known.

2.5.1 RESPONSE OF BEAMS TO BASE EXCITATION

Eq. (2.33) is a set of uncoupled differential equations for the solution of the response of a beam to external excitation forces. These forces are expressed using the generalized forces, which are obtained using either Eq. (2.31) for distributed loads or Eq. (2.32) for discrete loads.

When the base of a cantilever beam (the clamped edge) is excited by a time-dependent displacement $x_s(t)$, it can be shown (e.g., [10], [16]) that the differential equations of the excited beam are

$$M_r\ddot{\eta}_r + 2\varsigma_r\omega_r M_r\dot{\eta}_r + \omega_r^2 M_r\eta = -\ddot{x}_s(t) \cdot \int_0^L m(x)\phi_r(x)dx \qquad (2.34)$$

The r.h.s. of Eq. (2.34) contains, in fact, an equivalent generalized force. The quantity $-\ddot{x}_s \cdot m$ is a force per unit length acting in the opposite direction of the excitation displacement, and this force is weighted by the modal shape.

When $m(x)$ and the mode shape $\phi_r(x)$ are given by closed form expressions, it is easy to compute the equivalent generalized force. In cases where no closed form expressions are given, this equivalent generalized force should be computed numerically by performing a numerical integration. When using a finite element code, a small macro-program can be written for post-processing the numerical results of a modal analysis, by lumping the beam mass at the finite elements' model nodes.

2.5.2 RESPONSE OF A CANTILEVER BEAM TO HARMONIC TIP FORCE

The basic data of the beam is given in Table 2.1. Suppose that the free tip of the beam is excited by a force of amplitude $F_0 = 1$ kgf and a variable frequency Ω. The first three modes were calculated in Section 2.3 (Figure 2.1) and were normalized so that the tip displacement of all the three modes is $\phi(x = L) = 1$. The generalized excitation force, according to Eq. (2.32) is

$$N_r = F_0 \phi(x = L) \sin(\Omega t) = 1 \sin(\Omega t) \text{ (kgf)} \qquad (2.35)$$

Assume that all the nodal damping coefficients are $\zeta_1 = \zeta_2 = \zeta_3 = 0.02 = 2\%$.

Usually, higher modes are more damped than lower modes. Typical damping coefficients for metallic structural elements in the lower modes are in the order of magnitude of 1–2%. Higher modes, in which many nodal points exist, are in the order of magnitude of 5–10%. Rarely, higher values of damping coefficients exist for metallic structures.

The generalized masses for all the three resonance modes, as calculated in Eq. (2.23), are

$$M_1 = M_2 = M_3 = 0.25 \cdot m \cdot L = 0.25 \, W/g \qquad (2.36)$$

where m is the mass per unit length of the uniform beam and W is the total weight of the beam.

The amplitude of the generalized coordinates can be computed using the SDOF differential equation:

$$|\eta_r| = \frac{N_r}{M_r \omega_r^2} \cdot \frac{1}{\left[\left(1 - \left(\frac{\Omega}{\omega_r}\right)^2\right)^2 + 4\left(\frac{\Omega}{\omega_r}\right)^2 \zeta_r^2\right]^{1/2}} \qquad r = 1, 2, 3 \qquad (2.37)$$

The maximum value of the DLF for each mode is $\frac{1}{2\zeta_r} = 25$, as all damping coefficients are equal, $\zeta_r = 0.02$.

The maximal values of the generalized coordinate η_r are 9.9847, 0.2512, and 0.03345 for the first, second, and third mode, respectively. As all the generalized amplitudes have the same equation (differing only in the substituted numerical values), and the generalized masses, generalized forces, and the damping coefficients for all three resonances are equal, the following relationship must exist:

$$\frac{|\eta_2|_{\max}}{|\eta_1|_{\max}} = \frac{\omega_1^2}{\omega_2^2} = 0.02546$$

$$\frac{|\eta_3|_{\max}}{|\eta_1|_{\max}} = \frac{\omega_1^2}{\omega_3^2} = 0.00325$$

(2.38)

Within an acceptable computational error, these ratios agree with the preceding written maximum values.

The maximum response of the second mode, when excited in the second resonance frequency, is about 2.5% of the response of the first mode when excited in resonance. The third mode response is about 0.33% of the response of the first mode.

The fact that the generalized coordinate is proportional to the inverse of the frequency squared is the justification to use, in many practical cases, only the first resonance and mode shape in the analysis of beams to harmonic excitation. It will be shown later that the same justification also exists in the analysis of the response to many types of random excitation. Nevertheless, this should be done carefully, as will be shown in the next example.

The expression for the generalized coordinate η_r should include the phase angle between the response and the excitation force. This phase angle is also obtained from the SDOF system Eq. (1.10):

$$\theta_r = \arctan\left[2\zeta_r\left(\frac{\Omega}{\omega_0}\right) \bigg/ \left(1 - \left(\frac{\Omega}{\omega_0}\right)^2\right)\right]$$

(2.39)

Thus

$$\eta_r = |\eta_r| \cdot \sin\left(\Omega t + \theta_r\right) \tag{2.40}$$

and the total lateral displacement (which includes the first three DOFs) is

$$w(x,t) = \phi_1(x) \cdot \eta_1(t) + \phi_2(x) \cdot \eta_2(t) + \phi_3(x) \cdot \eta_3(t) \tag{2.41}$$

In the previous pages, stress modes for the cantilever beam were calculated for modes that were normalized to a unity at the beam tip. The actual stresses at the clamped edge of the beam can be obtained by multiplying the values of the stress modes at that edge with the true deflection of the beam tip. The values of the stresses contributed by the three modes are shown in Figure 2.3.

The same problem was solved using the ANSYS program. The ANSYS file (**beamharm.txt**, see Appendix) for a harmonic response analysis is included on the accompanying CD-ROM. In Figure 2.5, the absolute tip amplitude response as computed numerically is shown. In Figure 2.6, the absolute bending stress at the clamped edge is also shown.

In Table 2.6, results from the analytical and ANSYS finite element computations for the tip displacement and the clamped edge bending stresses are compared.

As can be seen, the agreement is excellent.

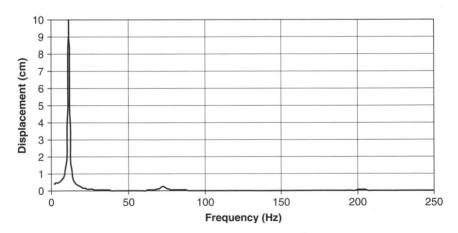

FIGURE 2.5 Generalized coordinates for three modes at the beam tip, as calculated by ANSYS. (The response of the third mode is too small to be seen in this scale).

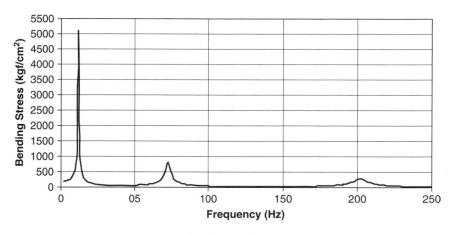

FIGURE 2.6 Bending stresses at the clamp, as calculated by ANSYS.

TABLE 2.6 Comparison of analytical and ANSYS solutions.

Mode	Tip Displacement (cm)		Bending Stress at Clamp (Kgf/cm²)	
	Analytical	ANSYS	Analytical	ANSYS
1	9.9843	9.9847	5119.44	5118.28
2	0.2542	0.2512	816.8	814.9
3	0.03243	0.03345	291.8	291.5

2.5.3 RESPONSE OF A CANTILEVER BEAM TO HARMONIC BASE EXCITATION

In this example, the clamped edge of the beam is moved with a harmonic displacement excitation of amplitude $x_0 = 1 \, \text{cm}$ and a frequency Ω. The acceleration of the base movement is

$$\ddot{x}_s = -x_0 \Omega^2 \sin(\Omega t) \qquad (2.42)$$

The following calculation is done for the first mode only. The generalized force, according to Eq. (2.31) is

$$N_1 = x_0 \Omega^2 \sin(\Omega t) mL \int_{\xi=0}^{\xi=1} \phi_1(\xi) \cdot d(\xi) \qquad (2.43)$$

where m is the mass per unit length and $\xi = \frac{x}{L}$. The integral on the r.h.s. of Eq. (2.43) is equal to 0.391496. Substituting the numerical values, one gets

$$N_1 = 3.921468\,\text{kgf} \tag{2.44}$$

The maximum absolute value of the generalized coordinate in the first resonance, when $\Omega = 2\pi\,(11.525\,\text{Hz}) = 72.41\,\text{rad/sec}$ and the damping coefficient is $\zeta_1 = 0.02$, is $|\eta_1|_{\text{max}} = 39.1496\,\text{cm}$. Note that this value is well above the assumption of small deflections relative to the beam dimensions. Nevertheless, this value will be further used in the example. Stress in the clamped edge can be calculated by the knowledge of the maximum tip deflection calculated for the previous example. This ratio is $39.1496/9.9843 = 3.9211$, thus the maximum stress at the clamp is $5118.28 \cdot 3.9211 = 20069.73\,\text{kgf/cm}^2$.

The same example can be computed using the ANSYS model in file **beamharm.txt**, by changing the command (in the/solution phase) $f,11,fy,1$ (unit harmonic force at the tip, node 11) with $d,1,dy,1$ (unit harmonic displacement at the tip, node 11), which means an imposed harmonic displacement of 1 at node 11, and computing only for the frequency $f_1 = 11.525\,\text{Hz}$. The result shows that the tip amplitude is $39.1605\,\text{cm}$, and the bending stress at the clamp is $20078.88\,\text{kgf/cm}^2$, values that agree well with the analytical solution.

When displaying the results of an ANSYS computation in the general-purpose post-processor/post1, special attention must be taken to compute the absolute value for the result. The ANSYS computation presents the results in the real and the imaginary planes, because a phase angle exists between the excitation and the response. The presentation of the results in one plane only may lead to erroneous conclusions!

An interesting result and an educational example is obtained when the behavior of the cantilever beam for which the free tip is excited by a given harmonic tip deflection. The method of solution was described previously, using the ANSYS file **beamharm.txt** (see Appendix). Of course, the displacement at the tip, node 11, is always 1. In Figure 2.7, the maximum amplitude at node 8 ($x/L = 0.7$) as obtained from the ANSYS solution is shown. In Figure 2.8, the maximum values of the response at the first resonance are shown. These values are not necessarily occurring at the same time, because of phase differences. The amplitude at the tip is, of course, 1. The different modal deflections along the beam are mostly negative, which means that while the

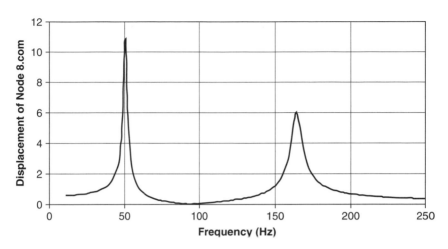

FIGURE 2.7 Maximum displacement at node 8, tip deflection excitation.

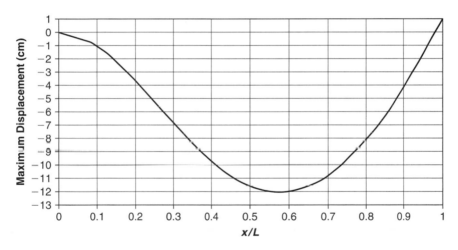

FIGURE 2.8 Maximum values of displacement along the beam (not necessarily the shape of the beam).

tip is moving in one direction, most of the beam moves in the opposite direction.

It can be seen that the two resonance frequencies shown in the solved frequency range are in the vicinity of $f_1 = 50.5$ Hz and $f_2 = 164$ Hz. These are certainly not the resonance frequencies of the calculated cantilever beam. A check will show that these frequencies are the two first resonance frequencies of a beam that is clamped at one end and simply supported on the other, as calculated in Table 2.4.

This leads to a very important conclusion: When a structure is excited by displacement "loads"—imposed deflection (or velocities or accelerations)—it responds in resonance frequencies and mode shapes that are those of a structure supported in these excitation points! This is a very important result that should be memorized when testing of a structure in laboratory conditions is scheduled, as will be shown later in this book.

2.5.4 TWO EXTERNAL FORCES

Suppose there are two forces acting on the beam. The first, f_1, is acting at the tip of the beam, and the second, f_2, is acting at the middle of the beam, in the opposite direction to f_1. Each force has amplitude of 1 kgf. Thus

$$f_1 = f_{x=L} = 1 \cdot \sin(\Omega t) \text{ kgf}$$
$$f_2 = f_{x=L/2} = -1 \cdot \sin(\Omega t) \text{ kgf}$$

(2.45)

From the normal modes described earlier, the modal displacements in each of these two points and for the three first modes, it can be found that

$$\phi_1(x = L) = 1 \qquad \phi_2(x = L) = 1 \qquad \phi_3(x = L) = 1$$
$$\phi_1(x = L/2) = 0.33952 \quad \phi_2(x = L/2) = -0.71366 \quad \phi_3(x = L/2) = 0.01969$$

(2.46)

The generalized forces are, in this case

$$N_1 = 1 \cdot 1 + (-1) \cdot 0.33952 = 0.660477$$
$$N_2 = 1 \cdot 1 + (-1) \cdot (-0.71366) = 1.71366$$
$$N_3 = 1 \cdot 1 + (-1) \cdot 0.01969 = 0.980312$$

(2.47)

Now the three equations (Eq. (2.37)) are no longer equal. The generalized force of the second mode is almost 2.6 times larger than the generalized force of the first mode. Thus, the response in this mode will be almost 2.6 times larger than the one computed in the previous example. In this case neglecting the second and third mode is still justified. However, if the force in the middle of the beam has amplitude of, say 100 kgf, neglecting the second mode will be an error.

This example shows again that neglecting terms in a solution of a problem should be done with care and with understanding of the problem and the data. It is clear to anyone driving on Highway 1 in California that, although the width of the road is much smaller than its length, the width of the road cannot be neglected. The larger response of the second mode happens because two forces, opposite in direction, are acting on the cantilever beam whose mid-point and tip vibrate in opposite directions in the second mode. Thus, examination of the properties of the loads relative to the beam mode shapes can help the designer in his analysis.

2.6 RESPONSE OF A STRUCTURE WITH MOUNTED MASS TO HARMONIC EXCITATION

The concept of mounted mass was introduced in Chapter 1 (1.5), where an uncoupled resonance frequency of the added mass and the rigidity of the connecting structure were mentioned. In the following section, an example of an attached ("mounted") mass with two different mounting rigidities is demonstrated by the following example.

Along the beam, at $x = 0.3\,L$, a mass with a weight of $W = 1$ kgf ($M = \frac{W}{g} = 1/980$ kgf $\cdot$ sec^2/cm) is attached to the beam through a spring with stiffness K_y. Two values of K_y are computed, $K_{y1} = 4$ kgf/cm (Casc (a)) and $K_{y2} = 65$ kgf/cm (Case (b)). The uncoupled resonance frequency for the mass itself on its mounting spring is 9.9646 Hz for Case (a) and 40.17 Hz for Case (b). The three first natural frequencies of the beam without the added mass and spring are 11.5245 Hz, 72.2228 Hz, and 202.2260 Hz, as shown in Table 2.3. Thus, Case (a) represents a mounting frequency very close to the first beam resonance, while in Case (b) the mounting frequency is in between the first and second resonance. The model is shown in Figure 2.9.

The ANSYS file **commass1.txt** (see Appendix) was prepared for computing both the resonance frequencies and modes of the system, and the response to a random tip force (shown in Figure 3.5 in Chapter 3). As the input PSD is between 5 to 250 Hz, only resonances within this range are of interest.

In Table 2.7, resonance frequencies of the four modes found in the relevant range are tabulated for both cases. Mode shapes are shown in Figure 2.10, where the diamond symbol represents the mounted mass modal displacement.

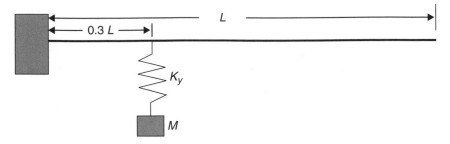

FIGURE 2.9 A mass elastically mounted on an elastic beam.

TABLE 2.7 Resonance frequencies: no added mass, Case (a) and Case (b).

	No Added Mass	*Case (a)* K_{y1}	*Case (b)* K_{y2}
1st Resonance (Hz)	11.525	9.4725	11.278
2nd Resonance (Hz)	72.212	12.034	35.995
3rd Resonance (Hz)	202.24	72.631	80.030
4th Resonance (Hz)	>250	202.54	207.33

It can be seen that the 3rd resonance of Case (a) system is close to the 2nd resonance of the original system, and the 4th resonance of this system is close to the 3rd resonance of the original system.

Analysis of the mode shapes shown in Figure 2.10 (it should be noted that for plotting convenience the vertical scales in the subfigures are not identical) shows that, for Case (a):

(a) The nodal displacement of the mounted mass in the first mode is maximal. This mode originates from the mass spring system. As the mass vibrates on the spring, its inertial forces excite the beam with a shape similar to the beam's first mode, but the modal displacements are smaller than those of the beam's first mode.

(b) The second mode originates from the beam's first mode. The mounted mass vibrates in a direction opposite to the beam.

(c) In the beam's second and third modes (the system's third and fourth modes), the mounted mass does not move, thus it does not participate in the system's movement in these modes.

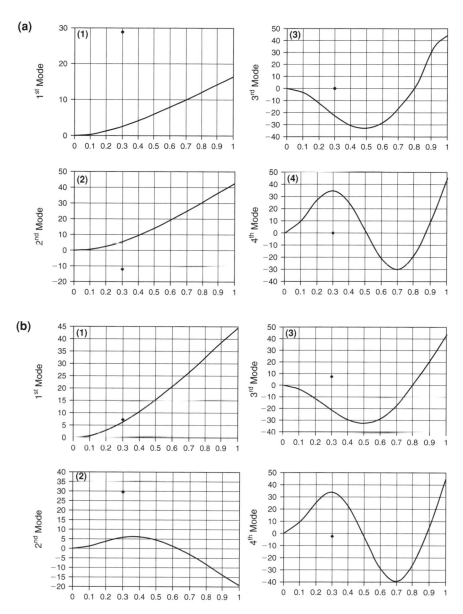

FIGURE 2.10 Mode shapes. (a) Case (a); (b) Case (b); diamond equals the mounted mass. Note that not all subfigures have the same *y* scale.

(d) The distance between the mounted mass and the relevant point on the beam expresses the elongation of the mounting spring. In other words, the stresses in the elastic mount are relative to this distance.

For Case (b):

(a) In the first mode (which is very close to the beam's first mode), the mass moves almost together with the beam. No stresses are built in the elastic mount (the spring).

(b) In the second system's mode, the modal displacements of the beam are small compared to their values in the other modes. The mounted mass has a maximal modal displacement, as the frequency is close to the mounting resonance frequency.

(c) There is a significant participation of the mounted mass in the third mode, which is mainly the beam's second mode. This participation is responsible to the change of the third frequency from the beam's second resonance frequency.

Such observations of the mode shapes can give the designer more insight into the behavior of the structural system, and may lead to some conclusions about the response behavior of the system, and to direct design changes, if required.

In order to show the different behavior of the cases of the beam with the mounted mass, the beam was subjected to a unit force at the tip, so that the excitation force at the tip is given by

$$F = F_0 \sin (\Omega t) = 1 \, \text{kgf} \cdot \sin (\Omega t) \tag{2.48}$$

where Ω is the excitation frequency, which was varied between 2 Hz to 250 Hz. This range covers the range of the four resonances, listed in Table 2.7.

The ANSYS file **commass1.txt** (see Appendix) includes the solution of the response of the system to the harmonic force excitation as well as the computation of the natural frequencies and the response to random excitation, which is described in Chapter 3. For simplicity, but without any loss of generality, the element connecting the mounted mass to the main structure comprises of only a spring, and no damper, which can be introduced by the real constant of the element combin14 that was used in the ANSYS file.

In Figure 2.11, the displacement response of the tip node is shown for Cases (a) and (b). The results are shown on a log-log scale so that resolution of the frequency values is clearer, as well as the difference in amplitude values. This

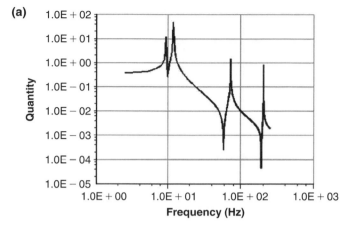

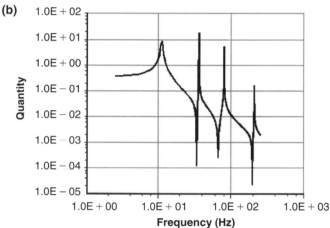

FIGURE 2.11 Displacement of the tip of the structure—(a) Case (a); (b) Case (b).

form of display is familiar to engineers who analyze results from vibration tests. It should be remembered that in some cases, displaying results in a logarithmic scale may be misleading.

In Figure 2.12, the displacement of the main structure (beam) at the point where the connecting spring is attached is shown, for Case (a) and Case (b), with the same log-log display.

In Figure 2.13, the displacement of the attached mass is shown for Case (a) and Case (b).

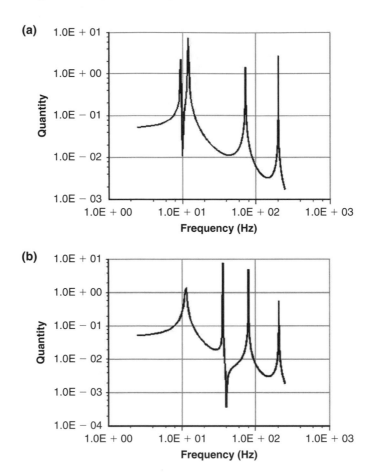

FIGURE 2.12 Displacement of node 4 (below the added mass) of the structure—(a) Case (a); (b) Case (b).

For Case (a), the two close peaks at the vicinity of 10 Hz are demonstrated in all three figures. For Case (b), the peak at the vicinity of 36 Hz is also demonstrated.

The reader should remember that the quantities shown in the figures are the real part of the computed variables. The true response comprises of in-phase (with the excitation force) and out-of-phase components. Usually, real and imaginary components of the results can be displayed after the computation with the finite element program. The effect of out-of-phase components is more dominant when damping is included.

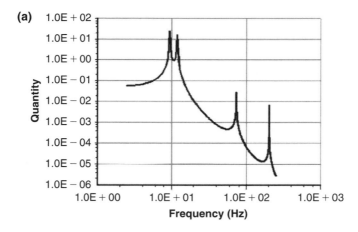

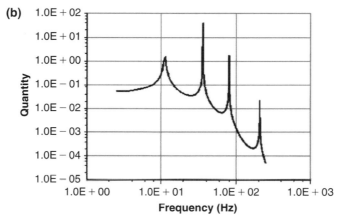

FIGURE 2.13 Displacement of node 14 (the mounted mass) of the structure—(a) Case (a); (b) Case (b).

In addition, the reader should bear in mind that by using the demonstrated procedure (harmonic response along the frequency axis), transfer functions between any two quantities can be computed. The use of transfer function is very helpful when response to random excitation is computed. This is demonstrated in more detail in Chapter 3, as well as the response of such a structure to random excitation.

In many cases of an attached mass, such as the one described by the preceding example, the design parameter on which the designer can influence at a given time is the rigidity of the mounted mass support. Usually, the major design of the main structure is a given fact, and the mass of the required attached mass (usually an equipment "black box") is also defined. By performing a coupled modal analysis of the main structure plus the attached mass, more insight into the problem can be gained. A better selection of the mounting rigidity (and therefore the coupled resonance frequencies) can be recommended, and thus a better design can be performed.

2.7 SYMMETRIC AND ANTI-SYMMETRIC MODES AND LOADS

In some structural elements such as the simply supported beam as well as other more practical structural systems, symmetric and anti-symmetric modes can be found. In order to demonstrate such modes and to present their relationship with symmetric and anti-symmetric loads (which may also occur in practical cases), a finite element solution is presented, based on the file **ssbeam.txt** listed in the Appendix. The material and geometric properties of the simply supported beams are identical to those of the cantilever beam presented in Table 2.1.

Three resonance frequencies and modes were computed. The frequencies of these modes are

$$f_1 = 32.349\,\text{Hz}$$
$$f_2 = 129.39\,\text{Hz} \tag{2.49}$$
$$f_3 = 291.08\,\text{Hz}$$

In Figure 2.14, the first three normal modes are depicted. One can see that the normal mode shapes are comprised of harmonic (sine) half-waves—the first mode having one half sine wave, the second mode has two half sine waves, and the third mode has three sine waves. This is also the outcome of the analytical analysis, described in most of the basic vibration textbooks. The analytical expression for the normal modes of a simply supported beam is

$$\phi_n\left(\frac{x}{L}\right) = \sin\left(n\pi\frac{x}{L}\right) \tag{2.50}$$

where n is the mode number.

It can also be seen that the first and the third modes are symmetric with respect to the beam mid-section, while the second mode is anti-symmetric.

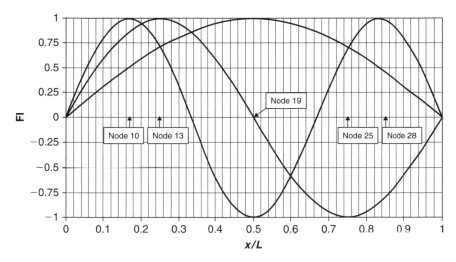

FIGURE 2.14 Three first modes of the simply supported beam.

Of course, when analyzing more than three modes, it can be concluded that for $n = 1, 3, 5, 7, \ldots$ symmetric modes are obtained, while $n = 2, 4, 6, 8, \ldots$ produce anti-symmetric modes.

As already stated, the final response of a linear structure is a weighted linear combination of the normal modes, and it is interesting to see the amount of participation of certain modes as a function of the symmetry or anti symmetry of the excitation forces.

To demonstrate these effects, the first load of the simply supported beam is loaded with a single harmonic force at the mid-beam cross-section (node 19 in file **ssbeam.txt**). This is a symmetric loading. Without a loss of generality, the magnitude of the force is selected as 1, and the frequency is varied between 20 Hz to 350 Hz, a range that includes the three computed resonances.

The amplitude at the middle cross-section (node 19) is shown in Figure 2.15, in a log-log scale. It can clearly be seen that there is no participation of the second, anti-symmetric, mode.

Next, the loading is set to 2 symmetric forces, at the quarter of the beam's length (node 10) and at 3/4 of the length of the beam (node 28). Each load has a magnitude of 1, and they both act in one direction. In Figure 2.16 the amplitude of node 10 (quarter length) is shown.

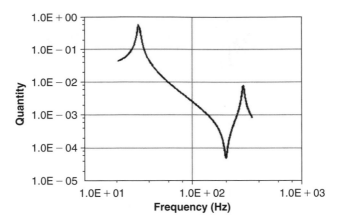

FIGURE 2.15 Response at node 19 (mid-length) to excitation of one force at node 19 (symmetric loading).

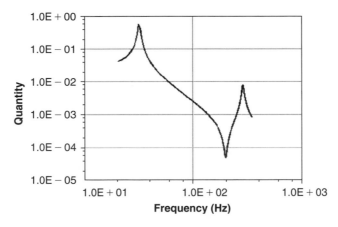

FIGURE 2.16 Response of node 10 (1/4 of beam's length) to excitation at nodes 10 and 28, same directions (symmetric loading).

Again, due to the symmetric loading, there is no participation of the second (anti-symmetric mode) in the response of the beam.

Suppose the same two forces are applied to nodes 10 and 28 in opposite directions. Thus, when one force points upward, the other one points downward. This is an anti-symmetric loading. The response of node 10 to this loading is shown in Figure 2.17.

It can be seen that the response is only in the second (anti-symmetric) mode, due to the anti-symmetry in the loading forces.

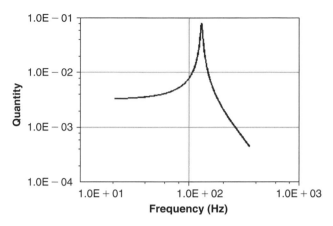

FIGURE 2.17 Response of node 10 (1/4 of beam's length) to excitation at nodes 10 and 28, opposite directions (anti-symmetric loading).

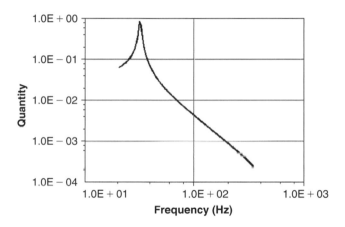

FIGURE 2.18 Response of node 13 (1/3 of the beam's length) to excitation at nodes 13 and 25, same directions (symmetric loading).

Another symmetric loading is applied at 1/3 of the beam's length (node 13, which is at a node of the third resonance) and at 2/3 of the beam's length (node 25, which is also a nodal point of the third mode). Both forces are acting in the same direction, thus the loading is symmetric. The amplitude of node 13 is shown in Figure 2.18.

Now, only the first (symmetric) mode is included in the response. There is no participation of the third (also symmetric) mode, because the locations of the symmetric loading are in the nodal lines of this mode.

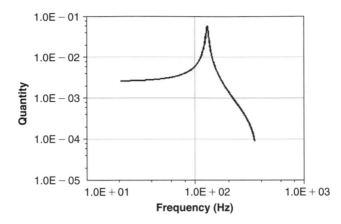

FIGURE 2.19 Response of node 13 (1/3 of the beam's length) to excitation at nodes 13 and 25, opposite directions (anti-symmetric loading).

Next, the same two forces are applied with "anti-phase"; i.e., when one is acting upward, the second is acting downward. This is an anti-symmetric loading. The amplitude of node 10 is depicted in Figure 2.19.

Only the second (anti-symmetric) mode is now included in the dynamic response.

Sometimes, a combination of substructures creates a system that has both symmetric and anti-symmetric modes, which has to be taken into account in the dynamic response analysis of the whole system. A well-known case is the symmetric and anti-symmetric response of an aircraft fuselage with its two attached wings. In the symmetric response, shown in Figure 2.20(a), the fuselage vibrates up and down, with the wings performing a symmetric bending. In this mode, the mass of the fuselage is one of the important parameters of the problem. In the anti-symmetric response, shown in Figure 2.20(b), the fuselage is doing a rolling movement, with anti-symmetric bending of the wings. In this response, the fuselage mass moment of inertia is the important parameter.

In general, it can be stated that symmetric loads do not excite anti-symmetric modes of the structure, while anti-symmetric loads do not excite symmetric modes. In addition, loads applied at a nodal line of a certain mode do not excite that node. Thus, a good knowledge of the complete set of modes of a structure can lead to an educated guess about its behavior under sets of external loads.

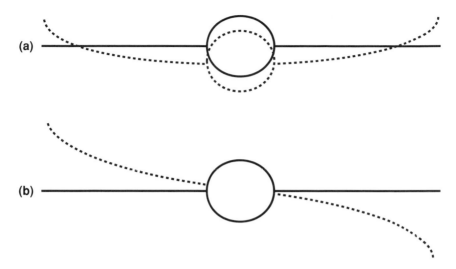

FIGURE 2.20 Symmetric (a) and anti-symmetric (b) modes of an aircraft fuselage-wings combination.

2.8 RESPONSE OF A SIMPLY SUPPORTED PLATE TO HARMONIC EXCITATION

In the previous section, many examples of beams' structures were demonstrated. Although the treatment of other structures, such as plates and shells, is completely similar, an example of the response of a plate structure to harmonic excitation is described in this section, and the response of the structure to random excitation is described in Chapter 3.

Without a loss of generality, the demonstrated structure is a rectangular plate, simply supported along all the four edges. The length of the plate is $a = 40$ cm in the x direction, the width of the plate is $b = 30$ cm in the y direction, and its thickness is $h = 0.5$ cm. The plate is made of steel, so that $E = 2.1 \cdot 10^6$ kgf/cm^2 and the mass density is $\rho = 7.959 \cdot 10^{-6}$ kgf $\cdot$ sec^2/cm^4. Note that because the numerical analysis is done in units of kgf, cm, and seconds, rather strange units are obtained for the mass density, which is not the weight density (not the "specific gravity"). The poison ratio of the material is assumed $\nu = 0.3$, and the damping ratio of all the relevant modes is assumed $\zeta = 2\% = 0.02$. The vibration of a simply supported rectangular plate can be computed analytically. The modes of such a plate are given by (e.g., [4])

$$\phi_{m,n} = \sin\left(m\pi\frac{x}{a}\right) \cdot \sin\left(n\pi\frac{y}{b}\right) \qquad (2.51)$$

In Eq. (2.51), m is the number of half sine waves in the x (length) direction, and n is the number of half sine waves in the y (width) direction.

It was also shown (e.g., [26]) that the resonance frequencies of a simply supported rectangular plate are given by

$$f_{m,n} = \frac{K_{m,n}}{2\pi}\sqrt{\frac{D}{\rho h b^4}} \qquad (2.52)$$

where

$$D = \frac{Eh^3}{12\left(1 - v^2\right)}$$

$$K_{m,n} = \pi^2\left(m^2 + \left(\frac{b}{a}\right)^2 \cdot n^2\right) \qquad (2.53)$$

b is the shorter edge of the plate

For the given data, the analytically computed first frequency is

$$f_{1,1} = 211.95\,\text{Hz} \qquad (2.54)$$

From Eq. (2.51), it can be seen that the modes of a simply supported plate comprises half sine waves in both x and y direction. If $m = 1$, it means that in the x direction there is only one half sine wave. If $m = 2$, there are two half waves, which means that at the center the modal displacement is zero. Same arguments can be evaluated for the y direction. Thus, for each normal mode (a combination of m and n), except for $m = 1$, $n = 1$, there are lines of zero modal displacement called nodal lines. These are analogous to nodal points in a beam structure.

The plate's data is introduced into an ANSYS file **ssplate.txt** (see the Appendix). A modal analysis was performed. In Table 2.8, the resonance frequencies of eight modes are listed, together with the m and n numbers of each of the calculated modes.

Note that the modes are arranged according to the increasing resonance frequencies. The relevant m's and n's are not necessarily in any order. Their

TABLE **2.8** Resonance frequencies and modal index of the simply supported plate.

	Frequency (Hz)	*m*	*n*
1	211.76	1	1
2	440.08	2	1
3	617.87	1	2
4	820.47	3	1
5	844.81	2	2
6	1222.8	3	2
7	1293	1	3
8	1352	4	1

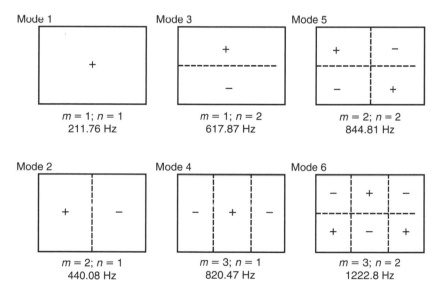

FIGURE **2.21** Six first modes and resonance frequencies.

order is determined according to the case data, and may be different (as well as the frequencies) when dimensions and material properties are different from those of the present example.

In Figure 2.21, six first normal modes are described. In the figure, nodal lines are shown in dotted lines. The (+) and (−) signs describe regions where the normal nodal displacements are upward and downward, respectively. In Table 2.9, four harmonic loadings are listed and described. The excitation is between 200 Hz and 1250 Hz (without any loss of generality), so that only

TABLE 2.9 Harmonic loading cases—location, magnitude, and direction of forces.

	Details			Remarks
Case 1	$x/a = 1/2$;	$y/b = 1/2$,	$F = +1$, Node 811	Symmetric Load
Case 2	$x/a = 3/4$;	$y/b = 1/4$,	$F = +1$, Node 416	Nonsymmetric Load
Case 3	$x/a = 3/4$;	$y/b = 1/4$,	$F = +1$, Node 416	Symmetric Load
	$x/a = 1/4$;	$y/b = 3/4$,	$F = +1$, Node 1206	
Case 4	$x/a = 3/4$;	$y/b = 1/4$,	$F = +1$, Node 416	Anti-symmetric Load
	$x/a = 1/4$;	$y/b = 3/4$,	$F = -1$, Node 1206	

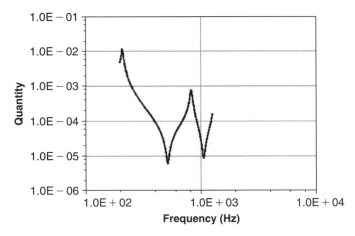

FIGURE 2.22 Response of mid-plate node (811) to Case 1 excitation.

the six modes depicted in Figure 2.21 are responding to this excitation. Note (from Table 2.8) that the seventh resonance (1293 Hz) is pretty close to the sixth frequency (1222.8 Hz), and the eighth frequency (1352 Hz) is also not far away.

As the amplitude of the excitation force is selected as 1, the response curves shown in the following figures describe also the transfer function between the excitation force at the relevant location and the response at the relevant nodes. The locations of these specific nodes are listed in Table 2.9.

In Figure 2.22, the response of the mid-plate node (node 811) to Case 1 excitation is shown. Because this location lies on nodal lines for all modes except the first and the fourth, there are peaks in the response only at the first

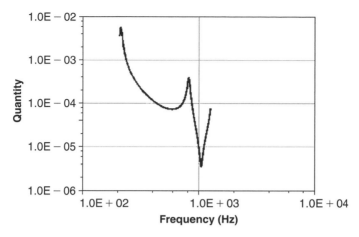

FIGURE 2.23 Response of mid-plate (node 811) to excitation at node 416 (Case 2).

and the fourth modes. The "rise" in the response at the end of the frequency range is due to the participation of the seventh mode (see Table 2.8), which is outside the range of the selected excitation.

In Figure 2.23, the response of the mid-plate (node 811) to the nonsymmetric excitation at node 416 is shown (see Table 2.9). Having a nodal line passing through this node to all but two modes results in an increased response only in the first and fourth modes.

In Figure 2.24, response to Case 2 excitation at node 416 (at the excitation point) is shown. Response to all the normal modes is depicted, although it is difficult to distinguish between the fourth and the fifth resonances (820 Hz and 845 Hz).

In Figure 2.25, response to Case 2 excitation at node 1206 is shown. Again, response to all the normal modes is depicted, although it is difficult to distinguish between the forth and the fifth resonances (820 Hz and 845 Hz).

In Figure 2.26, the response of the mid-plate node 811 to Case 3 excitation is shown (Table 2.9), which comprises two excitation forces in the same direction. Note that only the first and the fourth modes respond in high amplitudes. In Figure 2.27, the response of node 416 to the same excitation is shown. Only modes in which both nodes 416 and 1206 are with the same sign (first and fifth modes, Figure 2.21) participate in the response. The response of node 1206

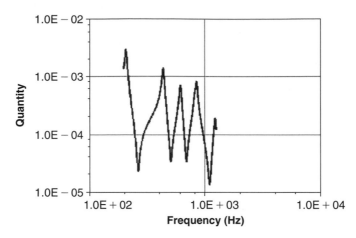

FIGURE 2.24   Response of node 416 to the excitation of Case 2 (Table 2.9).

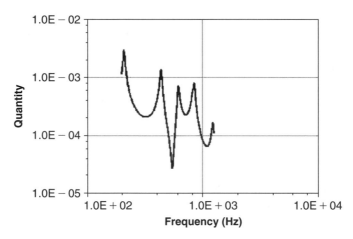

FIGURE 2.25   Response of node 1206 to the excitation of Case 2 (Table 2.9).

is exactly the same of that of node 416, due to the symmetry of the loading, and therefore is not plotted.

In Figure 2.28, the response of the mid-plate (node 811) to the anti-symmetric loading of Case 4 (Table 2.9) is shown. The response seems to be chaotic, but bearing in mind that the values of the response, as seen in the figure, are of the order of magnitude 10^{-17}, the chaotic behavior is really due to the numerical process, and practically the response is zero.

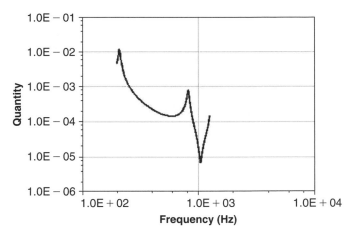

FIGURE 2.26 Response of mid-plate (node 811) to same direction excitations at nodes 416 and 1206 (Case 3 of Table 2.9).

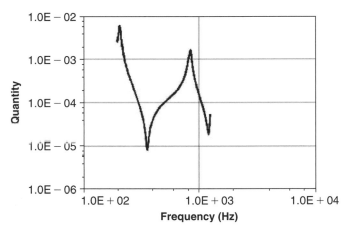

FIGURE 2.27 Response of node 416 to same direction excitations at nodes 416 and 1206 (Case 3 of Table 2.9).

In Figure 2.29, the response of node 416 to the opposite direction anti-symmetric loading (Case 4 of Table 2.9) is described. Only modes that have different direction modal response—the second, third, and sixth of Figure 2.21—respond to this excitation.

To demonstrate that the computing process is capable of also computing stresses in different locations of the plate, stresses in the x (longitudinal) and y (transversal) direction of one location in the plate—node 416—are also

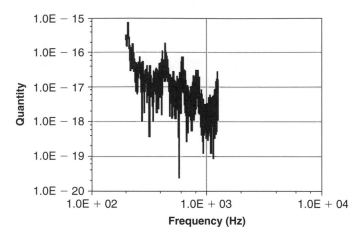

FIGURE 2.28 Response of mid-plate (node 811) to opposite direction excitations at nodes 416 and 1206 (Case 4 of Table 2.9).

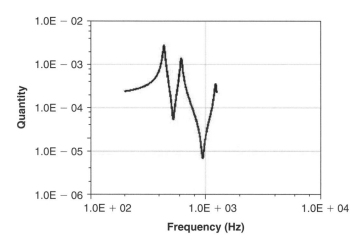

FIGURE 2.29 Response of node 416 to opposite direction excitations at nodes 416 and 1206 (Case 4 of Table 2.9).

included in the input file **ssplate.txt** (see Appendix). The stresses are computed for Case 2 of Table 2.9; i.e., a single force at node 416. In Figures 2.30 and 2.31, stresses in the x and y directions, respectively, are shown. As the excitation force in node 416 has a magnitude of 1, these two figures thus depict the transfer function between force at node 416 and stresses (x and y direction) in the same location. Stresses in any other location can be computed, using the correct commands that can be added to file **ssplate.txt**.

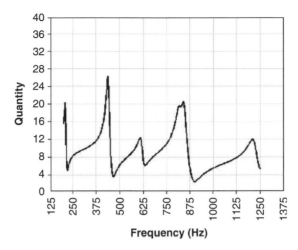

FIGURE 2.30 Stresses in the x direction at node 416, Case 2 loading.

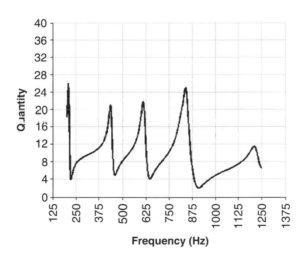

FIGURE 2.31 Stresses in the y direction at node 416, Case 2 loading.

Note that for the simply supported plate, the curvature of the modal-deflected plate is zero at the nodal lines, due to the sinusoidal characteristic of the solution (see Eq. (2.51)), and because lateral bending stresses of a plate are proportional to their curvature. This is not necessarily the case for other boundary conditions set on other plates. If, for instance, the plate is clamped in all or some of its boundaries, local curvature will cause bending stresses at these clamps. This is not demonstrated here, but can be computed by the reader by slightly changing the boundary conditions in the input file **ssplate.txt** listed in the Appendix.

2.9 VIBRATIONS OF SHELLS

Shells, especially thin walled, are an important part of any design of an aerospace structure. Vibrations of shells do not find their proper place in most of the textbooks of basic vibration analysis.

Once a numerical computational method is used to analyze a structure, it is not important whether the structure contains shells, as their inclusion becomes part of the structural model input file. Of course, there are considerations that influence what types of elements are selected for the computation, but these considerations are part of the finite elements code expertise, and do not necessarily belong to the structural analysis.

Nevertheless, it is important for the reader to be familiar with the characteristics of the modal analysis results of thin shells. As in plates, the modal analysis of shells is performed by solving the differential equations of the shells. The differential equations of shells are much more complex than those of plates, especially because an additional special dimension is introduced. There are many textbooks dealing with the differential equations of shells, and probably the best of them is [60]. The differential equations of shells can be extremely simplified (and still be much more complex than those of plates), and easier modal analysis can be performed on such simplified equations. Such simplifications and modal analysis can be found in [61], which is one of the few textbooks that include such analysis for thin shells of revolution.

When solving the differential equations of thin elastic shells, the typical solution contains a summation of modes that have the special form

$$\phi_{m,n} = \phi_m\left(\frac{x}{L}\right)\theta_n(\theta) \tag{2.55}$$

ϕ is a function of x—the shell axial direction—and contains the number m of half waves that the mode has in the axial direction, just as it is done in plates. This function has to meet the boundary conditions on both ends of the shell. θ is a function in the θ direction—the angle in the circumferential direction of the shell—and contains the number n of full circumferential waves (in difference from the plate's definitions). This function is periodic. For the simplest case where the shell is a cylinder of length L, and is simply supported in both ends, the functions ϕ_m are sine waves, as well as the θ_n functions. Eq. (2.55) takes the form

$$\phi_{m,n}\left(\frac{x}{L}, \theta\right) = \sin\left(m\pi\frac{x}{L}\right) \cdot \sin(n\theta) \tag{2.56}$$

The nodal lines (lines of zero modal displacements) of the modes described by Eq. (2.56) are circles and straight lines along the cylindrical shell. In Figure 2.32, nodal lines for the case $m = 3, n = 4$ are shown.

In Figure 2.33, cross sections of the modal displacements are shown for $n = 0$, $n = 1$, and $n = 2$. In the case of $n = 1$, the displacements of the cross sections are all in the same direction, in and out during the vibrations. These vibrations are like a breathing of the shell. In the case of $n = 1$, the displacements of the cross section forms one full wave (half of the circumference is out, the other half is in). This is a vibration in which the whole shell is vibrating like a beam with a ring cross section. Only when $n \geq 2$, one starts to see the "classical" cylinder's vibrations.

It is very difficult to find analytical solutions for the resonance frequencies of practical shells, which are part of a realistic practical structure. On the

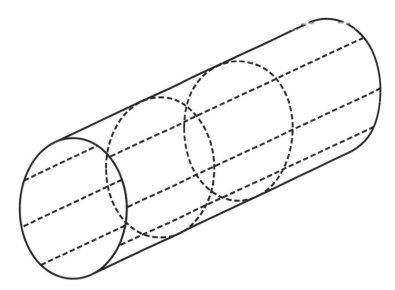

FIGURE 2.32   Nodal lines for simply supported cylindrical shell, $m = 3, n = 4$.

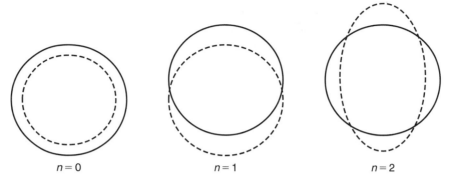

$n = 0$ $n = 1$ $n = 2$

FIGURE 2.33 Cross sections of modes with $n = 0$, 1, and 2.

other hand, there is no need for the design engineer to look for the analytical solution, as a numerical computational procedure based on finite element methods is always possible, with the advantage of performing a solution with the true structural boundary conditions. To demonstrate this, a finite element file **shell1.txt** (listed in the Appendix) was written as a very simple demonstration of the numerical procedure. Only 10 modes were requested in this numerical analysis. The data used in the computation is

$$r = 10 \, \text{cm, Radius of the Shell}$$
$$L = 60 \, \text{cm, Length of the Shell}$$
$$t = 0.1 \, \text{cm, Thickness of the Shell}$$
$$E = 2.1 \cdot 10^6 \frac{\text{kgf}}{\text{cm}^2}, \text{Young's Modulus} \tag{2.57}$$
$$\rho = 7.959 \cdot 10^{-6} \frac{\text{kgf} \cdot \text{sec}^2}{\text{cm}^4}, \text{Mass Density}$$
$$\nu = 0.3, \text{Poisson's Coefficient}$$

The results for the frequencies for the first 10 modes are summarized in Table 2.10.

A very interesting phenomenon is demonstrated by these results. There are five couples of modes, and for each couple the frequencies are identical. The reason for this is that for the same frequency, there are two possibilities for the shell to vibrate; these two are shifted by a rotation angle from each other. The results are better understood for the mode with $n = 1$ (which was not a solution for the present example). The $n = 1$ mode is a bending mode of the whole shell as a beam. This beam can vibrate either in the horizontal direction

TABLE 2.10 Resonance modes of the example.

Mode Number	Frequency (Hz)	m, Number of Axial Half-Waves	n, Number of Circumferential Waves
1	389.96	1	3
2	389.96	1	3
3	417.66	1	4
4	417.66	1	4
5	598.29	1	5
6	598.29	1	5
7	650.95	1	2
8	650.95	1	2
9	726.74	2	4
10	726.74	2	4

of the cross section, or in the vertical direction. These are two different modes even if their frequencies are equal, and should be taken into account when a response problem is solved.

For thin-walled shells, there are many possible modes of vibration, as the values of m and n and their combinations create a great number of modes. Some methods of shells response analysis (especially response to acoustic wide-band excitation) use the "modal density" concept; i.e., compute the response by assuming that in a given bandwidth of frequencies there is a certain (high) number of modes. These methods are not treated in this publication, and the interested reader should search for information on modal density.

Chapter 3 / Dynamic Response of a Structure to Random Excitation

3.1 RANDOM EXCITATION AND RESPONSE

The previous chapters demonstrated the response of a SDOF, MDOF, and continuous elastic systems to a deterministic time-dependent excitation, especially harmonic. A deterministic excitation is known (or can be calculated) at every time explicitly.

In practical structures, especially in aerospace and civil engineering designs, cases where the excitation is deterministic are not frequent. In most of the cases, the excitation is not known explicitly, and only some statistical properties of this excitation are known. This is the case, for instance, for aerodynamic forces acting on a wing, rough road excitation of a moving vehicle, earthquake excitation of buildings and bridges, wind excitation of tall buildings, waves excitation of marine structures, etc.

A random excitation includes a mixture of different levels of external forces or externally imposed displacements that contain components of many different frequencies. Boundary layer excitation over an aircraft wing is a typical example of external continuous random pressure excitation. Waves acting on an offshore oil rig are another example of random forces acting on a structure. Excitation of the wheels of a vehicle traveling on a rough road is an example of continuous imposed random displacements. An earthquake is a typical example of a transient randomly imposed displacement on the base of a building.

During many years, data on many kinds of excitations was collected and analyzed. Usually, the analysis determines the mean excitation, the dispersion in excitation level around the mean, and the frequency content of the excitation. Many specifications and standards were written, in which the random excitation is specified in terms of PSD (Power Spectral Density) of the required quantity (e.g., [29]).

The PSD is a function that describes the energy content distribution of a quantity over the frequency range. The reader who is interested in the mathematical definitions of the PSD and other characteristics of the terms involved in statistical dispersion should refer to one or more of the many textbooks and papers dealing with the subject (e.g., [10, 15, 16, 32–34]). In this chapter, only some basic terms will be explained more rigorously in order to enhance the physical understanding of the role of random excitations in structural dynamics—analysis and testing.

Random input of accelerations is a very common application in the environmental testing of a structural system, and of any designed system. The system is mounted on a shaker that applies acceleration to it. The control system of the shaker is set to input a given PSD of acceleration, according to a PSD curve obtained from formal specifications or data collected in field tests.

A typical input (obtained, say, from formal specifications) is described in Figure 3.1.

Usually, input PSD curves in formal specifications are given on a log-log scale.

It should be noticed that acceleration PSD, though most common in practical applications, is not the only possible random input. Input can be random forces, random displacements, random pressures, etc. The units of the PSD are always the relevant quantity squared divided by frequency. Usually, the frequency is given by Hz. Thus, acceleration PSD has units of g^2/Hz, force PSD—kgf^2/Hz, etc. In some analytical solutions, it is more convenient to use the angular frequency, in radian/sec, and it is very important to know how to transfer units from one system to another.

As can be seen later in this book, the act of imposing accelerations on the system, either from formal specifications or from prototype field tests, can lead to erroneous testing of the system. This will be shown later. Meanwhile, a better understanding of the types of possible random input is called for.

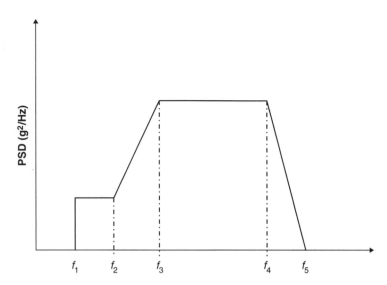

FIGURE 3.1 A typical acceleration input in a vibration test.

There are some special inputs, which are used extensively in the literature and in practice, which the reader should be familiar with. A "white noise" is an input that has a uniform value along the frequency axis, which means that it has equal energy in all the frequency range. An ideal mathematical white noise has frequency content from $-\infty$ to $+\infty$. This is really a mathematical definition, as in practice, a negative frequency is not defined. Thus, an "engineering white noise" can be defined as an input with a constant value of PSD from 0 to $+\infty$, thus the name "one sided white noise." The input may be a "band limited" when the frequency range of it is limited between two frequencies, ω_{low} to ω_{up}, close to each other. These three examples are depicted in Figure 3.2. The PSD described in Figure 3.1 is a practical case of a "wide band," nonuniform input PSD.

In Figure 3.3, a time signal with wide band frequency content is depicted. The artificially generated time signal for this example has a normal distribution with zero mean and a standard deviation (1σ) of 1. The amplitude is random, and seldom crosses the value of 3 (3σ). In fact, in the shown example, the signal touches and crosses the value of ± 3 only four times. The signal includes a large number of frequencies.

In Figure 3.4, a narrow band time signal is shown. The amplitude is random (and crosses the value of 3 only once), but a dominating frequency can be detected.

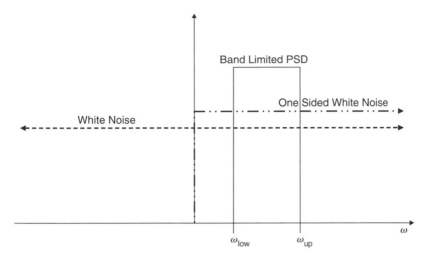

FIGURE 3.2 Three special PSD functions (based on [16]).

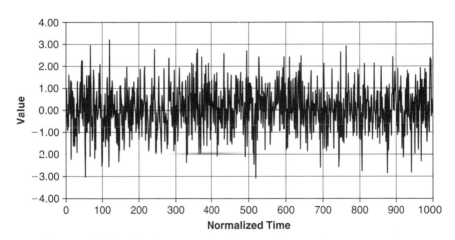

FIGURE 3.3 Wide band random signal (the time coordinate is insignificant).

A random variable has a probability density function (PDF) that describes its distribution. This distribution can be characterized by its statistical moments. The first moment is the mean value, which in fact is the center of gravity of the PDF curve. The second moment is the mean square (MS) value, which provides information about the dispersion of the random variable population around the mean. The MS is the second moment of inertia of the PDF curve around an axis at the mean value. The square root of the MS, the RMS (Root Mean Square) value, is also called the standard deviation (SD), and is usually called 1σ. The third moment is called skewness, and provides information about the

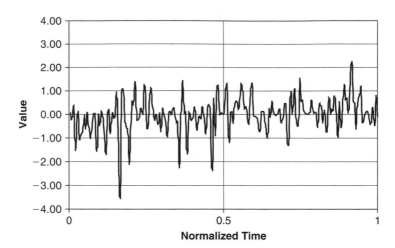

FIGURE 3.4 Narrow band random signal (the time coordinate is insignificant).

symmetry of the distribution of the random variable around the mean. There are also higher statistical moments. A normal (or Gaussian) distribution is fully described by its mean and MS values. It can be shown that, for a normal distribution, 84.13% of the values are smaller than 1σ, 97.72% are smaller than 2σ, 99.865% are smaller than 3σ, 68.27% of the random values are included in the range of $\pm1\sigma$, 95.45% are included in the range of $\pm2\sigma$, and 99.73% are included in the range of $\pm3\sigma$. Thus, when a range of $\pm3\sigma$ is selected, only 0.27% of the random values are out of this range.

More knowledge about random signals and their statistical parameters can be obtained from textbooks on statistics and probabilities (e.g., [35]).

Two important relations are extensively used in the analysis of the response of any system to random excitation (and not only in structural analysis):

1. The area under the PSD curve is the MS value of the random input. Thus, to obtain the MS value, the PSD curve is to be integrated with respect to the frequency. Note that when the PSD curve is depicted on a log-log scale (like, say, in Figure 3.1), the mean square is not the sum of the trapezoidal areas seen in the curve!

$$MS = \int_{\text{over all frequency range}} PSD \cdot d\omega$$

$$RMS = \sqrt{MS}$$

(3.1)

The square root of the MS, called the RMS, is the standard deviation of the given input, (1σ). The standard deviation is also a measure of the dispersion of the quantity around the mean value.

2. The PSD of the response of a system to an input PSD is given by the following expression:

$$\text{PSD}_{\text{response}}(\omega) = |H(\omega)|^2 \cdot \text{PSD}_{\text{input}}(\omega) \qquad (3.2)$$

In Eq. (3.2), the function $|H(\omega)|$ is the absolute value of the transfer function between the input and the output. When the input is in PSD of force, and $|H(\omega)|$ is the transfer function between force and displacement (as in Eq. (1.6), Chapter 1), then the PSD of the response is that of displacements. One can use any transfer function available from simulations or measured in experiments in Eq. (3.2); for example, a transfer function between a forced acceleration at a certain point on the structure, and the bending stress in another structural location. Once the PSD of the input is known (either by field tests, analytical or numerical considerations or from formal specifications) and the transfer function is known, the PSD of the desired response can be computed. This is the reason why the calculation or measurement of a transfer function is so essential. Transfer functions do not necessarily have to be computed. They also can be measured in well-planned experiments during a development phase or during field tests of prototypes.

Introducing Eq. (3.2) into Eq. (3.1), one obtains

$$\text{MS}_{\text{response}} = \int |H(\omega)|^2 \cdot \text{PSD}_{\text{input}}(\omega) \cdot d\omega$$
$$\text{RMS}_{\text{response}} = \sqrt{\int |H(\omega)|^2 \cdot \text{PSD}_{\text{input}}(\omega) \cdot d\omega} \qquad (3.3)$$

Eq. (3.3) is a basic relationship between input and output in a linear system. Except from the quantitative use of the equation, some important qualitative conclusions can be drawn. Suppose the input force PSD is a one-sided white noise. It was already shown that for a SDOF system, the transfer function between force input and displacement output is a function with very small values along most of the frequency range, and with a narrow peak (the "narrowness" of which depends on the damping coefficient) around the resonance frequency. Thus, multiplying the square of this transfer function with the constant value input PSD results in a function that also has small values along most of the frequency range, with a peak around the resonance frequency.

Thus, when a SDOF system is excited by a wide band force excitation the system responds with a random narrow band response, which means that the SDOF response has frequency components only around the resonance frequency, the amplitude of which is random. Thus, it is said that the SDOF is a "filter" to the wide band excitation. The contribution to the total MS of the response displacement comes mainly from the vicinity of the resonance frequencies.

When a MDOF system is treated, and this system has well separated resonance frequencies, the transfer function looks similar to that of the SDOF system, but has several (the number of which is the number of degrees of freedom) peaks. Thus, the response PSD curve also has several peaks. The response of the system has random amplitudes, but has several dominant frequencies, close to the resonance frequencies. When the resonance frequencies are not well separated, a "wider," and not a narrow, response PSD function is observed, for which all the described results are valid.

A continuous elastic structure behaves basically as a MDOF system. In fact, the usual solutions (like finite element computations) use discretization to replace the continuous system with a MDOF one. For a SDOF system, the absolute square of the transfer function between force input and displacement output is

$$|H(\omega)|^2 = \frac{1}{m^2 \omega_0^4} \cdot \frac{1}{\left[\left(1 - \left(\frac{\omega}{\omega_0}\right)^2\right)^2 + 4\left(\frac{\omega}{\omega_0}\right)^2 \zeta^2\right]} \tag{3.4}$$

Because of the argumentation described previously, the PSD of the response is almost totally from the component of excitation in the resonance frequency ω_0, no matter what the shape of $S_F(\omega)$—the PSD of the excitation force. $S_F(\omega)$ can be replaced by $S_F(\omega_0)$, the value of the PSD function at ω_0, and the PSD of the response displacement, according to Eq. (3.2), is

$$S_X(\omega) \cong \frac{1}{m^2 \omega_0^4} \cdot \frac{S_F(\omega_0)}{\left[\left(1 - \left(\frac{\omega}{\omega_0}\right)^2\right)^2 + 4\left(\frac{\omega}{\omega_0}\right)^2 \zeta^2\right]} \tag{3.5}$$

The MS of the displacement, according to Eq. (3.3) is

$$MS_X = S_F(\omega_0) \int_0^\infty |H(\omega)|^2 \, d\omega \tag{3.6}$$

Integration is done from 0 because negative frequencies are not considered. It can be shown that

$$\int\limits_{0}^{\infty} |H(\omega)|^2 \, d\omega = \frac{1}{m^2} \cdot \frac{\pi}{4\zeta\omega_0^3} \tag{3.7}$$

Thus,

$$MS_X \cong \frac{\pi S_F(\omega_0)}{4m^2\zeta m_0^3} \tag{3.8}$$

Eq. (3.8) is the exact solution for the MS of the displacement response of a SDOF system subjected to a "one-sided white noise" force excitation in the positive frequency range, and a very good approximation to the response to a random, nonconstant PSD force excitation.

Note that the MS value of the response is inversely proportional to the cube of the resonance frequency. This means that the RMS value is inversely proportional to the frequency to the power of 3/2. When a MDOF (or continuous) system is analyzed the contribution of each resonance to the MS decreases with the cube of the resonance frequencies. In addition, the MS is inversely proportional to the damping coefficient, and the RMS is inversely proportional to the square root of the damping coefficient.

Analytical solutions for the response of MDOF and continuous systems to random PSD excitation are not shown in this book. The reason is that these solutions usually contain integrals of the mode shapes and other parameters of the problem, which usually cannot be solved explicitly, and the user does a numerical integration. Therefore, it seems appropriate to solve the whole problem numerically from the beginning, usually by a finite element code. The interested reader can find analytical solutions in many technical and textbooks. Formulations and examples are included in [16], which contains practical material from many references in the literature. It should however be mentioned that when analytical solutions are performed for continuous elastic systems, the integrals in the MS expressions contain expressions of the transfer functions $|H_j^*(\omega)| \cdot |H_k(\omega)|$, where $|H_j^*(\omega)|$ is the conjugate of $|H_j(\omega)|$ and where j and k are the mode numbers. When $j = k$, the expressions obtained contain $|H_j(\omega)|^2$. When $j \neq k$, a mixed term is obtained. The integral of the terms that contain $|H_j(\omega)|^2$ gives the contribution of the mode j to the MS value. The integral of the mixed terms gives the inter-modal interaction between mode j and mode k. When the systems have well separated resonance frequencies (as is the case for the cantilever beam and other beams), the inter-modal interaction terms are very small compared to the "direct" contribution of each mode, and can usually be neglected. For structures in which the resonances are not well separated, the mixed terms may have a significant contribution to the result. In any case, the PSD module of one of the commercial finite element codes (like ANSYS® or NASTRAN®) takes care of the real case, and includes the inter-modal interactions inherently.

Sometimes, there is confusion between PSD functions given in (quantity)^2/Hz (i.e., along a circular frequency axis, cps or Hz) and (quantity)^2/(rad/sec) (i.e., along an angular frequency axis, radian/sec). Most of the closed form equations and solutions described or analyzed in the literature include ω, the angular frequency. On the other hand, in most specifications the PSD is given along the f axis. Special care should be given to these units, either by translating the given PSD to the required units (as shown below) or by introducing $2\pi f$ instead of ω in the relevant equations. The PSD is given in (quantity)^2/(rad/sec) and called PSD$_\omega$, then the PSD in (quantity)^2/Hz, called PSD$_f$. Suppose there is a uniform excitation of PSD$_\omega$((quantity)^2/(rad/sec)) between ω_1 and ω_2. The MS of this input is PSD$_\omega \cdot (\omega_2 - \omega_1)$. The equivalent PSD in (quantity)^2/Hz is PSD$_f$ and the MS of this input is PSD$_f \cdot (f_2 - f_1)$. The MS values in both cases must be equal, thus:

$$\text{PSD}_f \cdot (f_2 - f_1) = \text{PSD}_\omega \cdot (\omega_2 - \omega_1) = \text{PSD}_\omega \cdot 2\pi \left(f_2 - f_1\right) \qquad (3.9)$$

thus,

$$\text{PSD}_f = 2\pi \cdot \text{PSD}_\omega \qquad (3.10)$$

The finite elements codes (i.e., ANSYS, NASTRAN, etc.) are built in such a way that internal frequency computations are in Hz. Therefore, input PSD in these programs must always be given in (quantity)^2/Hz and not in angular frequency.

3.2 RESPONSE OF AN ELASTIC STRUCTURE TO RANDOM EXCITATION

The cantilever beam in the following demonstration is subjected to a random tip force, which has a power spectral density described in Figure 3.5. The

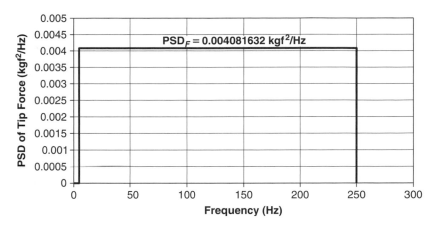

FIGURE 3.5 PSD of tip excitation force.

input was selected so it will contain the three first resonances of the beam. The value of the PSD of the force was selected so that the RMS (Root Mean Square) of the force is 1 kgf. Mean square value is obtained by an integration of the PSD function with respect to the frequency:

$$\text{Mean square of } F = \text{PSD}_F \Delta f = 0.004081632 \cdot (250 - 5) = 1 \, \text{kgf}^2$$
$$F_{\text{RMS}} = 1 \, \text{kgf}$$

(3.11)

First, a numerical solution is presented using the ANSYS finite element code. Then, a closed form approximate solution is presented and partially compared to the numerical analysis.

In file **beamrand_1.txt** (see Appendix), an ANSYS file for the solution of this problem is listed. When doing an ANSYS solution, there are two post-processors in which results can be analyzed and displayed. The first is /POST1 (the general post-processor), in which RMS values of displacements, velocities, accelerations, and stresses along the beam can be shown and listed. Remember that the RMS value is equal to 1σ of the results, and a value of 3σ contains 99.73% of the required quantity. The second post-processor is /POST26 (time domain post-processor) in which frequency domain results such as response PSD can be computed and displayed, and mathematical manipulations of response parameters, like integration, differentiation, multiplication, and much more can be performed.

In Figures 3.6, 3.7, and 3.8 the RMS values of the deflections, accelerations, and bending stress along the beam are shown, respectively. One

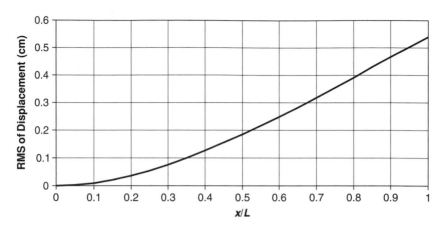

FIGURE 3.6 RMS values of displacements along the beam (/POST1).

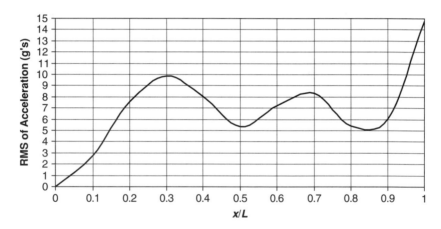

FIGURE 3.7 RMS values of the acceleration along the beam (/POST1).

should remember that the RMS values are statistical parameters, and two RMS values of, say, displacements in two locations, do not necessarily occur simultaneously.

In Figure 3.9, the PSD of the displacement at the tip of the beam (node 11) is shown. Note that almost all of the deflection response is due to the first mode. The magnitude of the responses in the second and third modes cannot be seen in the figure.

In Figure 3.10, the PSD of the acceleration at the tip of the beam (node 11) is shown. Note that the contribution of all the first three modes is almost equal.

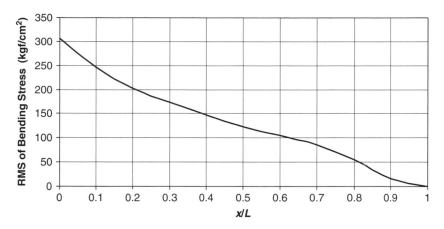

FIGURE 3.8 RMS values of the bending stress along the beam (/POST1).

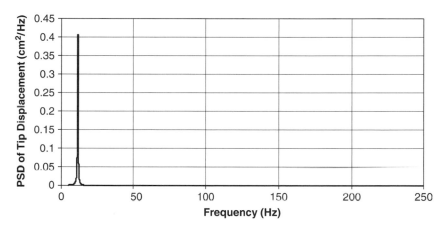

FIGURE 3.9 PSD of the deflection at the free tip (/POST26).

In Figure 3.11, the PSD of the bending stress at the clamped edge of the beam (node 1) is shown. Note that the contribution of the first second mode is very small relative to the first mode, and the contribution of the third mode is almost undistinguished.

When the PSD curves are integrated with respect to the frequency, the MS (Mean Square) values of the quantity are obtained. The square root of the MS is the RMS, which is equal to 1σ (one standard deviation) of the quantity. It is interesting to compare the RMS values obtained using the /POST1 and

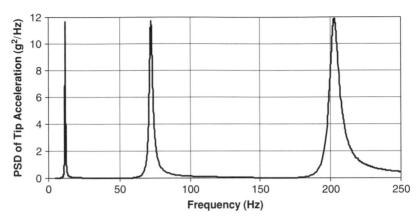

FIGURE 3.10 PSD of the acceleration at the free tip (/POST26).

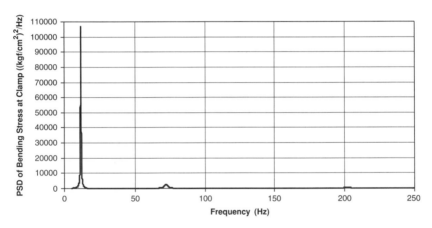

FIGURE 3.11 PSD of the bending stress at the clamped edge (/POST26).

the /POST26 post-processors. This integration can be performed directly in the /POST26 post-processor.

In Figure 3.12, the integral of the PSD of the tip displacement is shown. It can be seen that the contribution to the MS is mainly due to the first mode. The last value, at $f = 250\,\text{Hz}$, is $0.2922\,\text{cm}^2$, and therefore the RMS value is $w_{\text{RMS}}\,(\text{tip}) = \sqrt{0.2922} = 0.54055\,\text{cm}$, compared to $0.5403\,\text{cm}$ obtained in Figure 3.6.

In Figure 3.13, the integral of the PSD of the tip acceleration is shown. The last value at $f = 250\,\text{Hz}$ is $219.95\,\text{g}^2$, and therefore, the RMS value is $\ddot{w}_{\text{RMS}}\,(\text{tip}) = \sqrt{219.95} = 14.831\,\text{g}$, compared to $14.827\,\text{g}$ obtained in Figure 3.7.

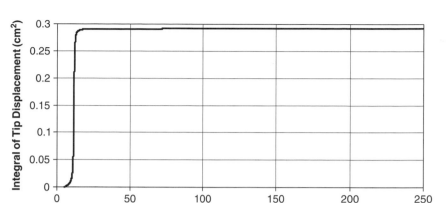

FIGURE 3.12 Integral of PSD of tip displacement.

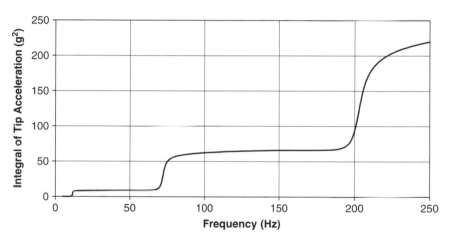

FIGURE 3.13 Integral of PSD of tip acceleration.

In Figure 3.14, the integral of the PSD of the bending stress in the clamped edge is shown. The last value at $f = 250\,\text{Hz}$ is $93402\,(\text{kgf}/\text{cm}^2)^2$, and therefore the RMS value is $\sigma_{RMS}(\text{clamp}) = \sqrt{93402} = 305.62\,\text{kgf}/\text{cm}^2$, compared to $305.6\,\text{kgf/cm}^2$ in Figure 3.8.

In addition, the contribution of each resonance to the total MS value can be found, by taking the values between two plateaus of the curves. In Table 3.1, the contribution of each resonance and comparison between total results of /POST1 and /POST26 post-processor are shown. The data is taken from the original ANSYS results file.

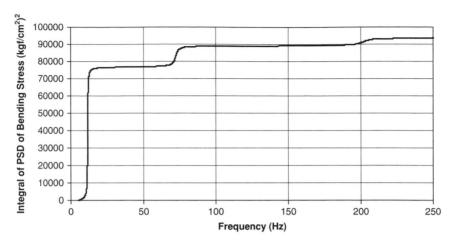

FIGURE 3.14 Integral of PSD of the clamped edge bending moments.

TABLE 3.1 Contribution of first three modes to the required quantities.

		Tip Deflection cm^2, *cm*	*Tip Acceleration* g^2, *g*	*Max. Bending Stress* $(kgf/cm^2)^2$, kgf/cm^2
1st Mode	MS	0.290998	8.8233	76745.3
	RMS	0.5394	2.97	277.03
2nd Mode	MS	0.00115	56.4969	12359.4
	RMS	0.0339	7.528	111.17
3rd Mode	MS	0.000052	154.4531	4297.3
	RMS	0.00721	12.428	65.55
Total /POST26	MS	0.2922	219.95	93402
	RMS	0.54055	14.831	305.62
Total /POST1	MS	0.2919	219.84	93391
	RMS	0.5403	14.827	305.6

It is interesting to analyze the results shown in Figures 3.9–3.14. It can be seen that although all the deflection is contributed by the first mode, with very small contributions of the second and third modes, the accelerations due to the different modes are significant. If an acceleration meter is put on the beam, the PSD of its measurements will look like Figure 3.10. The designer, looking at these "experimental" results, may conclude that there are three frequency ranges with which one has to be concerned, each at the vicinity of the resonance frequency. However, the fact is that the stresses, for which usually design criteria are set, do not exhibit the same results. About 85% of the MS value of these stresses is due to the first mode! The reason for this "pseudo-paradox" is that MS values of displacements are a function of $1/\omega_i^3$. Thus, as the resonance frequency is higher, the MS deflection response to that mode is decreased dramatically—by the cube of the frequency. As stresses are the results of the relative displacements between the different locations on the structure, the stresses too are less influenced by higher resonance frequencies.

These results demonstrate one of the most common errors performed by structural and mechanical design engineers during a design process. These engineers fear, for the wrong reasons, high accelerations, when they should fear high relative displacements (or more correct, curvatures) that contribute to the stresses. These existing stresses are usually the cause of structural failure. The fact that measurements (say in vibration tests) show high accelerations means usually that the structure is noisier, but is not necessarily going to fail.

In the past, the main measured quantities in vibration tests were of accelerations. This happened because acceleration is a convenient quantity to measure—it does not need an external reference, unlike displacement and/or velocities. Today, more and more environmental testing laboratories do a real-time signal processing, thus acceleration is measured directly, but displacement can be displayed on line during the tests. Nevertheless, it should be noted that as structural failure is usually due to excessive stresses in the structure, it is more desirable to measure stresses (or, as it can easier be done—strains) than acceleration. In fact, a reasonable "mixture" of many measurement devices tailored for the specific structure and its loads is the best possibility when a structural dynamic test is prepared.

3.2.1 CLOSED FORM SOLUTION

The solution follows the procedure outlined in ([16], Ch. 5), based on [15]. Suppose a structure is subjected to a cross-spectral density function (the cross-spectral density function is a more generalized form of the PSD function, to allow cross relation between excitation in one location to another location on the structure, and is described in many textbooks on signal processing).

The cross-spectral excitation function is $S_q(x_1, x_2, \omega)$, which means a cross-spectral excitation function of two locations x_1 and x_2 and a function of the angular frequency ω. This function can be evaluated in terms of the structural modes, with coefficients $S_{Q_j Q_k}$:

$$S_q(x_1, x_2, \omega) = \sum_{j=1}^{N} \sum_{k=1}^{N} S_{Q_j Q_k}(\omega) \cdot \phi_j(x_1) \cdot \phi_k(x_2) \tag{3.12}$$

It was shown in the literature ([15], [16]) that

$$S_{Q_j Q_k}(\omega) = \left[\int_{x_1} \int_{x_2} S_q(x_1, x_2, \omega) \cdot \phi_j(x_1) \cdot \phi_k(x_2) \, dx_1 dx_2 \right] \Big/ M_j M_k \tag{3.13}$$

where M_j is the jth generalized mass. $S_{Q_j Q_k}$ is a $N x N$ generalized cross-spectral density matrix for the N modes.

In many practical cases, the cross-spectrum can be separated into one of the following forms:

$$\begin{aligned} S_q(x_1, x_2, \omega) &= S_0(\omega) \cdot f(x_1, x_2) \\ S_q(x_1, x_2, \omega) &= S_0(\omega) \cdot f_1(x_1) \cdot f_2(x_2) \end{aligned} \tag{3.14}$$

and $S_0(\omega)$ can be taken out of the integral of Eq. (3.13).

When the matrix $S_{Q_j Q_k}$ is obtained, the cross-spectral density function of the response can be obtained [15] by

$$S_w(x_1, x_2, \omega) = \sum_{j=1}^{N} \sum_{k=1}^{N} S_{Q_j Q_k}(\omega) \cdot \overline{H}_j^*(\omega) \cdot \overline{H}_k(\omega) \cdot \phi_j(x_1) \cdot \phi_k(x_2) \tag{3.15}$$

where

$$\overline{H}_j(\omega) = \frac{1}{\omega_j^2 \left[1 - \left(\dfrac{\omega}{\omega_j} \right)^2 + i \cdot 2 \left(\dfrac{\omega}{\omega_j} \right) \cdot \zeta_j \right]} \tag{3.16}$$

and $\overline{H}_j^*(\omega)$ is the conjugate of $\overline{H}_j(\omega)$; i.e., the same expression with a minus sign preceding the complex term.

RMS values of the response are then obtained by integrating the cross-spectral density function of the response over the angular frequency range ω.

When the problem was solved using the ANSYS program, it was shown that the PSD of the tip displacement is obtained accurately using only the first mode. Therefore, in this closed form solution, and without any loss of generality, only the first mode is considered. Such approximations are very good engineering practice—the analytical solution is simplified using the consequence of a numerical computation.

The excitation is of a random tip force. Therefore, the cross-spectral density function is taken from the second Eq. (3.14):

$$S_q(x_1, x_2, \omega) = S_0(\omega) \cdot f_1(x_1) \cdot f_2(x_2)$$
$$f_1(x_1) = 1 \quad \text{when } x_1 = L, 0 \text{ elsewhere} \tag{3.17}$$
$$f_2(x_2) = 1 \quad \text{when } x_2 = L, 0 \text{ elsewhere}$$

As the mode shape displacement equal to 1 at the tip of the beam, the generalized PSD at the tip for the first mode is

$$S_{Q_1 Q_1} = \frac{S_0(\omega)}{M_1^2} \tag{3.18}$$

Note that in Eq. (3.18) the PSD is given as a function of ω, the angular frequency, while in Figure 3.5 it is given as a function of f, the circular frequency. It is advisable to do the calculation in the angular frequency range; therefore,

$$S_0(\omega) = \frac{1}{2\pi} S_f(f) = \frac{1}{2\pi} 0.004081632 \frac{\text{kgf}^2}{\text{Hz}} = 0.000649611 \frac{\text{kgf}^2}{\text{rad}/\text{sec}} \tag{3.19}$$

In addition, the generalized mass is given by (see Eq. (2.23))

$$M_1 = \frac{1}{4}\rho b h L = 0.00047754 \frac{\text{kgf} \cdot \text{sec}^2}{\text{cm}} \tag{3.20}$$

The product $\bar{H}_1(\omega) \cdot \bar{H}_1^*(\omega)$ that appears in Eq. (3.15) is

$$\bar{H}_1(\omega) \cdot \bar{H}_1^*(\omega) = \frac{1}{\omega_1^4} \cdot \frac{1}{\left\{ \left[1 - \left(\frac{\omega}{\omega_1}\right)^2 \right]^2 + 4\left(\frac{\omega}{\omega_1}\right)^2 \zeta_1^2 \right\}} \tag{3.21}$$

The maximum value of $S_w(\omega)$ at the tip of the cantilever beam is obtained at the first resonance frequency, when $\frac{\omega}{\omega_1} = 1$. The first resonance at the angular frequency is $\omega_1 = 2\pi \cdot (11.525 \text{ Hz}) = 72.4137 \text{ rad/sec}$, and for a damping coefficient $\zeta_1 = 0.02$ one obtains

$$\max S_w(\omega) = \frac{S_{Q_1 Q_1}}{M_1^2} \left(\max \overline{H}_1 H_1^* \right) = \frac{.000649611}{.00047754^2} \frac{625}{(72.4137)^2} = .06474 \frac{\text{cm}^2}{\text{rad/sec}}$$

(3.22)

To get the maximum PSD of the tip displacement response in circular frequency, this value should be multiplied by 2π, thus

$$\max S_w(f) = 2\pi \cdot 0.06474 = 0.40677 \frac{\text{cm}^2}{\text{Hz}} \tag{3.23}$$

This is the same as the maximum value obtained using the ANSYS code, shown in Figure 3.9. The MS value is obtained using

$$E\left(w^2 (x = L)\right) = \frac{\pi \cdot S_{Q_1 Q_1}}{4 \zeta_1 \omega_1^3} = 0.2946 \text{ cm}^2 \tag{3.24}$$

This value is compared to 0.2922 cm^2 obtained by the numerical computation. The RMS value is obtained by the square root of Eq. (3.24), and is 0.5482 cm, compared to 0.5406 cm obtained using the ANSYS code.

3.3 RESPONSE OF A CANTILEVER BEAM TO CLAMP DISPLACEMENT EXCITATION

Suppose the previous example is a real case of an engineering design. Assume that the results depicted in that example are real results measured on a prototype during a field test. Can we find another excitation, which can be performed in the laboratory using standard equipment to simulate the real measured conditions? Recall that the previous example was of a cantilever beam excited by a force at the tip. Standard commercial vibration test equipment usually excites accelerations (or displacements). Can the results of the previous example be simulated by a displacement excitation? The first thing to think of is to mount the cantilever beam, with a suitable clamp, on top of a shaker and excite the clamp with some given displacement. The following example shows that the problem is not so simple.

The beam is excited by a prescribed random displacement at the clamp, whose PSD is constant between 5 Hz to 250 Hz. The PSD of the clamp displacement is so selected to produce a tip displacement with RMS value equal to the

RMS of the previous example, 0.5403 cm, where random force was applied to the tip (the "experimental results"). This can be done by first applying a PSD that produces a RMS value of 1 cm at the tip, and then adjust it so that the MS value of the tip displacement is 0.2919 cm^2. A short calculation shows that the constant displacement PSD should be 0.000054128 cm^2/Hz (see file **beamrand_2.txt** in the Appendix).

In Figure 3.15, the RMS values of the relative displacements (that is, the displacement of the beam minus the displacement of the clamp) along the beam are shown. Comparing Figure 3.15 to Figure 3.6, it can be clearly seen that the deformation of the beam is quite different!

In Figures 3.16 and 3.17, RMS values of the accelerations and bending stresses along the beam are shown, respectively. Comparison to Figure 3.7 and Figure 3.8 reflects quite different behavior. The acceleration at the clamp is no longer zero (because it was excited by a given random displacement). The large curvature near the clamp is very high, which results in much higher bending stresses in the clamp. Results of the tip displacement, tip acceleration, and bending stress at the clamp presented in Figures 3.18, 3.19, and 3.20, respectively, demonstrate quite a different (from the previous example) spectral content of each quantity. A much more response component of the second and third mode is demonstrated. Performing the integrals of the PSD curves, this can be seen more explicitly in Figures 3.21, 3.22, and 3.23,

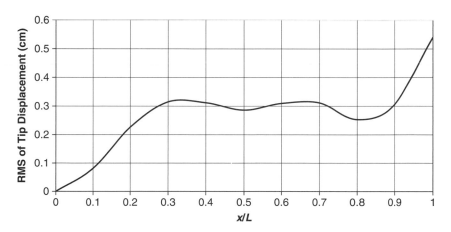

FIGURE 3.15 Mean square values of displacement along the beam (clamp excitation).

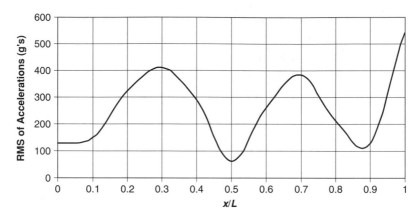

FIGURE 3.16 Mean square values of accelerations along the beam (clamp excitation).

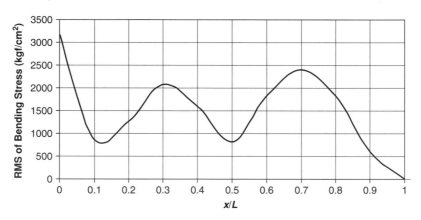

FIGURE 3.17 Mean square values of bending stress along the beam (clamp excitation).

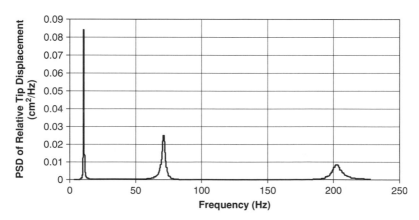

FIGURE 3.18 PSD of beam's tip displacement (clamp excitation).

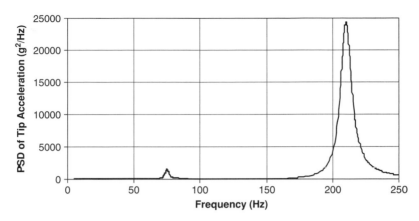

FIGURE 3.19 PSD of beam's tip acceleration (clamp excitation).

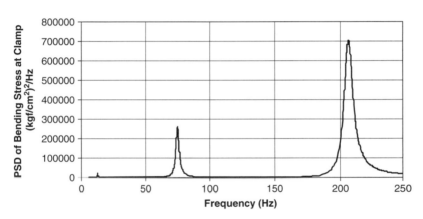

FIGURE 3.20 PSD of bending stress at the clamp (clamp excitation).

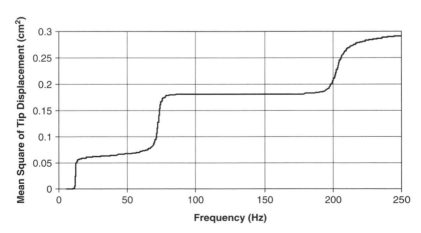

FIGURE 3.21 Mean square of tip displacement (clamp excitation).

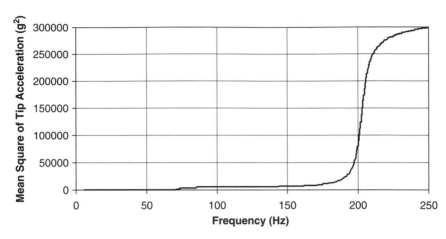

FIGURE 3.22 Mean square of tip acceleration (clamp excitation).

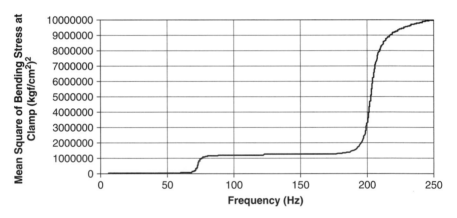

FIGURE 3.23 Mean square of bending stress at the clamp (clamp excitation).

where the integrals of the PSD functions are shown. Thus, a simulation of one loading condition (assumed to be the real experimental result) by a different kind of excitation is quite difficult. The problem is further discussed in Section 3.5.

3.4 RESPONSE OF A CANTILEVER BEAM TO TIP DISPLACEMENT EXCITATION

Suppose the cantilever beam, which is originally excited by a random tip force, is tested by applying a random displacement at the tip. This is equivalent to mounting the beam on its clamp, and attaching a shaker to the tip of the beam, with a control input that is adjusted to create a flat, constant displacement PSD in the frequency range of 5 Hz to 250 Hz. The ANSYS file for this example is **beamrand_3.txt** (see Appendix). As displacement is imposed on the tip of the beam, the modal analysis should be done with a constraint at the tip. Thus, the resonance frequencies are no longer those of the cantilever beam, but of a beam clamped at one end and simply supported on the other end. From this point of view, such a test will not be representative to the "real" behavior of the beam in a field test!

The input PSD of the tip displacement is such that a RMS value of 0.5403 cm is obtained at the tip, so that the RMS displacement is equal to the "real" field test result. To do so, the constant value of the tip displacement PSD should be $0.00119152\,\mathrm{cm}^2/\mathrm{Hz}$.

In Figures 3.24, 3.25, and 3.26, the RMS displacements, accelerations, and bending stresses along the beam are shown, respectively. These should be compared to Figures 3.6, 3.7, and 3.8 for the "real" results.

In Figures 3.27, 3.28, and 3.29, the PSD of the tip displacement, tip acceleration, and the bending stress at the clamp are shown, respectively. The beam has now three resonance frequencies at 50.5 Hz, 163.8 Hz, and 341.9 Hz. Thus, only the two first resonances participate in the response, as the excitation has a zero PSD above 250 Hz.

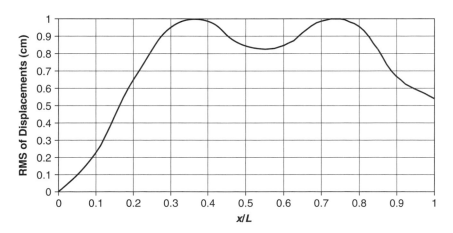

FIGURE 3.24 RMS of displacements along the beam.

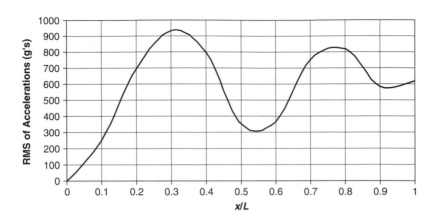

FIGURE 3.25 RMS of accelerations along the beam.

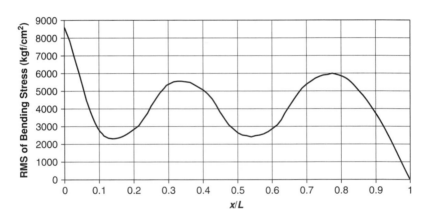

FIGURE 3.26 RMS of the bending stress along the beam.

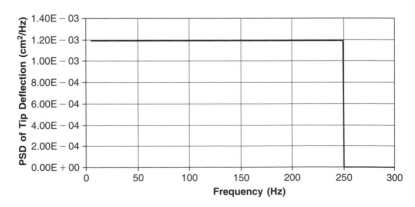

FIGURE 3.27 PSD of tip displacement (equal to the input).

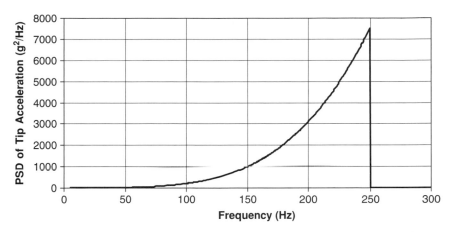

FIGURE 3.28 PSD of tip acceleration (equal PSD of displacement times ω^4).

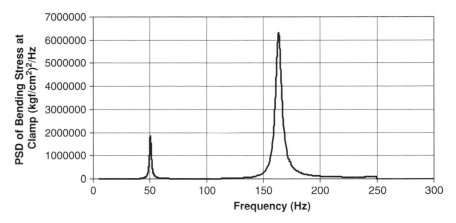

FIGURE 3.29 PSD of bending stress at the clamp.

It can clearly be seen that the computed PSD of the tip displacement is equal exactly to the imposed displacement, as it should be. The acceleration PSD has the form of a 4th order parabola. Acceleration equals displacement multiplied by $\omega^2 = (2\pi f)^2$, and the PSD, which includes a square of the quantity, has a factor of $\omega^4 = (2\pi f)^4$. In the PSD of the bending stress, two resonances can be seen; the third is not responding as the excitation ends at 250 Hz.

It is clearly seen that in this example, our search for an excitation, which is different from the original force excitation, led to a solution in which the magnitudes of the stresses and displacements are "erroneous," and resonance frequencies are not those demonstrated in the virtual "field test!"

3.5 SIMULATION OF AN IMPORTANT STRUCTURAL PARAMETER IN A VIBRATION TEST

In the previous sections, it was demonstrated that it was impossible to completely simulate results of one excitation (random force at the tip), called "the real" conditions, by other types of excitations, which are more easily obtained using standard environmental test laboratory equipment. The demonstrations included an excitation of the clamp by prescribed random displacements and excitation of the tip by a prescribed random displacement spectrum. In the latter, even the frequencies of the response were not the correct ones.

The main purpose of vibration laboratory tests is to discover failures in the design before a final structural design (product) is determined. Therefore, it is important for the designer to be able to analyze the performance of the structure, and to decide "what he is afraid of." This decision (the determination of failure criteria) is in fact one of the most important issues in the design process. If the failure criterion is based on a limit value of stress, it is important to test the structure in such a way that the stresses are simulated in the laboratory test. If the design criterion is based on a limit value of displacements or accelerations, it is important to check these values in the tests. When the design criterion includes several criteria, it is usually easier to design different tests for the different criteria, conditioned on the knowledge of the interference between those criteria. It is easier to design a test in which only part of the important parameters of the structural performance are tested, and then two or more kinds of tests are performed.

The importance of establishing failure criteria, or to determine failure modes of the structure, is again emphasized, not only as a routine procedure during the design process, but also as a must in the design of experimental testing of the designed structure. Designers should bear in mind that a complete laboratory simulation of the real field conditions is seldom possible, and should be ready to design several different tests for a given structure, so that each test may check for a different possible failure mode.

Assume, reasonably, that the failure criterion of the cantilever beam structural design is the stress at the clamped edge, for the "real" case where the tip is excited by a random force. The designer does not want this bending stress to be higher than a given value, say the material yield stress. Then, a test that involves base acceleration excitation (which can be performed using the standard environmental testing laboratory equipment) can be designed in such a way that the magnitude and the PSD of the bending stress at the clamp can be simulated, without any certainty that other parameters are correctly simulated in other locations. Thus, a reasonable test for one of the design failure criterion can be established, while, when more than one criterion exists, other kinds of tests may be required.

Sometimes a design criterion on acceleration is required, not necessarily because of structural aspects but for other reasons. For instance, if sensitive equipment is to be mounted at the tip of a cantilever structure, the acceleration at that tip is the input that this equipment should withstand. Thus, it is important sometime to simulate, in tests, the acceleration PSD at the tip.

When both criteria are required (i.e., simulation of the PSD of the clamp stress *and* simulation of the PSD of the beam's tip acceleration), two different tests may be required, as will be demonstrated in the following sections.

The method in which such tests can be designed is demonstrated for both the cantilever beam and a cantilever plate (which may simulate an aerodynamic stabilizer).

3.5.1 TWO EXAMPLES

The Cantilever Beam

The "Real" Behavior

The cantilever beam demonstrated here is the one for which the "real" case was solved in Section 3.3. The RMS values for the displacement, acceleration, and bending stress along the beam were depicted in Figures 3.6, 3.7, and 3.8, respectively. The PSD of the response to a tip random force were shown in Figures 3.9, 3.10, and 3.11 for the tip displacement, tip acceleration, and clamp stress, respectively. It may be clearer if the results shown in these figures are plotted on a log-log scale, as was done in Figures 3.30 (for the tip displacement), 3.31 (for tip acceleration), and 3.32 (for clamp stress). The representation of test results on a log-log scale is very common in the routine procedures of environmental testing laboratories. Such representation may have more resolution when close resonances exist, and better depict the results when values of response are of a different order of magnitude.

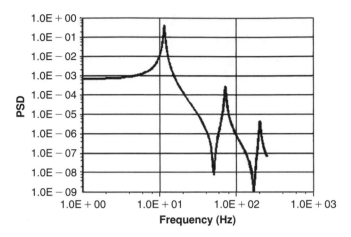

FIGURE 3.30 PSD of beam's tip displacement (identical to Figure 3.9).

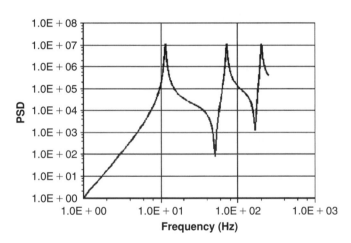

FIGURE 3.31 PSD of beam's tip acceleration (identical to Figure 3.10).

The procedures described in the following section deal with the determination of equivalent tests, and describe how, based on experimental results (such as flight tests), the spectrum of the excitation can be determined).

Determination of the Excitation Force

Based on Eq. (3.2), the following equation can be written:

$$\text{PSD}_{\text{output}} = |\text{TF}|^2_{\text{input-output}} g \text{PSD}_{\text{input}} \tag{3.25}$$

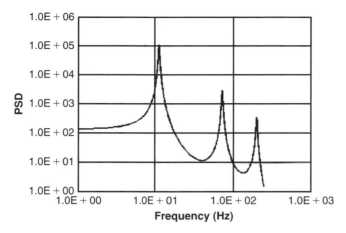

Figure 3.32 PSD of beam's clamp stress (identical to Figure 3.11).

where $|TF|^2$ (the Transfer Function Squared, TFS) has replaced $|H(\omega)|^2$, for convenience. It can be written, for the transfer function between the tip force and the tip displacement

$$PSDy_{tip} = |TF|^2_{F \to y_{tip}} \cdot PSD_F \qquad (3.26)$$

Using the structural model, one can calculate the TFS between the excitation force and the tip deflection by using the harmonic response option of the finite element program. A unit tip force of different frequencies is applied to the model, and the required output response is calculated. In the ANSYS program, the harmonic response module is used. Usually, the transfer function has a real part and an imaginary part (or an amplitude and a phase angle), but the square of the transfer function is a real quantity. In Figure 3.33, the square of the transfer function between a tip force and a tip displacement is shown. In Figure 3.34, the square of the transfer function between a tip force and the stress at the clamped edge is shown. These transfer functions can also be obtained experimentally, using a unit excitation force swept slowly in the frequency domain, measuring and plotting the tip displacement and the clamp stress.

When the transfer functions are known (either from a model or from experiments), the use of Eq. (3.26) and similar may be used to calculate the PSD of the excitation force:

$$PSD_F = PSDy_{tip} / |TF|^2_{F \to y_{tip}}$$
$$PSD_F = PSDS_{bend} / |TF|^2_{F \to S} \qquad (3.27)$$

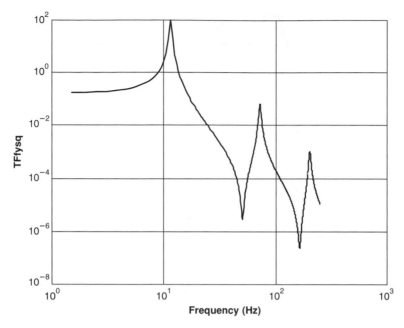

FIGURE 3.33 Squared transfer function between tip force and tip displacement.

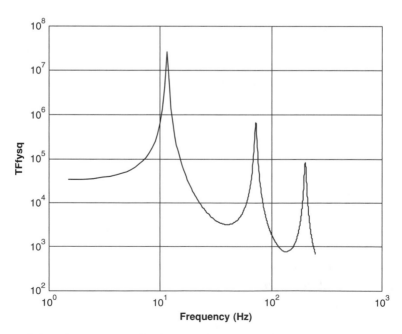

FIGURE 3.34 Squared transfer function between tip force and clamp stress.

The values of $PSDy_{tip}$ and $PSDS_{bend}$ are taken from experiments of the "real" system; for instance, flight tests. Using the results presented in Figures 3.30 and 3.31, or in Figures 3.32 and 3.33, the PSD of the excitation force (Figure 3.5) is obtained.

The square of the transfer function between a tip force and the tip acceleration $|TF|^2_{F \to a_{tip}}$ can also be computed or measured. In this case, the PSD of the excitation force is calculated using

$$PSD_F = PSDa_{tip} / |TF|^2_{F \to a_{tip}} \tag{3.28}$$

where $PSDa_{tip}$ is the PSD of the measured tip acceleration.

In the described case, only one excitation force exists, and therefore only one experimentally measured PSD is required in order to find its PSD function. In cases where more than one excitation force exists, the number of required measured quantities is equal to the number of unknown excitations (say n excitations), and a set of n linear equations must be solved.

The previous case was solved using ANSYS as the finite element program, and transferring the results into an EXCEL spreadsheet, and into a MATLAB® file (to calculate the squared transfer functions). Nevertheless, these calculations can be performed directly using the spreadsheet.

Equivalent Base Excitation

In the case described (which was called the "real" case), the cantilever beam is excited by a tip force. A true test of such a system is to excite the tip of the beam by a random force. This kind of test (force control) is not a standard procedure in the traditional equipment of vibration testing laboratories where the control is done by accelerometers. If one excites the tip of the beam by a random acceleration, the natural frequencies of the system are changed, because they are now those of a beam clamped in one side and simply supported in the excited tip. The only way to retain the true resonance frequencies is to excite the clamp of the beam—base excitation. As this excitation is not identical to the "true" excitation, the stresses at the clamp and the displacement of the tip will not represent the real behavior of the structure. There is a possibility to excite the clamp so that the PSD of the tip acceleration will be the "true" one, and there is a possibility to excite the clamp so that the PSD of the stress at the clamped edge will represent the "true" stresses. It is not possible to do both with the same base excitation. To decide which condition to fulfill—the clamped edge stress or the tip displacement—depends on the failure mode analysis of the structure. If the most dangerous failure mode is the stress at the clamped edge, the test should be designed to imitate the stress PSD. If the dangerous condition is the tip displacement (or the tip acceleration), the test should be designed to imitate that PSD. Sometimes, two dynamic tests should be designed in order to complete an experimental proof of the structure.

The PSD function of the relative displacement between the base and the tip is

$$\text{PSD}y_{\text{rel}} = |TF|^2_{y_0 \to y_{\text{rel}}} \cdot \text{PSD}y_{0,1} \tag{3.29}$$

where $\text{PSD}y_{\text{rel}}$ is the PSD of the relative displacement, and $\text{PSD}y_{0,1}$ is the PSD of the required base excitation. The squared transfer function in Eq. (3.29) can be obtained by either a model (say ANSYS) or an experiment. ANSYS computation is shown in Figure 3.35.

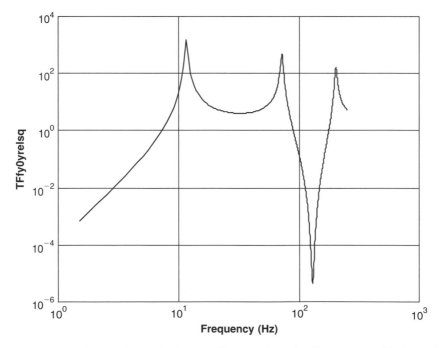

FIGURE 3.35 Squared transfer function between base displacement and (relative) tip displacement.

In Figure 3.36, the squared transfer function between base displacement and the clamped edge stress is shown:

$$PSDS = |TF|^2_{y_0 \to S} \cdot PSDy_{0,2} \qquad (3.30)$$

where PSDS is the PSD function of the stress at the clamped edge, and $PSDy_{0,2}$ is the required PSD of the base excitation.

Using Eq. (3.29), the squared transfer function of Figure 3.35 and the "true" PSD of relative displacement (Figure 3.30), the equivalent PSD of base excitation (already translated to PSD of base acceleration) is calculated and shown in Figure 3.37. Using Eq. (3.30), the squared transfer function of Figure 3.36 and the true PSD of the clamped edge (Figure 3.32), the equivalent base excitation (already translated to PSD of base acceleration) is calculated and shown in Figure 3.38.

The acceleration PSD functions shown in Figures 3.37 and 3.38 can be "digitized" and reintroduced to an ANSYS solution of the cantilever beam

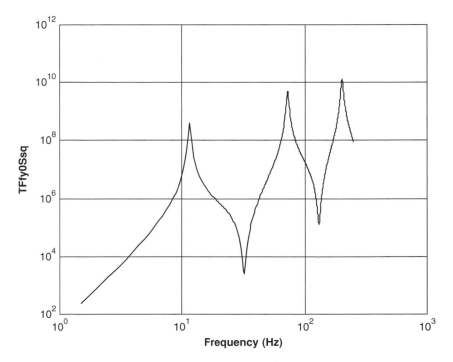

FIGURE 3.36   Squared transfer function between base displacement and clamped edge stress.

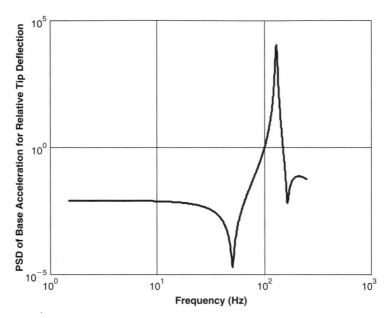

FIGURE 3.37 PSD of equivalent base acceleration to simulate "true" relative tip displacement.

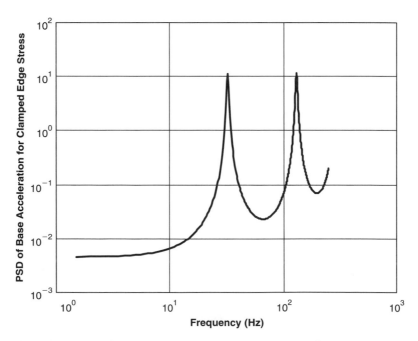

FIGURE 3.38 PSD of equivalent base acceleration to simulate "true" clamped edge stress.

problem, this time with base excitations. It can be shown that if the excitation of Figure 3.37 is used, the PSD of the tip relative displacement is the "true" one (while other quantities, such as stresses, are not reproduced). If the excitation of Figure 3.38 is used, the PSD of the stresses at the clamped edge is reproduced, while other quantities are not reproduced. Thus, the first case can be used to test the structure when tip deflection is important to its functionality, while the second excitation can be used to test the case where the stress at the clamp is the important feature. Note that in the ANSYS program, only 50 values of PSD input are possible; therefore, digitization must be done carefully, to represent the peaks and valleys of the digitized curve.

The problem of performing a test that will simulate the true behavior of a structure as measured in field tests is a major concern to environmental test designers. Until lately, the standard equipment of vibration environmental testing laboratories was controlled by accelerometers, and force excitation, which usually simulates better the true excitations, was impossible. In the last five years, with the improvement of digital control equipment and the miniaturization of load cells, capable of measuring forces, a different approach starts to emerge. In research performed by the JPL laboratory of NASA, a new method of controlling the equipment with force gages was developed. A new testing procedure was suggested [36], based on [37] and on JPL research published in the literature (e.g., [38]). The presented approach carries the test methods into a more adequate experimental representation, although further development, based on the transfer function approach described in the last pages, is still required.

A Cantilever Delta Plate

The "Real" Behavior

In the previous chapter, the process of "designing" an equivalent vibration test was demonstrated for a one-dimensional structure. In this chapter, a two-dimensional (plate) structure is presented.

In Figure 3.39, the cantilever delta steel plate finite element model is shown. This model is included in the file **wing1.txt** (see Appendix). The root chord is 40 cm long, the tip chord is 20 cm long, and the span is 30 cm long. The thickness of the plate is 0.5 cm. The node numbers are also shown in Figure 3.39.

In Figure 3.40, the modal shapes for the first four resonance frequencies are shown. The first mode is a first bending, the second mode is the first torsion mode, the third mode is a second bending, and the fourth mode is a second torsion mode. The resonance frequencies are

$$f_1 = 55.026 \, \text{Hz}$$
$$f_2 = 157.85 \, \text{Hz}$$
$$f_3 = 303.3 \, \text{Hz}$$
$$f_4 = 429.8 \, \text{Hz}$$

(3.31)

In the "real" case ("flight"), two random forces are acting on this structure. One force is acting on the trailing edge of the tip chord (node 89). It has a uniform PSD function between 20 Hz to 500 Hz, and its RMS value is 2 kgf. At the center of the tip chord (at node 49) is an additional random force with the same frequency content and an RMS value of 1 kgf. The "real results" are the response of the plate to these random forces, and as such can be treated as "flight test results." For simplicity and a clear understanding, force excitations

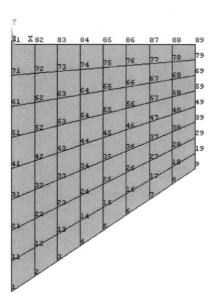

FIGURE 3.39 The structure, the elements, and the nodes.

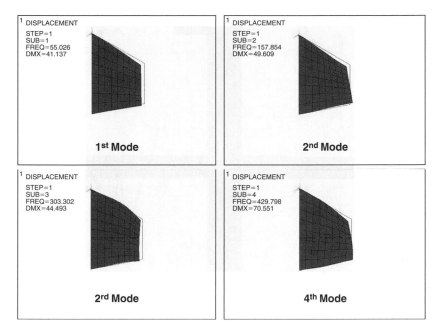

FIGURE 3.40   Four modal shapes of the delta plate.

are used. When, for instance, pressures rather than forces excite the plate, the same procedure can be used.

Two important quantities are considered in this example:

1. The real displacement of node 89 (trailing edge, tip chord).

2. The maximal equivalent stress (Von-Mises stress) in the plate. It is possible to show that this stress has a maximum in node 71 (one node before the last one in the root chord).

In Figure 3.41, the PSD function of the "real" displacement of node 89 is shown. In Figure 3.42, the PSD function of the equivalent stress at node 71 is depicted.

Four transfer functions (squared) are of interest, and calculated using ANSYS (but also can be obtained experimentally):

1. Squared transfer function between excitation force at node 89 and the displacement at that node, shown in Figure 3.43.

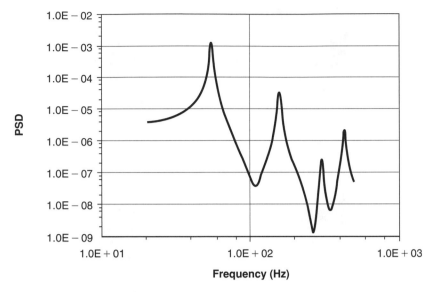

FIGURE 3.41   PSD function of displacement at node 89.

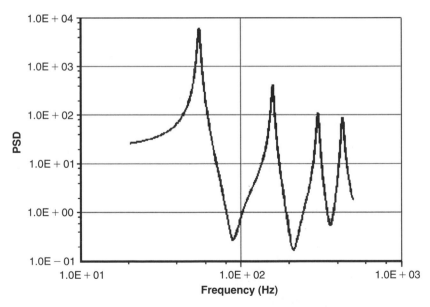

FIGURE 3.42   PSD function of equivalent stress at node 71.

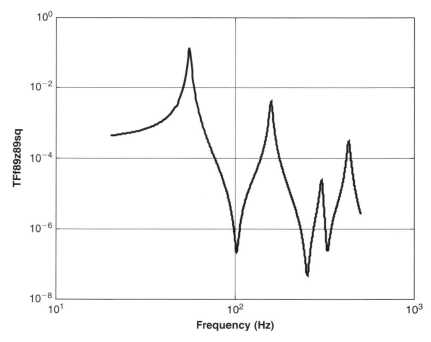

FIGURE 3.43   Squared transfer function, force at node 89, displacement at 89.

2. Squared transfer function between excitation force at node 89 and the equivalent stress at node 71, shown in Figure 3.44.

3. Squared transfer function between excitation force at node 49 and the displacement at node 89, shown in Figure 3.45.

4. Squared transfer function between excitation force at node 89 and the equivalent stress at node 71, shown in Figure 3.46.

The following equations can be written for the PSD functions of the displacement in node 89 and the stress at node 71:

$$\begin{aligned}
\text{PSD}_{Z89} &= |TF|^2_{f89 \to z89} \cdot \text{PSD}_{f89} + |TF|^2_{f49 \to z89} \cdot \text{PSD}_{f49} \\
\text{PSD}_{\sigma71} &= |TF|^2_{f89 \to \sigma71} \cdot \text{PSD}_{f89} + |TF|^2_{f49 \to \sigma71} \cdot \text{PSD}_{f49}
\end{aligned} \tag{3.32}$$

The left-side terms are the PSD functions that can be measured—for instance, in "flight tests"—and were calculated by the ANSYS model (Figures 3.41 and 3.42). They are also computed using the squared transfer

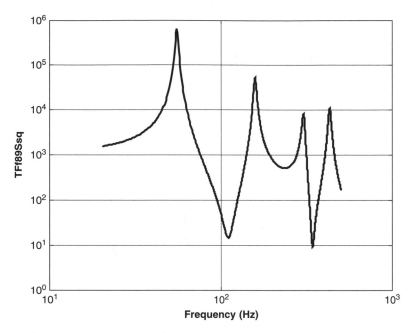

FIGURE 3.44 Squared transfer function, force at node 89, stress at 71.

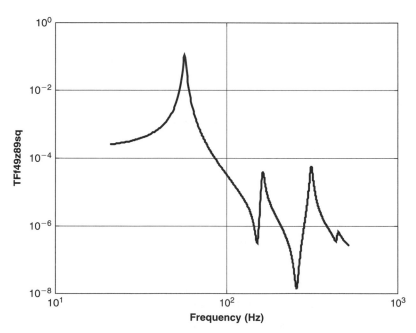

FIGURE 3.45 Squared transfer function, force at node 49, displacement at 89.

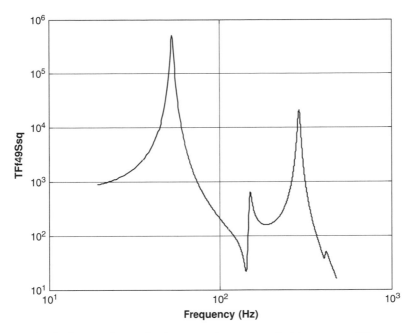

FIGURE 3.46   Squared transfer function, force at node 49, stress at 71.

functions and the PSD of the input forces. Figure 3.47 describes the results obtained using the first of Eq. (3.32), and is identical to Figure 3.41. Figure 3.48 was obtained using the second Eq. (3.32) and is identical to Figure 3.42.

Equivalent Base Excitation

The equivalent excitation is obtained by exciting the root chord with a random displacement excitation (that can easily be transformed to an acceleration base excitation). For this computation, two squared transfer functions are required, and were computed using the ANSYS:

1. Squared transfer function between root chord displacement z_0 and the displacement in node 89, shown in Figure 3.49.

2. Squared transfer function between root chord displacement z_0 and the stress in node 71, shown in Figure 3.50.

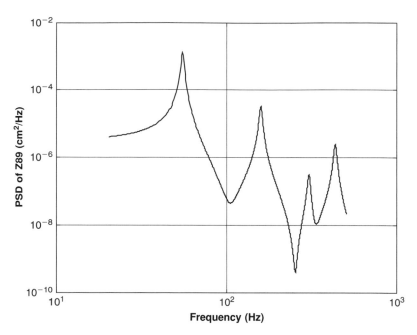

FIGURE 3.47 PSD function of displacement, node 89 (identical to Figure 3.41).

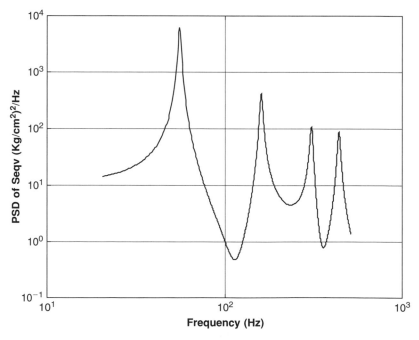

FIGURE 3.48 PSD function of stress, node 71 (identical to Figure 3.42).

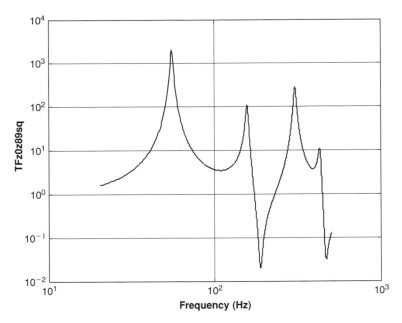

FIGURE 3.49 Squared transfer function between base excitation and displacement in node 89.

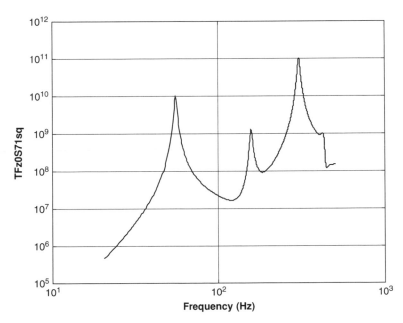

FIGURE 3.50 Squared transfer function between base excitation and stress in node 71.

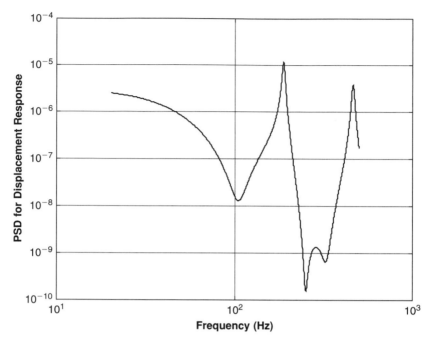

FIGURE 3.51 PSD of base excitation required to simulate displacement at node 89.

The following equations can be used for the computation of the PSD function of the equivalent excitations:

$$
\begin{aligned}
\text{PSD}_{Z89} &= |TF|^2_{z0 \to z89} \cdot \text{PSD}_{z0z} \\
\text{PSD}_{\sigma 71} &= |TF|^2_{z0 \to \sigma 71} \cdot \text{PSD}_{z0s}
\end{aligned}
\tag{3.33}
$$

where PSD_{z0z} is the PSD function of base displacement excitation that simulates displacement at node 89 (shown in Figure 3.51), and $\text{PSD}_{z0\sigma}$ is the PSD function of base displacement that simulates stress at node 71 (shown in Figure 3.52). If required, these functions can be modified to include PSD functions of acceleration.

The results of Figures 3.51 and 3.52 can be digitized and used to solve (using ANSYS) the response in two cases, the displacement at node 89 and the stress at node 71. Results are shown in Figures 3.53 and 3.54.

Comparing Figures 3.53 and 3.41, and Figures 3.54 and 3.42, it can be seen that the peak levels are identical, and the frequencies at which these peaks exist. There are some minor differences in the shape of the curves. These

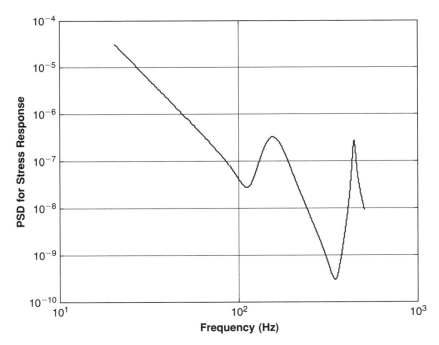

FIGURE 3.52 PSD of base excitation required to simulate stress at node 71.

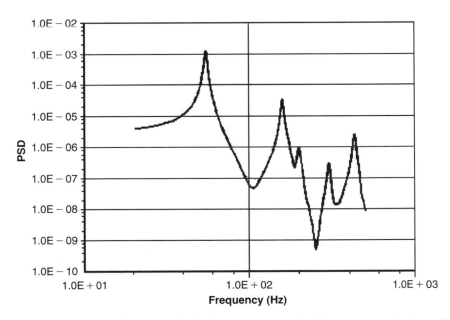

FIGURE 3.53 PSD function of displacement at node 89, using excitation of Figure 3.51 (compare to Figure 3.41).

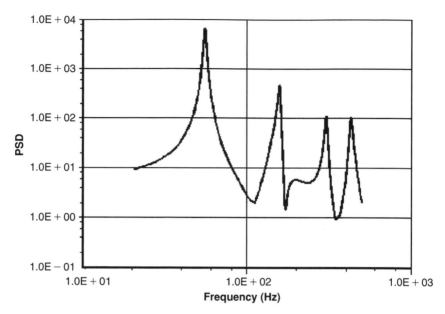

Figure 3.54 PSD function of displacement at node 89, using excitation of Figure 3.52 (compare to Figure 3.42).

changes are the result of the ANSYS computation method, which collocate frequency points around the resonance frequencies.

3.6 RESPONSE OF A STRUCTURE TO ACOUSTIC EXCITATION

Computation of the response of a structure to acoustic (random pressure) excitation using a finite element code like the ANSYS has some unique features that differ from the routine computation of a response to random forces or random displacements. Therefore, an example of such a response is presented here.

Acoustic excitation is a random load generated by pressure fluctuations acting on the structure. For flight vehicles, a major kind of excitation is the turbulent flow that exists around the structure—the flow causes pressure fluctuations on different surfaces of the structure. These pressure fluctuations are random in nature and cause a vibration response of the surfaces of flight vehicles. There are models that simulate such pressure fluctuations. One model is described in [16], based on [39] and [40]. Similar response to wind loads exists in civil engineering structures, and response to water flow in marine structures.

To solve the problem of the response of a structure to pressure fluctuations (either analytically or numerically), the power spectral density (PSD) of the excitation is required. In many Standards and Specifications, acoustic excitation (pressure fluctuation) is described not by direct PSD curves, but by acoustic decibels (dBs) (e.g., MIL-STD 810, procedure 515 [31]. These values are usually given in the specifications with values of acoustic dBs in a "1/3 octave bandwidth" and thus it is important to convert these values to a regular PSD function. A procedure to do such conversion is shown in this paragraph.

Some basic concepts are detailed here. An octave is a bandwidth limited by two frequencies so that the upper frequency is twice the lower frequency. Third octave (1/3 octave) bandwidths are obtained by dividing the octave into three "sub-bandwidths," which are all equal in a frequency logarithmic scale. Thus, an octave between 20 Hz and 40 Hz is divided into three log-equal bandwidths: 20–25.2 Hz, 25.2–31.7 Hz, and 31.7–40 Hz.

Acoustic sound pressure levels (SPL) is defined by

$$\text{SPL} = \text{Acoustic Decibel (dB)} = 20 \cdot \log \left(\frac{P_{\text{rms}}}{P_{\text{ref}}} \right) \tag{3.34}$$

P_{rms} is the root mean square value of the pressure, and P_{ref} is a reference pressure defined as

$$
\begin{aligned}
P_{\text{ref}} &= 2 \cdot 10^{-4} \, \text{dyne/cm}^2 = 2 \cdot 10^{-5} \, \text{Pascal} = \\
&= 2.0408 \cdot 10^{-6} \, \text{kgf/m}^2 = 2.0408 \cdot 10^{-10} \, \text{kgf/cm}^2 \\
&= 2.9 \cdot 10^{-9} \, \text{psi}
\end{aligned} \tag{3.35}
$$

For a continuous PSD function, the PSD function is given for bandwidth of 1 Hz wide, and the relation is

$$\text{PSD}(f) = (P_{\text{rms}})^2 / 1 \, \text{Hz} \tag{3.36}$$

To correct for bandwidths that are not 1 Hz, one defines

$$
\begin{aligned}
\text{LPS} &= \text{SPL} - \Delta L \\
\Delta L &= 10 \cdot \log (\Delta f)
\end{aligned} \tag{3.37}
$$

where Δf is the actual bandwidth. ΔL is a "trimming" factor, which takes care of the fact that the bandwidths are not 1 Hz, but Δf. Then, the P_{rms} is obtained from Eq. (3.34) as

$$P_{rms} = P_{ref} \cdot 10^{LPS/20} \tag{3.38}$$

For the reader's convenience, Table 3.2 describes 1/3 octave bandwidth between 20 Hz and 2000 Hz, a range that is of common use in aerospace applications. The center frequency, the individual bandwidth of each division, and the value of the corresponding ΔL are also shown. These are used later in the numerical example.

Using the values of ΔL given in Table 3.2 and Eqs. (3.37) and (3.38), one can find a relation between the PSD value $((P_{rms})^2/1\,Hz$, see Eq. (3.36)) and the sound pressure level spectrum defined in a given specification. In Table 3.3, SPL levels that define a uniform $PSD(f) = 1 \cdot 10^{-6}(kgf/cm^2)^2/Hz$ in the range 20–1000 Hz, together with the local P_{rms} are shown. A constant value

TABLE 3.2 1/3-octave divisions, 20–2032 Hz.

	Bandwidth Range, Hz	Bandwidth, Hz	Center Frequency, Hz	ΔL
1	20–25.2	5.2	22.4	7.16
2	25.2–31.7	6.5	28.3	8.129
3	31.7–40	8.3	35.6	9.191
4	40–50.4	10.4	44.9	10.170
5	50.4–63.5	13.1	56.6	11.173
6	63.5–80	16.5	71.3	12.175
7	80–100.8	20.8	89.8	13.181
8	100.8–127	26.2	113.1	14.183
9	127–160	33	142.5	15.185
10	160–201.6	41.6	179.6	16.191
11	201.6–254	52.4	226.3	17.193
12	254–320	66	285.1	18.195
13	320–403.2	83.2	360	19.201
14	403.2–508	104.8	453	20.204
15	508–640	132	570	21.206
16	640–806	166.3	718	22.209
17	806–1016	209.7	905	23.216
18	1016–1280	264	1140	24.216
19	1280–1613	333	1437	25.224
20	1613–2032	419	1810	26.222

TABLE 3.3 SPL level for a range of 20–1000 Hz.

	SPL, dBs	PSD, $\left(kgf/cm^2\right)^2/Hz$	P_{rms}, $\left(kgf/cm^2\right)$
1	141	$1 \cdot 10^{-6}$	$1 \cdot 10^{-3}$
2	142	$1 \cdot 10^{-6}$	$1 \cdot 10^{-3}$
3	143	$1 \cdot 10^{-6}$	$1 \cdot 10^{-3}$
4	144	$1 \cdot 10^{-6}$	$1 \cdot 10^{-3}$
5	145	$1 \cdot 10^{-6}$	$1 \cdot 10^{-3}$
6	146	$1 \cdot 10^{-6}$	$1 \cdot 10^{-3}$
7	147	$1 \cdot 10^{-6}$	$1 \cdot 10^{-3}$
8	148	$1 \cdot 10^{-6}$	$1 \cdot 10^{-3}$
9	149	$1 \cdot 10^{-6}$	$1 \cdot 10^{-3}$
10	150	$1 \cdot 10^{-6}$	$1 \cdot 10^{-3}$
11	151	$1 \cdot 10^{-6}$	$1 \cdot 10^{-3}$
12	152	$1 \cdot 10^{-6}$	$1 \cdot 10^{-3}$
13	153	$1 \cdot 10^{-6}$	$1 \cdot 10^{-3}$
14	154	$1 \cdot 10^{-6}$	$1 \cdot 10^{-3}$
15	155	$1 \cdot 10^{-6}$	$1 \cdot 10^{-3}$
16	156	$1 \cdot 10^{-6}$	$1 \cdot 10^{-3}$
17	157	$1 \cdot 10^{-6}$	$1 \cdot 10^{-3}$

PSD spectrum was selected arbitrarily, since it is convenient for the numerical example. This case was done without a loss of generality for the general case.

The values of the PSD and the P_{rms} in Table 3.3 are identical only because a uniform pressure spectrum was selected. This is not general, and in other cases, different values may fill the table. The SPL curve as a function of frequency is linear in a logarithmic scale of the frequency, and the PSD curve has a uniform value over the frequency range. Specific nonlinear curves from given specifications can be treated using the same approach. In such a case, the PSD curve is not necessarily linear.

The total root mean square value $P_{RMSTotal}$ of the pressure acting on the structure can be obtained by integrating the PSD(f) function over the given frequency range.

$$P_{RMSTotal} = \sqrt{\int_{f1}^{f2} PSD(f) \cdot df} \qquad (3.39)$$

Suppose the PSD range is 20–1000 Hz. For the uniform values of PSD given in Table 3.3, the P_{RMST} is computed by

$$P_{RMSTotal} = \sqrt{1 \cdot 10^{-6} \cdot (1000 - 20)} = 0.03130 \, ^{kgf}\!\big/_{cm^2} \qquad (3.40)$$

The process of computing the response of a structure to acoustic excitation is now demonstrated. The structure is a simply supported wide beam, shown in Figure 3.55. Dimensions and material properties are shown in the figure. The beam was modeled by plate elements. An acoustic pressure excites one face of the beam, with PSD function shown in Figure 3.56. This excitation is corresponding to the PSD calculated and presented in Table 3.3, where SPL levels are shown in the second column of the table.

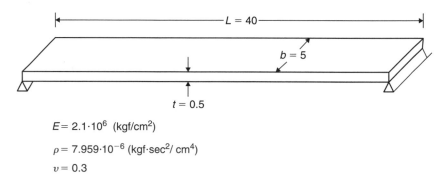

$E = 2.1 \cdot 10^6$ (kgf/cm²)

$\rho = 7.959 \cdot 10^{-6}$ (kgf·sec²/ cm⁴)

$v = 0.3$

FIGURE 3.55 Simply supported steel wide beam.

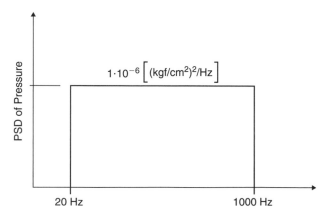

FIGURE 3.56 PSD function of excitation pressure.

The problem was solved using ANSYS. As shown in previous examples, the response of a structure to spectral excitation is performed in ANSYS by running two analyses the modal analysis and the spectral analysis.

The file for the numerical computation, **bbplate.txt**, is listed in the Appendix. Running a spectral response analysis to pressure excitation requires that in the modal analysis, the pressure should be introduced into the solution commands (this is different from response to force excitation). The introduction of pressures into the modal analysis does not change the modal results, but a certain matrix that is required in the spectral analysis for scaling the results is created and saved in the working files of the finite element program. This command is shown in the **bbplate.txt** file.

In order to compare the response of the beam to equivalent force excitation (each pressure on an element can be replaced by four nodal random forces), the **bbplate.txt** file also contains a solution procedure for the response to equivalent force excitation. As forces on the edge simply supported nodes do not contribute to the response, the equivalent forces on the other nodes were corrected by a factor so that the static deflection of the beam to these forces will yield the same mid-beam deflection as the pressure excited beam. This factor is found by short trial and error computations. This is also reflected in the **bbplate.txt** file (see Appendix).

The wide beam model is shown in Figure 3.57. The resonance frequencies of the beam are

$$f_1 = 73.05 \, \text{Hz}$$
$$f_2 = 295.01 \, \text{Hz}$$
$$f_3 = 671.4 \, \text{Hz} \qquad (3.41)$$
$$f_4 = 805.2 \, \text{Hz}$$
$$f_5 = 1086.3 \, \text{Hz}$$

Only the first four resonances are included in the range of the excitation input, 20–1000 Hz.

In Figure 3.58, the RMS values of the beam's lateral displacements are shown. It is qualitatively seen that these displacements contain mainly the first mode response.

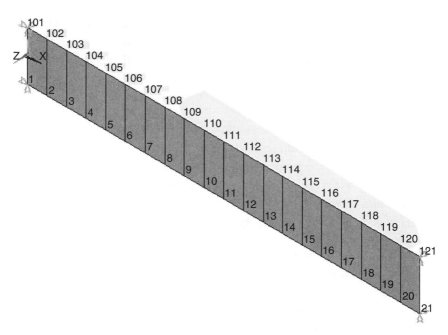

FIGURE 3.57 Finite element model with boundary conditions.

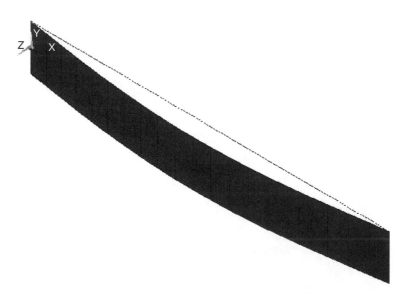

FIGURE 3.58 RMS values of the beam's response. Only the first mode participates.

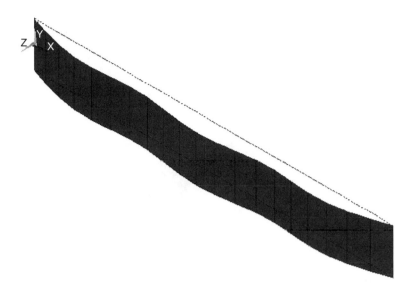

Figure 3.59 RMS values of acceleration response. More modes participate.

In Figure 3.59, the RMS values of the accelerations are depicted. It can be seen that although the displacement response is comprised mainly of the first mode response, there are acceleration responses in other modes, too. Higher modes in the displacement response are attenuated because displacements are inversely proportional to the resonance frequencies squared, as was also demonstrated by the previous numerical examples.

In Figure 3.60, RMS values of bending stresses in the simply supported beam are shown. As stress is proportional to the relative displacements of the structure, the behavior of the stress distribution is similar to the displacement distribution. Thus, the fact that higher modes participate in the acceleration response, this does not reflect on the stress distribution.

In Figure 3.61, the PSD function of the mid-beam displacement is shown. As there is no response in higher modes, only the frequency region around the first resonance is shown. This is also shown in Figure 3.62, where the integral of the mid-beam displacement PSD (MS value of the displacement) is shown.

In Figure 3.63, the integral of the mid-beam bending stress PSD is shown. It can be seen that although the major part of the MS value is contributed by the first mode, there is a small contribution of the third mode, 671.4 Hz, which is a symmetric mode.

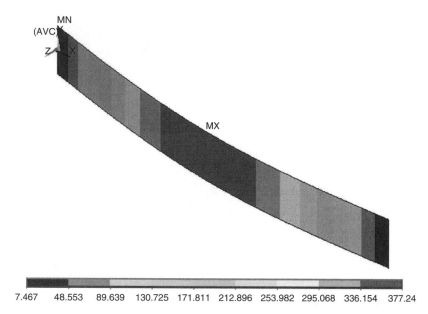

| 7.467 | 48.553 | 89.639 | 130.725 | 171.811 | 212.896 | 253.982 | 295.068 | 336.154 | 377.24 |

FIGURE 3.60 RMS values of bending stress; the first mode governs the response. (Note that when ANSYS analysis is performed, the RMS values appear in color.)

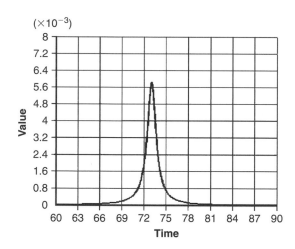

FIGURE 3.61 PSD of mid-beam response; higher frequencies are not shown.

When the equivalent random force excitation, also described in file **bbplate.txt** (see Appendix), was run, identical results were obtained. These results are not shown here. It is suggested that the reader will complete these runs and compare them to the random pressure excitation response.

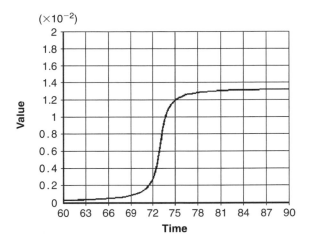

FIGURE 3.62 Integral of mid-beam PSD of deflection.

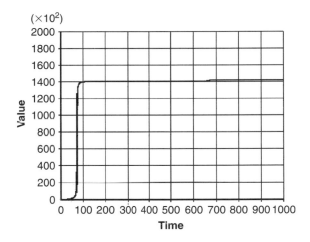

FIGURE 3.63 Integral of PSD function of mid-beam bending stress; some contribution of the 3rd mode can be traced.

3.7 AN EXAMPLE OF A FRAME STRUCTURE

In the preceding paragraphs, examples for beam elements were demonstrated. In Section 3.5, an example of a plate was shown. Here, an example of a frame structure is depicted. This is done in spite of the fact that the described process does not depend on the type of the structure. Once a numerical input file of a structure is prepared—be it a beam, a frame, a plate, a shell, or a structural combination of different structural elements—the treatment is identical to all the structures.

In addition, analytical solution is not included in these examples. The two examples can be solved analytically, using the methods described in Chapter 5 of [16]. The solution process is long and includes calculation of resonance frequencies, modal shapes, generalized masses, generalized forces, and integrals of the transfer functions. This can be avoided if a finite element solution can be obtained very quickly, in a time schedule that is compatible with the needs of a real practical project.

The frame treated here has three members—lengths L_1, L_2, and L_3. Each of these members has a width b_1, b_2, and b_3, and thickness of h_1, h_2, and h_3. Material properties can differ from member to member. A schematic description is shown in Figure 3.64. The left vertical member is clamped at the bottom, while the right member is simply supported at the bottom node. The frame is excited by random force acting horizontally at the tip of the left member. The PSD of this force is described in Figure 3.65. The PSD is uniform, and the RMS of the force was selected to be 1 kgf.

In the demonstrated example, without a loss of generality, all three members are of equal length, width, and thickness, and are made of the same material. In the input file **frame1.txt** (see Appendix), general geometry and materials can easily be included.

The data for the demonstrated frame is

$$
\begin{aligned}
L_1 &= L_2 = L_3 = 60\,\text{cm} \\
b_1 &= b_2 = b_3 = 8\,\text{cm} \\
h_1 &= h_2 = h_3 = 1\,\text{cm} \\
E_1 &= E_2 = E_3 = 2100000\,\text{kgf/cm}^2 \\
\rho_1 &= \rho_2 = \rho_3 = 7.959 \times 10 - 6\,\text{kgf sec}^2/\text{cm}^4 \\
\nu_1 &= \nu_2 = \nu_3 = 0.3
\end{aligned}
\tag{3.42}
$$

This data was introduced to file **frame1.txt** in a parametric way, so changes in these parameters can be introduced easily, changing the data parameters in the input file. This form of input file should be preferred when parametric studies are required.

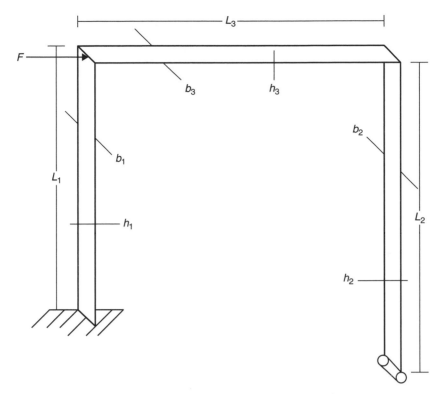

FIGURE 3.64 Geometry of frame.

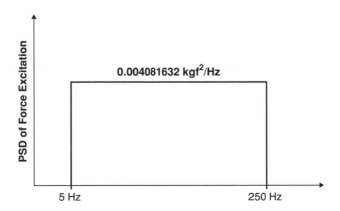

FIGURE 3.65 PSD of excitation force RMS of the force is 1 kgf.

Six resonance frequencies were calculated. Results for the preceding data are

$$f_1 = 15.712 \, \text{Hz}$$
$$f_2 = 72.025 \, \text{Hz}$$
$$f_3 = 109.75 \, \text{Hz}$$
$$f_4 = 142.24 \, \text{Hz} \tag{3.43}$$
$$f_5 = 273.09 \, \text{Hz}$$
$$f_6 = 327.18 \, \text{Hz}$$

Only the four first frequencies are within the bandwidth of the excitation; therefore, there is no response in the fifth and sixth frequencies. These four resonance and mode shapes are shown in Figure 3.66.

For demonstration purposes, different damping coefficients were selected to the four modes:

$$\zeta_1 = 0.03$$
$$\zeta_2 = 0.01$$
$$\zeta_3 = 0.005 \tag{3.44}$$
$$\zeta_4 = 0.01$$

The damping coefficients of modes 2, 3, and 4 were specially selected low values, in order to enhance response in these modes (the lower the damping, the higher the response).

In Figure 3.67, the RMS value of the displacements along the frame, in the horizontal (X) direction, is shown. It can be seen that the displacement response is composed mainly of the first mode.

This is not the case for the RMS values of the X accelerations along the frame structure, as can be seen in Figure 3.68. Again, it is demonstrated that acceleration may comprise higher modes. The acceleration RMS values due to higher modes are the result of the higher frequencies, and not due to higher amplitudes!

When an element table is created, the RMS value of the bending stress in node 1—the clamped end of the left member—is found to be 28.276 kgf/cm^2.

In Figure 3.69, the PSD function of the X displacement of node 11—the tip node of the left member of the frame—is shown. It can be seen that the response is mainly due to the first mode of this member, with no participation of the higher modes.

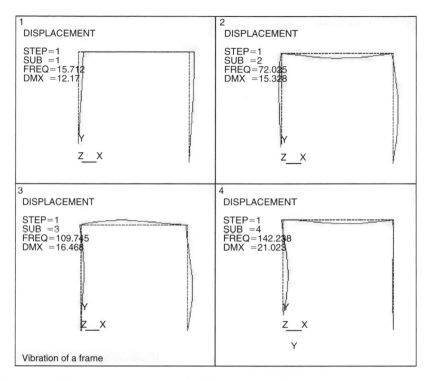

FIGURE 3.66   Four resonances of the demonstrated frame.

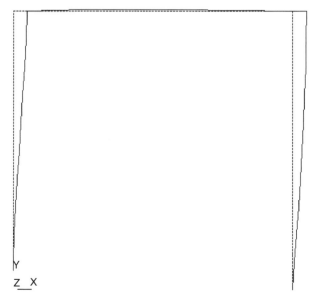

FIGURE 3.67   RMS values of X displacements along the frame.

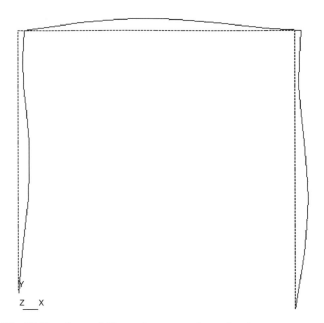

Figure 3.68 RMS values of X accelerations along the frame.

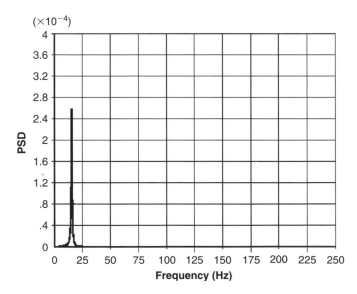

Figure 3.69 PSD of X displacement of node 11.

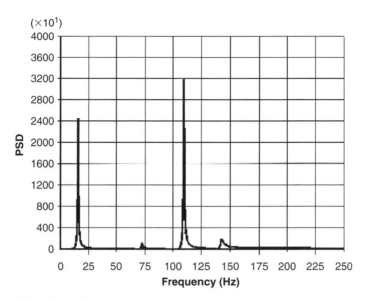

FIGURE 3.70   PSD of X acceleration of node 11.

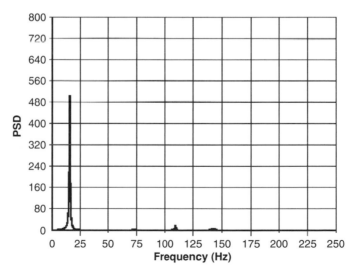

FIGURE 3.71   PSD of bending stress at the clamp (node 1).

In Figure 3.70, the PSD of the X acceleration of node 11 is depicted. The first and the third modes exhibit large PSD values, while modes 2 and 4 are contributing much less response values to the PSD function.

In Figure 3.71, the PSD of the bending stress at the clamped edge is shown. Although some responses of higher modes are seen, the PSD is mainly comprised of the first mode.

Many other results can be plotted while analyzing this case, as well as all other examples. The main conclusion from the present example is that the displacements are influenced mainly by the first mode, and so are the stresses. On the other hand, the acceleration shows response in higher modes too, due to the higher values of frequencies.

3.8 RESPONSE OF A STRUCTURE WITH MOUNTED MASS TO RANDOM EXCITATION

A random force at the tip excites the beam, with the input force PSD given in Figure 3.5. This is also a part of the file **comass1.txt** (see Appendix) that was prepared for ANSYS. Results for RMS values along the beam for displacements, accelerations, and bending stresses are shown in Figures 3.72, 3.73, and 3.74, respectively.

The PSD functions of several quantities are shown in the following figures for Cases (a) and (b). Figure 3.75 describes the PSD of the tip deflection of the beam. Figure 3.76 describes the PSD of the displacement of the mounted mass. Figure 3.77 describes the PSD of the tip acceleration. In Figure 3.78, the PSD of the acceleration of the mounted mass, and in Figure 3.79 the PSD of the bending stress at the clamp are shown. For clarity, only the frequency ranges where significant values can be detected in the graphs are included.

As was already stated in Chapter 2, in many cases of an attached mass, such as the one described by the preceding example, the design parameter on which the designer can influence at a given time is the rigidity of the mounted mass support. Usually, the major design of the main structure is a given fact, and

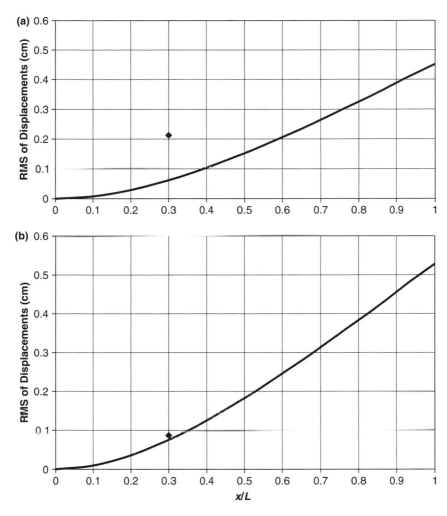

FIGURE 3.72 RMS of displacements along the beam: (a)—Case (a); (b)—Case (b). Diamonds represent the mounted mass.

the mass of the required attached mass (usually an equipment "black box") is also defined. By performing a coupled modal analysis of the main structure plus the attached mass and a response to random excitation, more insight into the problem can be gained. A better selection of the mounting rigidity (and therefore the coupled resonance frequencies) can be recommended, and thus a better design can be performed.

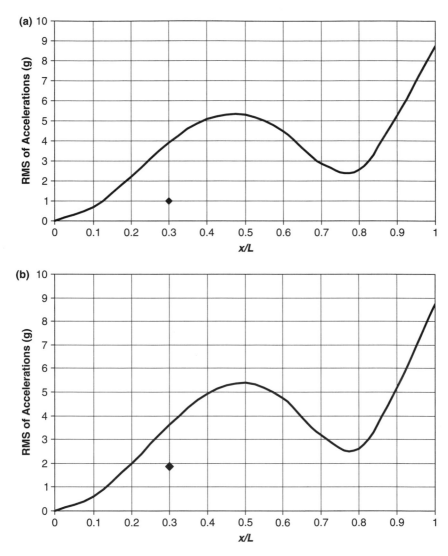

FIGURE 3.73 RMS of accelerations along the beam: (a)—Case (a); (b)—Case (b). Diamonds represent the mounted mass.

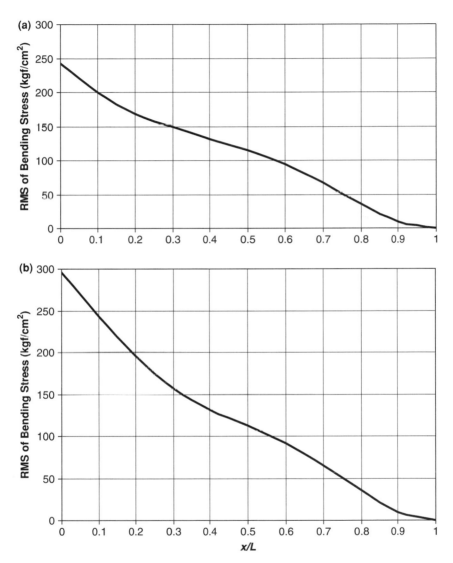

FIGURE 3.74 RMS of bending stresses along the beam: (a)—Case (a); (b)—Case (b).

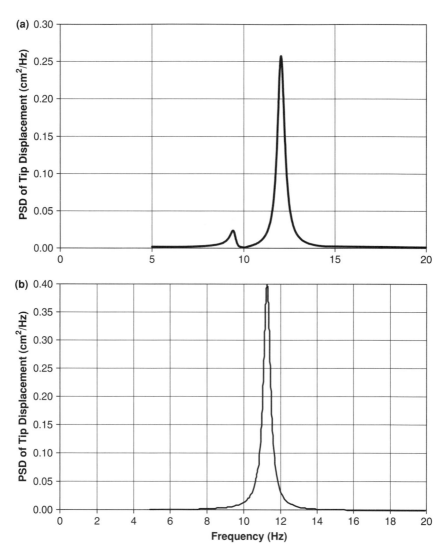

FIGURE 3.75   PSD of beam's tip displacements: (a)—Case (a); (b)—Case (b).

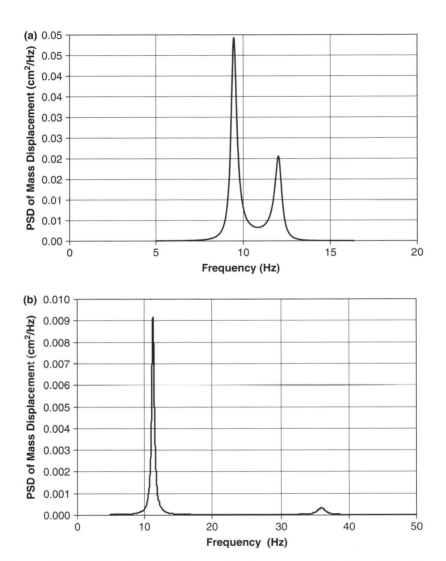

FIGURE 3.76 PSD of mounted mass displacements: (a)—Case (a); (b)—Case (b).

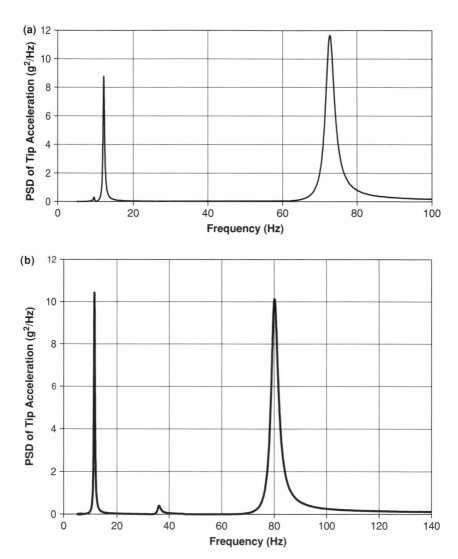

FIGURE 3.77 PSD of beam's tip acceleration: (a)—Case (a); (b)—Case (b).

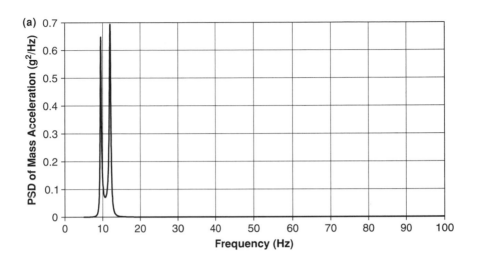

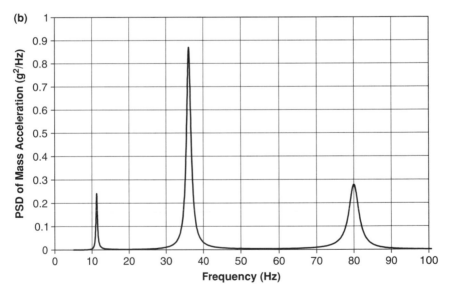

FIGURE 3.78   PSD of mounted mass acceleration: (a)—Case (a); (b)—Case (b).

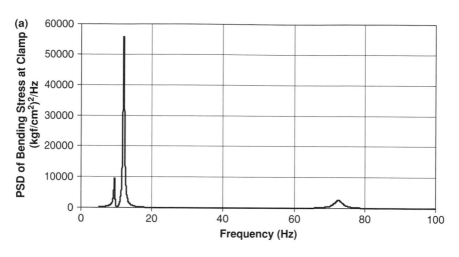

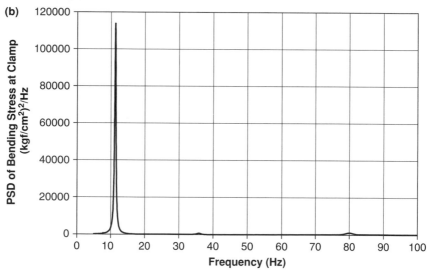

FIGURE 3.79   PSD of bending stress at clamp: (a)—Case (a); (b)—Case (b).

3.9 RESPONSE OF A SIMPLY SUPPORTED PLATE TO RANDOM EXCITATION

The simply supported plate, whose response to harmonic excitations was demonstrated in Chapter 2, Section 2.8, is now treated numerically, using ANSYS, for its response to random excitation. The plate's data is identical to the data described in Chapter 2, Section 2.8.

For this analysis, a file **ssplaterand.txt** (which is listed in the Appendix) was written. The post-processing part of this file is identical to **ssplate.txt**, also listed in the Appendix.

In order to use only the first six modes of the plate, described in Chapter 2, Section 2.8, and without any loss of generality, the simply supported plate is now subjected to random excitation force acting at node 416 (location of this node is described in Table 2.9). The PSD function of the excitation force was selected as a uniform PSD between the frequencies 200 Hz and 1250 Hz. This range ensures that only the first six modes are participating in the response. Any other PSD functions can be selected and replace the input in file **ssplaterand.txt**. The PSD excitation function is shown in Figure 3.80.

When a computation of the response to random excitation is run with a finite elements code, computation of normal modes is required. This is not so for a harmonic response analysis demonstrated in Chapter 2. The modal analysis was introduced into file **ssplate.txt** just for the purpose of computing and demonstrating the normal modes, but is not required for the harmonic response analysis.

When response to random excitation described in Figure 3.80 is run, the regular post-processor of the finite elements code is used to find the RMS

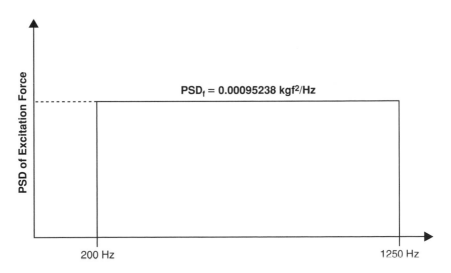

FIGURE 3.80 PSD of force excitation at node 416.

values of displacements, stresses, velocities, accelerations, etc. It was found that the RMS value of the mid-plate displacement is

$$Z_{RMS} \text{ (node 811)} = 0.00062941\text{cm} \tag{3.45}$$

The RMS of the displacement at node 416 (under the acting random force) is

$$Z_{RMS} \text{ (node 416)} = 0.00043744\,\text{cm} \tag{3.46}$$

The values are very small because the excitation force is very small. Computing the RMS value of the excitation force, by integrating the PSD function in Figure 3.80 yields

$$F_{RMS} = 1\,\text{kgf} \tag{3.47}$$

The acceleration of node 811 (mid-plate) is

$$A_{RMS} \text{ (node 811)} = 2582.6\,\text{cm/sec}^2 = 2.6353\,\text{g} \tag{3.48}$$

and the acceleration of node 416 (at the force location) is

$$A_{RMS} \text{ (node 416)} = 6441.8\,\text{cm/sec}^2 = 6.573\,\text{g} \tag{3.49}$$

The PSD functions of different parameters are computed using the time and frequency domain post-processor of the finite elements code. Some of the results are shown in the following pages.

In Figure 3.81, the PSD function of the displacement of the mid-plate (node 811) is shown. PSD function of the displacement of node 416 (at the force location) is shown in Figure 3.82. In Figure 3.83, the PSD function of σ_x at node 416 is shown, and that of σ_y (at the same node) in Figure 3.84. PSD function of the acceleration of node 416 is shown (in cm/sec^2) in Figure 3.85, and in g's in Figure 3.86.

When the integral of a PSD function is performed, the Mean Square (MS) value of that parameter is obtained. Taking the square root of this value, the Root Mean Square (RMS) value is obtained. In Figure 3.87, the integral of the PSD function of the acceleration of node 416 (in g's, Figure 3.86) is performed with the finite elements code. From the figure, the contribution of each of the participating modes to the total MS value can also be concluded.

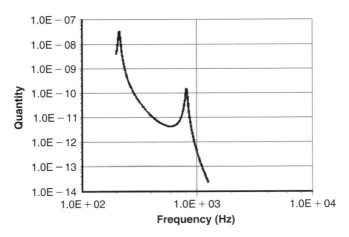

FIGURE 3.81 PSD function of displacement at node 811 (mid-plate).

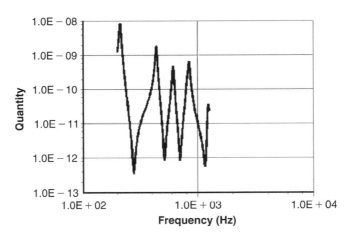

FIGURE 3.82 PSD function of displacement at node 416 (at the force location).

The RMS value obtained by this integration is

$$A_{\text{RMS}} \, (\text{node } 416) = 6.575 \, \text{g} \tag{3.50}$$

which is in very good agreement with the result obtained in Eq. (3.19).

Similar integrations were performed for the RMS value of the displacements in nodes 811 and 416. The following results were obtained:

$$
\begin{aligned}
Z_{\text{RMS}} \, (\text{node } 811) &= 0.0006296 \, \text{cm} \\
Z_{\text{RMS}} \, (\text{node } 416) &= 0.0004376 \, \text{cm}
\end{aligned}
\tag{3.51}
$$

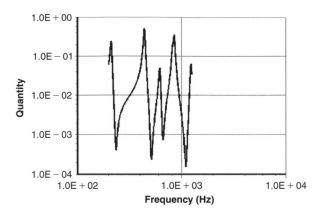

FIGURE 3.83   PSD function of the σ_x stress at node 416.

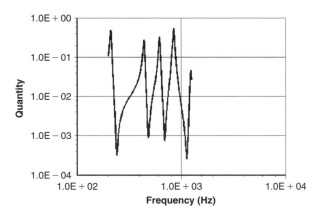

FIGURE 3.84   PSD function of the σ_y stress at node 416.

which are in very good agreement with Eqs. (3.15) and (3.16).

From Figures 3.81 through 3.87, the designer can gain a lot of knowledge about the behavior of the structural element under analysis—the amount of participation of different modes in the dynamics of this element, the contribution of each mode to the total behavior of the structural element, the different behavior of different locations on the structure, and much more. There is practically no limit to the number of structural parameters that can be computed using the capabilities of the spectral analysis module and the time-frequency domain post-processors of the commercially available finite elements codes.

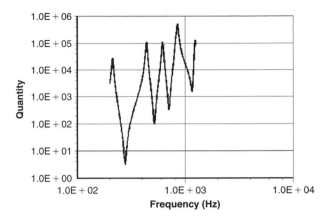

FIGURE 3.85 PSD function of acceleration at node 416 (in cm/sec^2).

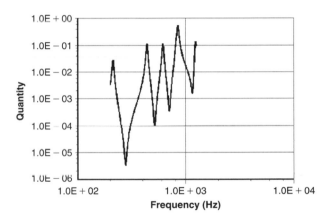

FIGURE 3.86 PSD function of acceleration at node 416 (in g's).

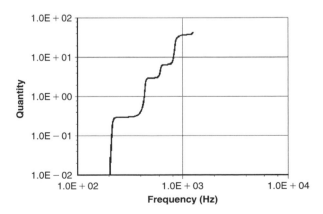

FIGURE 3.87 The integral of the PSD function shown in Figure 3.86.

Chapter 4 / Contacts in Structural Systems

4.1 STATIC CONTACT

In many cases, a structural element makes contact with another part of the structure during its response to external loading. If the loading is slow—static or quasi-static—the contact creates a new situation, in which boundary conditions of the system are changed. If the loading is dynamic, there are new boundary conditions for a short period of time, and there is an impact on the system that may, in some cases, create excessive stresses and other undesirable effects. Examples for contact problems can be found in many engineering applications. The response of a packaged equipment in a box subjected to external vibrations may exhibit contact problems when the equipment is not pre-loaded in its packaging. The fins of a missile carried externally by a combat aircraft may vibrate and hit the fuselage due to the airflow beneath the plane. A leaf spring in a relay switch may respond with vibrations that cause it to contact its envelope, etc.

Contact problems involve many undesirable structural effects. They certainly create a geometric nonlinearity, as the boundary conditions of the problem are changed during the vibrations. They are associated with impacts that give rise to stress wave propagation phenomena. When impacts occur, there is also a rebound, associated with the law of conservation of impulse at the impact time.

Practical solutions to contact problems are usually difficult to be generalized, and in most cases, such problems have to be analyzed with different tools for each problem. Sometimes, a closed-form formulation can be formulated, but usually the solution for these formulations must be performed using numerical algorithms. In many cases, the use of contact elements that exist in the commercially available finite elements computer codes are the best way to obtain a solution to these problems. Use of contact elements in these programs is often tricky, and calls for a lot of experience from the design engineer. In many cases, the designer or the analysts spend more time solving numerical convergence problems rather than concentrating on the physical, structural behavior.

In this chapter, an attempt is made to familiarize the designer with the expected phenomena and the expected difficulties involved. When the solution is done using a finite element code, the ANSYS® is used and the relevant file is quoted in the Appendix.

4.1.1 AN EXAMPLE OF A STATIC CONTACT PROBLEM

At a distance of DEL = 3 cm above the tip of the cantilever beam, in the positive Y direction, there is an anvil with rigidity much higher than the rigidity of the beam. This is achieved by a Young modulus 10,000 larger than the beam's Young modulus. As stresses at the tip are required when the beam touches the anvil, beam elements are not suitable for this purpose, and the beam is modeled by solid (plate) elements. A force of 20 kgf is applied at the tip. In order to avoid tip cross-section distortion, this force is divided equally to five nodes at the tip. The computed configuration is shown in Figure 4.1. The input data text file for the ANSYS is given by **plate2.txt** (see Appendix).

The force is increased from 0 to 20 kgf; thus, the distance between the tip and the anvil decreases. From the static beam formulas, the force required to get a 3 cm deflection can be calculated explicitly [26]. A value of force of 7.3 kgf is obtained. Thus, when the external total force is 7.3 kgf, contact between the beam and the tip is obtained. For the larger values of force (up to 20 kgf), there is no movement of the tip, and local stresses are developed in the tip elements, particularly in the Y direction. The ANSYS is run with the file **plate2.txt** described in the Appendix, and the tip deflection as a function of the external total load is shown in Figure 4.2.

The contact elements in ANSYS are quite tricky. A contact stiffness is to be applied (KN in ANSYS contact 48 element), which influences the convergence

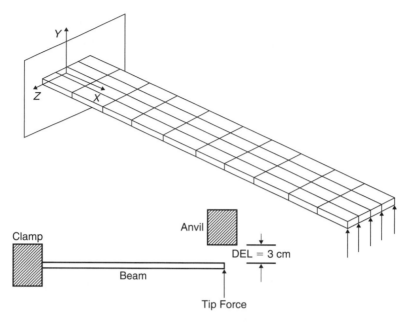

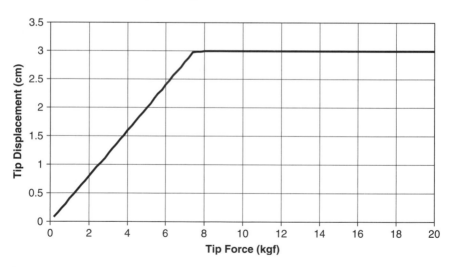

FIGURE 4.1 Static example configuration.

FIGURE 4.2 Tip deflection as a function of the external load.

of the numerical nonlinear analysis. In addition, a tolerance on the contact "depth" is required. These are described in the real constants of the contact elements, as depicted in file **plate2.txt** (see Appendix). Larger contact stiffness is more realistic, but this sometimes causes the solution not to converge.

Sometimes, several attempts are to be run in order to obtain a converged solution. Therefore, the point of initial contact may present some fluctuations, as seen (not so clearly) in Figure 4.2.

It is of interest to plot the stresses in the Y direction at the tip of the beam. In Figure 4.3, Y stresses (in the direction of the thickness of the beam) are shown for two nodes. Node 51 is on the lower side of the tip cross-section, where 1/5 of the total force is applied. Node 151 is on the upper side of the tip cross-section, where the beam contacts the anvil. Because of the total stress field in the beam (described with plate elements, including Poisson ratio effects), there are stresses in the Y direction even when the beam does not reach contact with the anvil. Then, the stresses on the upper face start to decrease toward negative values (compression due to the contact), and the stresses on the lower face start to increase toward tension, due to the total behavior of the tip cross-section. The noncontinuous behavior is due to the load steps and the iterations for convergence of the numerical procedure.

In Figure 4.4, the bending stresses at the lower and upper faces of the clamped cross-section are shown as a function of the tip applied force. This cross-section is far enough from the tip where contact occurs, and only some numerical effect is observed at the value of the external force in which contact begins.

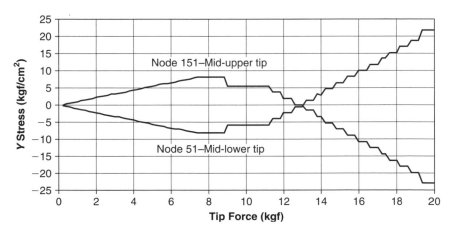

FIGURE 4.3 Stresses in Y direction at the center of the tip cross-section.

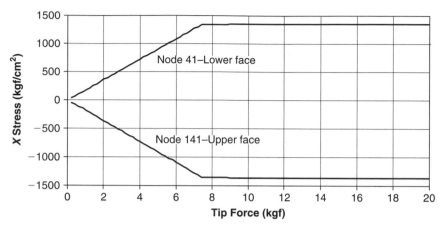

FIGURE 4.4 Bending stresses at the clamped cross-section.

4.2 ANALYTICAL SOLUTION FOR A DYNAMIC CONTACT PROBLEM

Dynamic contact problems appear when a structural subsystem is packed inside another structural system. In many cases, the structural subsystem is pre-pressed by some elastic elements toward the enveloping structure. When the whole system is excited by vibrations, gaps can be opened between the two elements in part of the process, while collisions occur in other instances. These collisions create local dynamic stresses against which the structural elements should be designed for.

The phenomenon is better understood if a simple model is applied and solved. The best simple models can be created, solved, and understood if a SDOF or MDOF system replaces the actual structure. Although the solution of such a "replacement" system may not be rigorous, it provides a much better insight into the physical interpretation of the phenomena that occur. This statement is good for any dynamic problem, and not only to the described dynamic contact issue.

The following example provides insight to the dynamic contact problem. Analytical expressions can be written, and any of the mathematical solver computer programs (i.e., MATLAB®, TK Solver™) can be used to numerically solve these expressions. These solutions usually take only seconds.

In Figure 4.5, the demonstrated system is shown. The mass m is connected through a spring k to an external rigid frame. The spring k is initially compressed by a length δ_0 so that an initial force F_{in} is formed in the spring. The movement of the base is x_b, and the movement of the mass is x_s. The whole system—the rigid frame and the base—is acted by a static acceleration field a_0. The "base"—the rigid frame—is moved by an external excitation so that

$$x_b = \frac{1}{2}a_0 \cdot t^2 + x_{b_0} \sin(\omega_i t) \tag{4.1}$$

ω_i is the excitation frequency, and b_0 is an amplitude selected so that the base has a given value of harmonic acceleration amplitude. The movement of the mass is

$$x_m = x_{m_0} + x_{mv} = \frac{1}{2}a_0 \cdot t^2 + x_{mv} \tag{4.2}$$

x_{m_0} is the movement due to the static acceleration a_0, and x_{mv} is the movement due to the harmonic acceleration. Using this model, it can be concluded that the mass moves relative to the base only when $x_m > x_b$. If $x_m < x_b$, there is no relative movement between the mass and the base. The force applied on the mass by the spring k is $\downarrow F_k = k\delta_0 + k(x_m - x_b)$, considered positive when acting downward. Force equilibrium on the mass requires

$$m\ddot{x}_m + k\delta_0 + k(x_m - x_b) = 0 \tag{4.3}$$

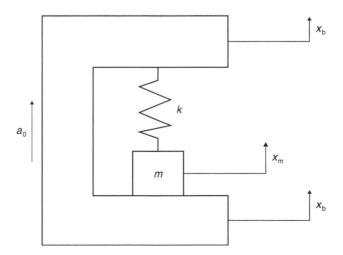

FIGURE 4.5 The model.

Using Eqs. (4.1), (4.2), and (4.3), the following differential equation is obtained for x_{mv}:

$$m \cdot \ddot{x}_{mv} + k \cdot x_{mv} = k \cdot x_{b_0} \cdot \sin(\omega_i t) - k \cdot \delta_0 - m \cdot a_0 \qquad (4.4)$$

This equation has a general solution comprised of a homogenous solution and three particular solutions:

$$x_{mv} = A \sin(\omega_r t) + B \cos(\omega_r t) + \frac{\omega_r^2}{\omega_r^2 - \omega_i^2} x_{b_0} \sin(\omega_i t) + \delta_1 \cos(\omega_r t) - \delta_1 \qquad (4.5)$$

$$\delta_1 = \delta_0 + \frac{a_0}{\omega_r^2}; \qquad \omega_r = \sqrt{\frac{k}{m}}$$

The model does not include damping. The resulting movement comprise many "one cycle" excitations, where the damping effect is negligible. To also allow excitation when $\omega_i = \omega_r$, the DLF (which include the damping effect) is introduced, so that

$$x_{mv} = A \sin(\omega_r t) + B \cos(\omega_r t) + \text{DLF} \cdot x_{b_0} \sin(\omega_i t) + \delta_1 \cos(\omega_r t) - \delta_1 \quad (4.6)$$

where

$$\text{DLF} = + \frac{1}{\left[\left(1 - \frac{\omega_i^2}{\omega_r^2} \right)^2 + 4\zeta^2 \frac{\omega_i^2}{\omega_r^2} \right]^{1/2}} \qquad \text{when } \omega_i \leq \omega_r$$

$$\text{DLF} = - \frac{1}{\left[\left(1 - \frac{\omega_i^2}{\omega_r^2} \right)^2 + 4\zeta^2 \frac{\omega_i^2}{\omega_r^2} \right]^{1/2}} \qquad \text{when } \omega_i > \omega_r \qquad (4.7)$$

Substituting Eqs. (4.6), (4.1), into (4.2), the following solution is obtained:

$$x_m = \frac{1}{2} a_0 t^2 + A \sin(\omega_r t) + B \cos(\omega_r t) + \text{DLF} \cdot x_{b_0} \sin(\omega_i t) + \delta_1 \cos(\omega_r t) - \delta_1 \qquad (4.8)$$

The velocity is

$$\dot{x}_m = a_0 t + A \omega_r \cos(\omega_r t) + B \omega_r \sin(\omega_r t)$$
$$+ \text{DLF2} \cdot x_{b_0} \omega_i \cos(\omega_i t) + \delta_1 \omega_r \sin(\omega_r t) \qquad (4.9)$$

When, in any initial stage, the mass has a displacement x_{m_0} and a velocity $\dot{x}_{m_0}$, the coefficients A and B can be calculated using

$$A = \frac{\begin{vmatrix} x_{m_0} - \text{DLF2} \cdot x_{b_0} \sin(\omega_i t) + \delta_1 - \delta_1 \cos(\omega_r t) - \frac{1}{2}a_0 t^2 & \cos(\omega_r t) \\ \dot{x}_{m_0} - \text{DLF2} \cdot \omega_i \cdot x_{b_0} \cos(\omega_i t) + \delta_1 \omega_r \sin(\omega_r t) - a_0 t & -\omega_r \sin(\omega_r t) \end{vmatrix}}{-\omega_r}$$

$$B = \frac{\begin{vmatrix} \sin(\omega_r t) & x_{m_0} - \text{DLF2} \cdot x_{b_0} \sin(\omega_i t) + \delta_1 - \delta_1 \cos(\omega_r t) - \frac{1}{2}a_0 t^2 \\ \omega_r \cos(\omega_r t) & \dot{x}_{m_0} - \text{DLF2} \cdot \omega_i \cdot x_{b_0} \cos(\omega_i t) + \delta_1 \omega_r \sin(\omega_r t) - a_0 t \end{vmatrix}}{-\omega_r}$$

$$(4.10)$$

where $|\cdot|$ represents a determinant.

The algorithm for solution of the mass movement is as follows:

1. At time $t = 0$, $\dot{x}_{m_0} = x_{b_0}\omega_i$, $x_{m_0} = 0$ and A and B are calculated using Eq. (4.10).

2. For given values of Δt, a new value of x_m is computed. If $x_m < x_b$, then $\dot{x}_m = \dot{x}_b$, $x_m = x_b$, and recalculation of A and B is performed. When $x_m > x_b$, the mass leaves the rigid support and a gap is formed. This gap increases and decreases (with the addition of time steps Δt) until $x_m = x_b$. This is the first collision.

3. Denoting v_{m-}, v_{b-} the velocity of the mass and the base just before this collision, respectively, and v_{m+}, v_{b+} the velocity after collision, the following relation exists (due to conservation of impulse): $(v_{m+} - v_{b+}) = -\varepsilon \cdot (v_{m-} - v_{b-})$, where ε is the coefficient of resilience. From this expression, v_{m+} can be calculated.

4. With the new initial velocity v_{m+} and the known mass displacement, new values of A and B can be calculated, and the computation is repeated until the next collision.

5. The process is repeated until a given time and/or number of collisions.

This algorithm was programmed with the TK Solver mathematical solver file **contact7.tkw** (see the Appendix). The capabilities of the TK Solver program are described in Chapter 8. More details about the program can be found in [41]. The following example describes some of the results. The data used for the example is

$$W = 20\,\text{kgf; Weight of the mass}$$

$$k = 72.5\,\text{kgf/mm; Spring constant}$$

$$\zeta = 0.02;\,\text{Damping coefficient}$$

$$nv = 1\,\text{g; Input acceleration amplitude}$$

$$f_{\text{in}} = 40\,\text{Hz; 60\,Hz; Input frequency} \qquad (4.11)$$

$$\varepsilon = 1; 0.4;\,\text{Coefficient of resilience}$$

$$nf = 0; 0.25; 0.5; 0.75;\,\text{Ratio of initial force}$$

$$E = 21000\,\text{kgf/mm}^2;\,\text{Young's modulus (steel)}$$

$$\gamma = 7.8 \cdot 10^{-6}\,\text{kgf/mm}^3\,\text{(steel)}$$

The mass and the spring were selected so that the resonance frequency is $f_{\text{r}} = 30\,\text{Hz}$. Two values of excitation frequencies, two values of coefficient of resilience, and four values of the initial pressing force were selected. In Figure 4.6, the base and the mass movement for an harmonic input of 1 g, with a frequency of 40 Hz and a coefficient of resilience $= 1$, is shown. In Figure 4.7, the same data is applied, but the input frequency is 60 Hz, which is twice the resonance frequency. In Figure 4.8, the data applied for Figure 4.7, with a coefficient of resilience $= 0.4$, is shown.

It is interesting to note the divergence of the mass movement in Figure 4.7, where the excitation frequency is twice the resonance frequency and the coefficient of resilience is 1 (perfectly elastic collision). This occurs because

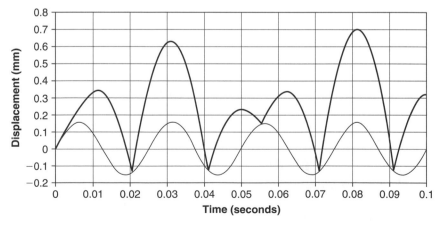

FIGURE 4.6 Displacement of the base (light line) and the mass (heavy line).

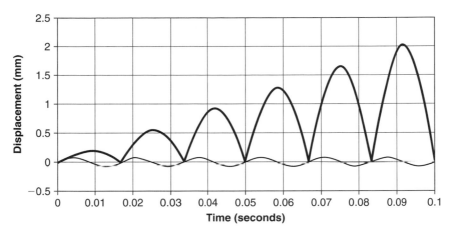

FIGURE 4.7 Displacement of the base (light line) and the mass (heavy line). (Coefficient of Resilience = 1)

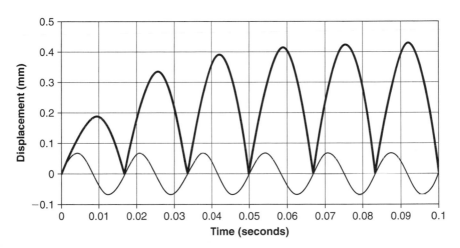

FIGURE 4.8 Displacement of the base (light line) and the mass (heavy line). (Coefficient of Resilience = 0.4)

in each collision, the mass is "falling" on the base while the latter is in the maximum upward velocity, adding momentum to the mass in a perfectly elastic collision. This is the "worst case" for the mass displacement. When the collision is partially plastic (coefficient of resilience smaller than 1, Figure 4.8), there is a stabilization of the amplitude toward a finite one.

It can be shown that for a perfectly elastic collision, with a frequency of excitation twice the resonance frequency, the local peaks of the mass movement lie on a straight line so that

$$x_{mv_{max}}(n) = \frac{8 + 16(n-1)}{3} \cdot x_{b0} \qquad (4.12)$$

When the collision is not perfectly elastic, the maximal amplitude tends to

$$x_{mv_{max}}(n \to \infty) = \frac{8}{3} \cdot \frac{1+\varepsilon}{1-\varepsilon} \cdot x_{b0} \qquad (4.13)$$

which is obtained after a relatively small number of collisions.

When the initial force is equal or larger than the amplitude of the input acceleration, multiplied by the weight of the mass, no gap (and therefore no collision) occurs. When this force is less than this, "partial gaps" (and therefore less intense collisions) occur. In Figure 4.9, the effects of forces 25%, 50%, and 75% of the "no gaps" force are demonstrated.

In a point mass, there are no stresses. In the spring, there are three kinds of stresses: the initial stress due to the initial force (the shortening of the spring in δ_0), the "vibrational" stress due to the changes in the distance of the mass from the base, and collision stresses that create stress waves in the spring. The order of magnitude of these stresses can be estimated using the one-dimensional stress wave relation [42]

$$\sigma = \rho c \Delta V = \rho \sqrt{E/\rho} \cdot \Delta V = \sqrt{\rho E} \cdot \Delta V \qquad (4.14)$$

where $c = \sqrt{E/\rho}$ is the speed of sound in the material, E is the Elastic (Young's) modulus, ρ is the mass density of the material, and ΔV is the relative velocity between the two colliding surfaces, which can be calculated from the

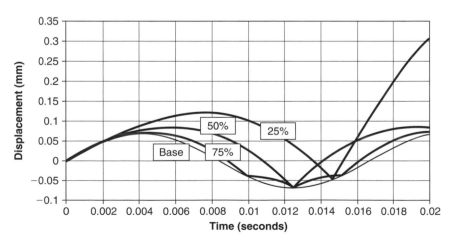

FIGURE 4.9   Effect of initial force on the collisions.

velocities of the mass and the base. It can be shown that when $\varepsilon < 1$ (partially plastic collisions), the maximum value of this one-dimensional stress is:

$$\sigma_{max}(n \to \infty) = \rho \cdot c \cdot \frac{8}{3} \frac{1}{1 - \varepsilon} \cdot x_{b_0} \cdot \omega_i \qquad (4.15)$$

The described example was also solved using ANSYS file with contact elements. The contact elements are nonlinear, and therefore the use of the harmonic excitation solutions in ANSYS cannot be performed. In addition, response to random excitation cannot be performed, as these two modules accept only linear elements. The only way to solve such a problem with the finite element code is to use the transient module, in which many transient load steps replace the harmonic excitation. Such a procedure is easily susceptible to numerical divergence. The results obtained are similar, but not exactly identical. The reason is that the numerical contact elements of the ANSYS have a finite rigidity, and finite contact accuracy must be defined. Thus, small penetration exists. When the contact parameters are changed, many times a numerical divergence occurs. Solutions of contact problems with a finite elements program may therefore be quite tricky and demands much experience, and much patience.

4.3 THE TWO DOF CONTACT PROBLEM

In the previous section, a very simple case (described in Figure 4.5) was demonstrated. The analytical solution performed by building the differential equations for this case and solved numerically seems quite long and tiresome. When more complex structural systems are involved, it is impossible to form the differential equations of the case, and therefore an analytical solution is impossible. In Section 4.4, a more complex case, although relatively simple, is demonstrated. In this section, numerical solution of the system described in Figure 4.5 is demonstrated, using a two DOF equivalent system.

It should be noted that in many practical cases the use of an equivalent MDOF system to practical complex cases should be encouraged. Performing an analysis on a simplified MDOF system can give the designer insight on the general behavior of the analyzed system. Such models are suitable for use in the preliminary design phase, where the influence of relevant parameters can be investigated without performing a complex analysis, and with a possibility of parametric quick checks.

The system described in Figure 4.5 is replaced by the one described in Figure 4.10. Although three masses are described in the figure, there are practically two masses, as the upper half and lower half of the "external" mass is considered as one in the numerical analysis, by forcing the two halves to move completely together. The file describing this case is named **two2.txt**, and is listed in the Appendix.

The spring element between the central mass and the upper half mass is replaced, in the numerical analysis, by a rod of given length and cross-section, in such a way that the spring constant is simulated by the rod. The use of a rod element in ANSYS allows initial preloading easily. A point-to-point contact element is introduced between the lower half of the external mass and the internal mass.

The following values were assumed for the parameters of the problem:

$$W_1 = 10\,\text{kgf}; \quad W_2 = 5\,\text{kgf}; \quad W_3 = 10\,\text{kgf}$$
$$K = 100\,\text{kgf/cm}; \quad E = 2.1 \cdot 10^6\,\text{kgf/cm}^2; \quad L = 10\,\text{cm} \tag{4.16}$$

Values of W_n ($n = 1, 2, 3$) are the weight of the masses, K is the rigidity of the spring, E is the Young's modulus of the spring material, and L is the initial length of the rod-spring. The area of the rod cross-section is calculated within the ANSYS file using

$$A = \frac{K \cdot L}{E} \tag{4.17}$$

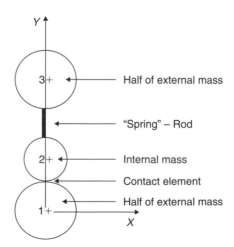

FIGURE 4.10   An equivalent two DOF system to the one described in Figure 4.5.

It can be found that the frequency of the internal mass on the spring is 22.28 Hz. As was shown in the previous section, the maximal response of the mass is obtained when the excitation is double this value. An excitation of the external mass by an imposed harmonic acceleration of 5g's in a frequency of 44.56 Hz was first calculated.

In Figure 4.11, the displacements of both the external and the internal masses are shown. Note the nature of the results, which are identical to the nature of the response shown in Section 4.2. In Figure 4.12, the gap opened between the external and internal masses during excitation is shown. Because the acceleration input was assumed to start with positive values, the displacements start at negative values; therefore, there is no gap in the beginning of the process.

As described in Section 4.2, the initial force required for the internal mass not to move is

$$\text{Minimal } F_{\text{initial}} = n \cdot W_2 \tag{4.18}$$

For the given data, this force is 25 kgf. An initial force of 20 kgf was applied, with the same forcing parameters of the previous case. In Figure 4.13, the displacements of the two masses are shown, and Figure 4.14 depicts the gap opened between the masses. It can be seen that as the initial force is smaller than the minimum required, there is still a relative movement between the

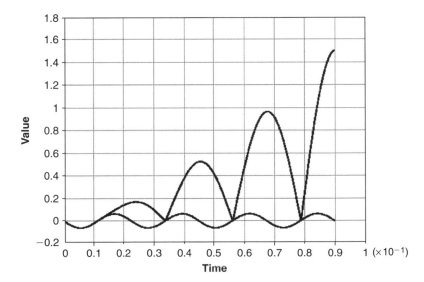

FIGURE 4.11 Displacements of the two masses; input acceleration 5 g's at 44.56 Hz.

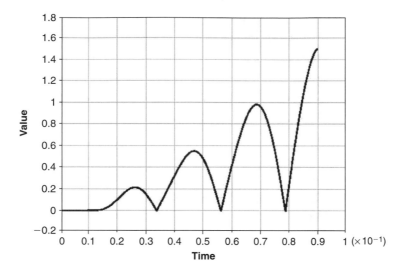

FIGURE 4.12 Gap generated between the masses, same input as in Figure 4.11.

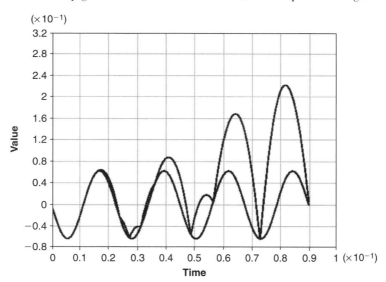

FIGURE 4.13 Displacements of the two masses; input acceleration 5 g's at 44.56 Hz, initial force is 20 kgf.

masses. When an initial force of 25 kgf and larger was applied, there was no relative movement between the masses.

To complete the numerical example, a harmonic acceleration input of 5 g's in a frequency of 100 Hz was applied, without an initial force. Displacements are shown in Figure 4.15, and the gap in Figure 4.16.

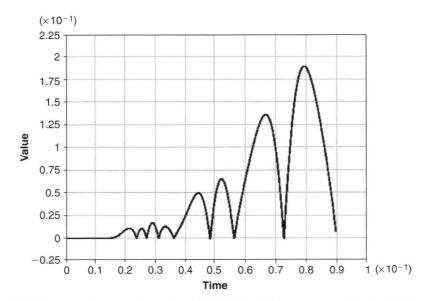

FIGURE 4.14 Gap generated between the masses; same input as in Figure 4.13, initial force is 20 kgf.

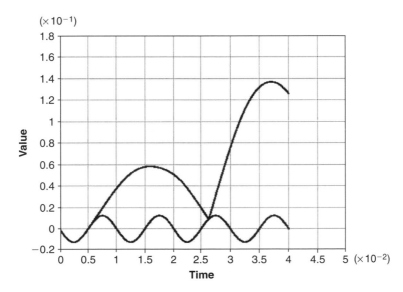

FIGURE 4.15 Displacements of the two masses; input acceleration 5 g's at 100 Hz.

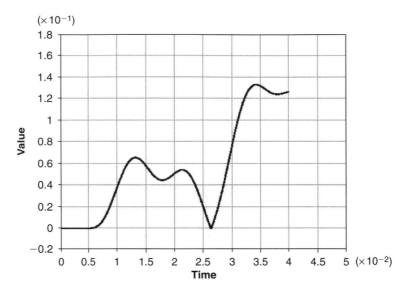

FIGURE 4.16 Gap generated between the masses; same input as in Figure 4.15.

4.4 NUMERICAL SOLUTION OF A DYNAMIC CONTACT PROBLEM—FORCE EXCITATION

As already stated, the contact element in the finite elements codes is nonlinear. Therefore, all the modules that are capable of only linear analysis cannot be used. Thus, the response of a structural model, which has contact elements to harmonic excitation, cannot be solved by using either the harmonic response module or the spectral analysis module. The way to solve such problems is to "translate" the harmonic excitation into a transient one, by solving numerous load steps whose magnitudes follow the harmonic time behavior of the external excitation. This is done easily by a "do-loop" inside the finite element program.

When doing so, much experience is required from the designer in order to select the appropriate time step between solution points. This is added to the experience required to select the (arbitrary) contact rigidity. In addition, the time step should allow the description of the harmonic excitation. The ANSYS manual recommends that the selected time step will be such as to get at least 30 points in a cycle of the excitation, thus

$$\Delta t \leq \frac{1}{30 \cdot f} \tag{4.19}$$

where f is the frequency of the excitation.

Another point to consider when doing a transient analysis with contact elements is the wave propagation phenomena that can be captured with the finite elements solution, as long as the time step is small enough. In order to capture the stress wave propagation, the time step size multiplied by the speed of sound in the material must be much smaller than the selected element size. The speed of sound can be obtained by

$$c = \sqrt{\frac{E}{\rho}} \tag{4.20}$$

for steel, this speed is approximately 5140 m/sec.

Usually, these effects are not required, and the very small time step required to capture those will lead to a very long and time-consuming computation.

In the following example, the response of the cantilever beam with an harmonic tip force excitation of 20 kgf (divided between five tip nodes), with a frequency of the first resonance of the beam ($f = 11.525$ Hz, the period is 0.087 sec), is computed using the ANSYS program. The input file **plate3.txt** is described in the Appendix. In the solution phase of the file, note the introduction of an initial time $t0$, the time step dt, the expression for the applied "transient" force, and the "do-loops" for which N steps are given. These can be changed between computation runs to check for more precise behavior.

In Figure 4.17, the vertical displacement of the beam's tip is shown.

The tip amplitude is increasing, until the beam touches the anvil. At this moment, there are two opposing effects. There is a rebound because the beam hits the anvil with a finite velocity, but the external force is still increasing. Several rebounds can be traced in the figure. When the contact occurs the second time, the rebounds are more distinguished. It should be emphasized that the accuracy of the behavior at times closed to the contact is influenced by the resolution of the solution; i.e., the time step selected.

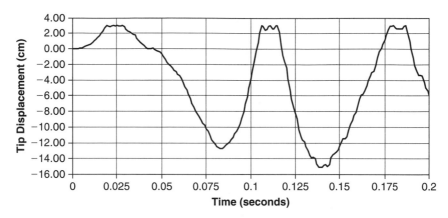

FIGURE 4.17 Tip deflection of the beam (contact side).

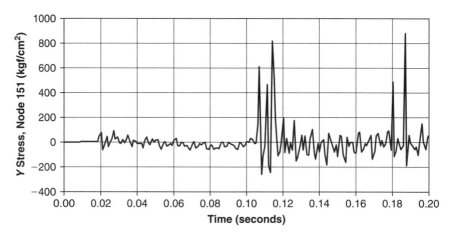

FIGURE 4.18 *Y* stress at the upper center of the tip cross-section (time scale differs from Figure 4.17).

In Figure 4.18, the *Y* stress (in the direction of the thickness) at node 151—the node at the center of the tip—is shown. In Figure 4.19, the *X* stress (bending) at node 141 (upper center of the clamped cross-section) is shown.

A better insight into the behavior of the beam at contact can be obtained when the time step is decreased to $\Delta t = 0.000125$ sec. In Figure 4.20, the tip deflection at the first contact is described. Tip rebounds and the effects of wave propagation can be seen.

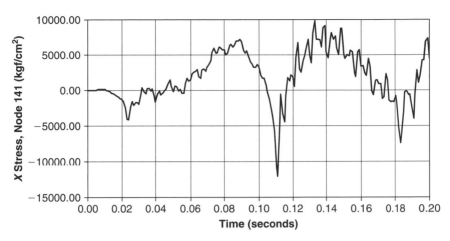

FIGURE 4.19 Bending stress at the upper center of the clamped cross-section.

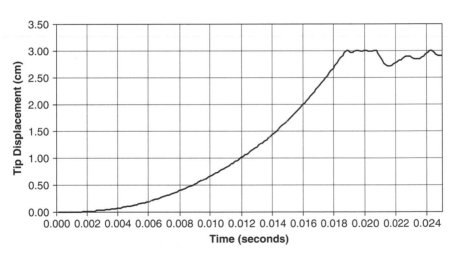

FIGURE 4.20 Tip displacement, node 51.

In Figure 4.21, the Y stress (the direction of the thickness of the beam) is shown at node 151 (the contact side). In Figure 4.22, the bending X stress at the upper side of the clamped cross-section is shown. It is interesting to note the "ripple" in the bending stress. The period of this ripple is the time required for a stress wave to go from the tip to the clamped edge, thus it represents the behavior of the stress waves reaching the clamped edge. Each time there is a contact, or an impact of the tip, there is an effect on the bending stress of the clamped edge.

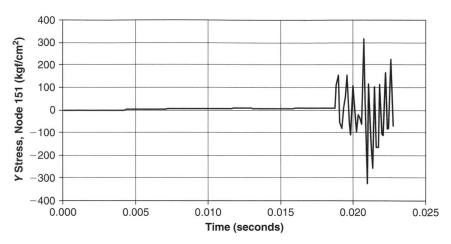

FIGURE 4.21 Y stress, node 151, tip of the beam.

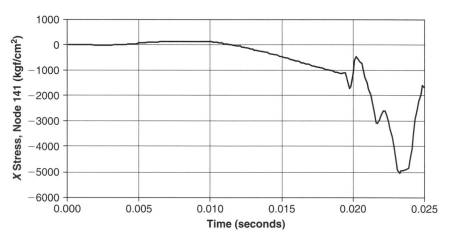

FIGURE 4.22 X stress (bending stress), node 141, clamped cross-section.

Use of a "regular" finite element program is not recommended for tracing the stress wave propagation in structures. There are special-purpose programs to apply when the stress wave's analysis is the main issue of a structural program. These programs are outside the scope of this book. These effects were demonstrated here just to emphasize that the commercially available finite element programs are capable of describing the phenomena, and sometimes it is required to check the dynamic effects of contacts and impact during a design.

4.5 NUMERICAL SOLUTION OF A DYNAMIC CONTACT PROBLEM—BASE EXCITATION

When the imposed excitation on the structure is a displacement, velocity, or acceleration, solution with the finite element program can be done by converting acceleration or velocity excitation to a displacement, and using this displacement as imposed at the relevant nodes. Usually, imposed excitations in realistic structures are given by an acceleration excitation. Such problems are encountered when a substructure is "packed" in another (external) structure, and both are subjected to acceleration input described by an external acceleration.

As contact elements are nonlinear, harmonic and spectral response modules cannot be used in the finite element programs. A harmonic signal can then be described as a "harmonic transient," using the transient analysis module of these codes.

In the example demonstrated in this chapter, the external structure is excited by a harmonic acceleration

$$a = a_0 \sin(\Omega t) \tag{4.21}$$

Ω is the angular frequency of the excitation, a_0 is the amplitude of the acceleration, sometimes given in units of g (the gravitational acceleration), thus

$$a_0 = n_g \, g \tag{4.22}$$

where n_g is the number of g's of the amplitude of the acceleration. When the excitation is harmonic, the displacement is given by

$$x = x_0 \sin(\Omega t) = -\frac{a_0}{\Omega^2} \sin(\Omega t) = -\frac{n_g g}{\Omega^2} \sin(\Omega t) \tag{4.23}$$

When the acceleration excitation is not harmonic, the acceleration signal should be transformed into a displacement signal by double integration, in order to use it in the finite element programs.

The cantilever beam described in the previous section (by solid plate elements) is clamped on one side. Above it, there is an anvil. Suppose that the anvil and the clamp are parts of an external fixture and are connected. For example, there is not a real connection, but in the excitation input, both the anvil and the clamp are excited by the same displacement excitation. Such connection means that the external fixture is very rigid compared to the beam. This is not a required demand, as the external fixture can also be modeled by finite elements of any rigidity. In order to avoid long periods of the time variable, the distance between the anvil and the beam is set to DEL = 1 cm, instead of the value DEL = 3 cm in the previous example.

The frequency of the harmonic excitation is the first resonance frequency of the beam; thus, $\Omega = 11.525$ Hz. The fixture is excited by an acceleration whose amplitude is $-3g$. The minus sign means that the displacement of the fixture starts toward the positive direction of the Y-axis. In Figure 4.23, the modified model is shown.

In the Appendix, the ANSYS file (**plate4.txt**) for the solution is described. The model file is very similar to the file **plate3.txt**, except for a few more parameters that were introduced in order to describe the excitation. In the solution phase, first an initial displacement is introduced in an initial, very small time value, as suggested by the ANSYS manual. Then, a "do-loop" for other times in the "harmonic transient" is run, with a time step also described in the file. The smaller the time step, the more wave propagation phenomena are included in the solution.

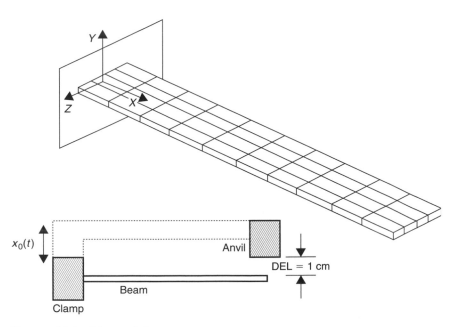

FIGURE 4.23   The model.

In Figure 4.24, the displacement excitation is shown. Of course, it is a harmonic sine.

In Figure 4.25, the absolute displacement of the mid-point of the tip cross-section is shown. The quantity of interest is the relative displacement between the clamp and the beam's tip. This is obtained in the "time history post-processor," by subtracting the clamp displacement from the tip displacement. The result is shown in Figure 4.26.

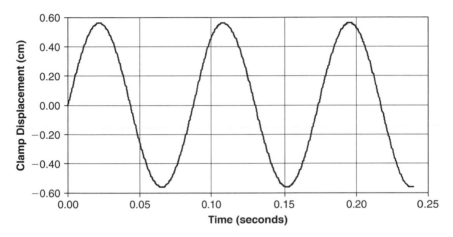

FIGURE 4.24 The displacement of the fixture (the excitation).

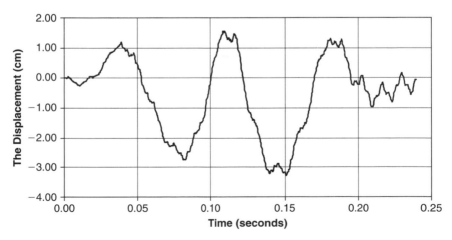

FIGURE 4.25 The displacement of the tip cross-section (node 51).

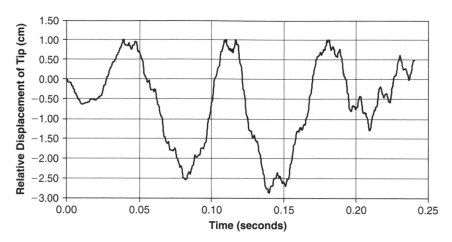

FIGURE **4.26** The relative displacement of the tip.

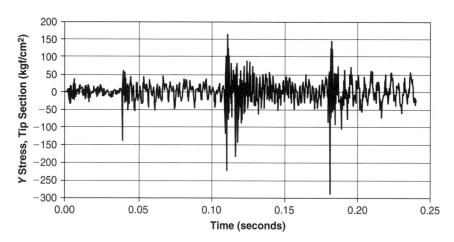

FIGURE **4.27** Y stress at the middle of the tip cross-section.

Contact between the beam and the tip can be seen at times: 0.04 seconds, 0.11 seconds, and 0.18 seconds, when the relative displacement is equal to DEL = 1 cm. Rebounds and wave effects can also be traced, although the time step selected is 0.0003 seconds. Whenever there is a contact (or contacts), there is an impact that influences the stresses which show spikes as can be seen in Figure 4.27 (vertical stress at the middle of the tip cross-section) and with less severity in Figure 4.28 (bending stress at the upper side of the clamped cross-section, this cross-section being far from the points of impact).

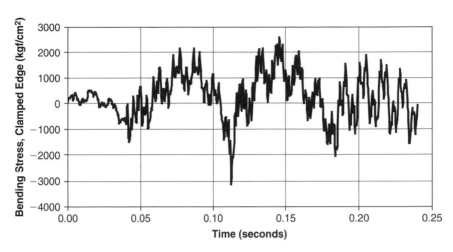

FIGURE 4.28 *X* (bending) stress at the middle upper face of the clamped cross-section.

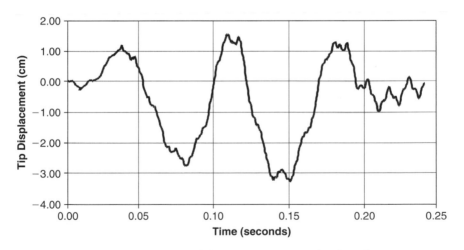

FIGURE 4.29 The displacement of the tip cross-section (node 51).

The example was run for an excitation frequency equal to the first resonance frequency of the beam, but the excitation frequency is a parameter that can be changed in the model file **plate4.txt** shown in the Appendix.

In Figure 4.29, the absolute displacement of the mid-point of the tip cross-section is shown. The quantity of interest is the relative displacement between the clamp and the beam's tip. This is obtained in the "time history

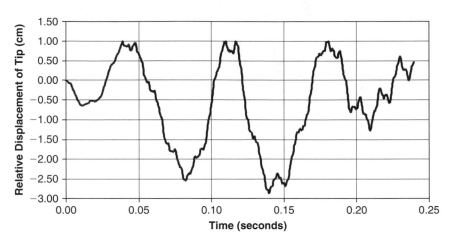

FIGURE **4.30**   The relative displacement of the tip.

post-processor," by subtracting the clamp displacement from the tip displace-ment. The results are shown in Figure 4.30.

Contact phenomena are presented in a practical design. Therefore, they should be analyzed in the design procedure. The use of contact elements in a finite elements code enables the designer to take into account the possi-bility of collisions during the operational life of the designed structure. When blast loads exist, it is recommended not to use finite elements codes, but other programs that are more suitable for the description of such phenomena. These programs are available, but are outside the scope of this publication.

Chapter 5 / Nondeterministic Behavior of Structures

5.1 Probabilistic Analysis of Structures

An engineering design is a process of decision-making under constraints of uncertainty. This uncertainty is the result of the lack of deterministic knowledge of different physical parameters and the uncertainty in the models with which the design is performed. Such uncertainties exist in all engineering disciplines such as electronics, mechanics, aerodynamics, as well as structural design.

The uncertainty approach to the design of systems and subsystems was advanced by the engineering and scientific communities by introducing the concept of reliability. Every system is now supposed to be analyzed for possible failure processes and failure criteria, probability of occurrence, reliability of basic components used, redundancy, possibilities of human errors in the production, and other uncertainties. Using this approach, a total required reliability of a given design is defined (usually it is a part of the project contractual or market demands), with proper reliability appropriations for subsystems. The required reliability certainly influences both the design cost and the product cost.

Nevertheless, in most cases structural analysts are still required to produce a structural design with absolute reliability, and most structural designs are performed using deterministic solutions, using a factor of safety to cover for the uncertainties. The use of a safety factor is a *de facto* recognition of the random characters of many design parameters. Another approach is to use a worst-case analysis in order to determine the design parameters.

During the last two decades, the need for application of probabilistic methods in structural design started to gain recognition and acceptance within the design communities. Structural designers started to use the stochastic approach and the concepts of structural reliability, thus incorporating the structural design into the whole system design. By the beginning of the 21st century, these approaches start to dominate many structural analysis procedures, incorporating the structural design in the total system design. In the last 30 years, theoretical and applied researches were developed to the extent that the probabilistic analysis of structures can use design tools (methods, procedures, computing codes) to apply the probabilistic design to all practical industrial designs. The development of commercially available computer programs, which can be incorporated with the "traditional" tools, like the finite elements computer codes, has contributed vastly to this practical progress. Dozens of textbooks dealing with the probabilistic approach are now available, hundreds of new papers on the subject are published every year, and conferences that include nondeterministic approaches are held. Thus, the field has reached a maturity, which justifies its routine use in the design process.

It is not the purpose of this book to describe the theoretical and practical advances of probabilistic structural analysis. Many basic concepts and procedures can be found in the list of references (e.g., [15, 43–47]). In the present publication, practical aspects of the use of these methods are presented and demonstrated in order to encourage the users to use these methods. Most of the demonstrations are simple, and use one of the commercially available probabilistic analysis computer codes. It is assumed that the reader can find at least a demonstration version of one of these codes. In fact, the ANSYS® program, as well as the NASTRAN®, has, in its latest versions, a probabilistic module, which can solve these problems in one of two basic methods—the Monte Carlo simulation and the response surface methods. Methods for closed form solutions of simple problems (with a small number of random variables) can be found in [16]. Many computer programs are now commercially available; i.e., NESSUS® [48] (developed and maintained by Southwest Research Institute in San Antonio, TX), ProFES® [49] (developed and maintained by Applied Research Associates, in Raleigh, NC), and PROBAN® [50] (developed and maintained by Det Norske Veritas in Oslo, Norway).

The safety factor and worst-case approaches are demonstrated in the following simple example. Suppose that the cantilever beam, with the data given in Table 2.1 of Chapter 2 has three parameters that have tolerance ranges.

TABLE 5.1   Parameters with tolerances (gray rows).

Parameter	Value	Range
Length	60 cm(deterministic)	N/A
Width	8 cm(deterministic)	N/A
Thickness	0.5 ± 0.06 cm	0.44–0.56
Young modulus	2100000 ± 150000 kgf/cm^2	1950000–2250000
Applied force	10 ± 1 kgf	9–11
Max. allowed displacement	7 ± 1 cm	6–8

They are listed in Table 5.1. Two other parameters (L and b) are considered deterministic, without any loss of generality, just in order to simplify the example. The beam has to be designed so that its tip deflection is not larger than a prescribed value, δ_0, and the required factor of safety is (at least) 1.2.

Note that the applied force is usually part of the project requirements, included usually in the technical specifications defined by a customer, or computed by, say, the aerodynamics group. In addition, the maximum allowed displacements depends usually on contractual requirements.

The tip deflection for the cantilever beam is obtained by [26]:

$$\delta_{tip} = \frac{4PL^3}{Ebh^3} \tag{5.1}$$

The nominal value is obtained using the nominal values of the parameters, thus

$$\delta_{tip,nominal} = \frac{4 \cdot 10 \cdot 60^3}{2.1 \cdot 10^6 \cdot 8 \cdot 0.5^3} = 4.114 \text{ cm} \tag{5.2}$$

The ratio between the allowed displacement and the nominal result is the nominal factor of safety of the nominal design:

$$\text{Nominal factor of safety} = \frac{6}{4.114} = 1.46 \tag{5.3}$$

which fulfill the requirement of a factor of safety larger than 1.2.

The worst-case design is obtained by using the maximum values of the parameters in the nominator of Eq. (5.1) and the minimum values for the parameters in the denominator, thus:

$$\delta_{tip,worst\ case} = \frac{4 \cdot 11 \cdot 60^3}{1.95 \cdot 10^6 \cdot 8 \cdot 0.44^3} = 7.152 \text{ cm} \tag{5.4}$$

The factor of safety is now

$$\text{Worst-case factor of safety} = \frac{6}{7.152} = 0.839 \qquad (5.5)$$

This value clearly does not meet the requirement (it even means a sure failure, as it is smaller than 1), and the designer has to change the design parameters. There are many ways to change the design, to list a few:

1. The nominal thickness of the beam can be increased. The result is a heavier and more expensive structure.

2. The nominal Young's modulus can be increased by selecting a better material. The result is usually a more expensive structure.

3. The nominal external force can be decreased. This means a better definition of the loads, and usually involves negotiations with the customer, and more efforts in defining the external loads (say by the aerodynamicists). The cost of the project is increased.

4. Tolerances on all or part of the parameters can be decreased. This means a more expensive production and quality control, resulting in a more expensive product.

5. A combination of part or all the above measures.

The preceding nominal and worst-case designs have demonstrated that two different design approaches resulted in different conclusions. According to the nominal approach, the nominal design is safe and has the required factor of safety, while the worst-case design (which seems more appropriate as it incorporates knowledge on the expected tolerances) results in a design that has to be modified. Nevertheless, the described worst-case design is based on a major pessimistic assumption that all "worst values" occur simultaneously. This is a rare event (and therefore has some probability of occurrence), so it seems that a better procedure is to use a probabilistic approach. We assume that each parameter of Table 5.1 that has tolerances is a random variable, with a given distribution. We also assume (for this simple example, without any loss of generality) that the distribution is normal (Gaussian) and that the tolerance range is $\pm 3\sigma$. Thus, the standard deviation of each parameter is a third of the tolerance described. Table 5.1 is modified to form Table 5.2.

Using this data, the problem was solved using the Lagrange multiplier method, described in [16, Ch. 9], for the probability of failure. This explicit

TABLE 5.2 Random variables, means, and standard deviations.

Parameter	Dist.	Mean	Standard Deviation
Length	N/A	60 cm	N/A
Width	N/A	8 cm	N/A
Thickness	normal	0.5 cm	0.02
Young modulus	normal	2100000 kgf/cm²	50000
Applied force	normal	10 kgf	0.33333
Max. allowed displacement	normal	7 cm	0.33333

method can be used when a small number of random variables (four, in this case) are involved. The result is

$$P_f = \text{Probability of failure} = \Pr(\delta_{\text{tip}} > \delta_0) = 0.000108637 = 0.010864\%$$

(5.6)

Thus, the reliability of this structure (to the defined failure criterion) is

$$\text{Reliability} = 1 - P_f = 99.989136\%$$

(5.7)

These numbers mean that there is an expected failure of one structure (spec-imen) out of approximately 9200 cases. Such a result should be studied carefully. Does it fulfill the requirement? A direct answer is not possible, as the requirement is not expressed in a probabilistic form, but with a determin-istic demand—the traditional factor of safety. Until now, only a small number of official codes and requirements specifications are based on the probabilis-tic approach, and most of these are in the civil engineering community. The adaptation of probabilistic requirements is a process that will be enhanced in the near future, as the benefits gained by such an approach, which ensure for better-optimized designs, are significant. The standardization official system has a very large "inertia" and is very conservative, especially when legal con-siderations are involved. Meanwhile, it is recommended to use a bridging approach, which incorporates the traditional factor of safety definitions with the probabilistic approach, as described in Chapter 7, Section 7.2.

5.1.1 THE BASIC STRESS-STRENGTH CASE

In probabilistic analysis of structures, frequent use is made of the basic "stress-strength" model, which is used extensively in reliability analysis. Suppose there is a component or a system with strength R, on which stresses S are applied. Then, a failure surface can be defined, which divides the (R, S) space into two parts—"safe" and "fail" regions. A failure function is defined as

$$\text{Failure function} \equiv g(R, S) = R - S = 0 \tag{5.8}$$

The region $g(R, S) \leq 0$ is defined as the failure range, and $g(R, S) > 0$ is the safe range. The words "strength" and "stress," although originating from structural analysis, do not necessarily belong to this discipline. The "strength" is in fact the resistance capability of the system, and the "stress" is the existing conditions. For instance, "strength" may be the highest voltage V_{max} an electronic component may take, and the "stress" may be the actual voltage V_{actual} on this component. The failure function is then $g(V_{max}, V_{actual}) = V_{max} - V_{actual}$, and when this quantity is equal or smaller than zero, the component fails. The "strength" term may be the income of a business, and the "stress" the expenses. When (income-expenses) are equal or smaller than zero, the business loses money ("fails").

The failure function in Eq. (5.8) is a straight line in the (R, S) plane. Practically, R and S may both be functions of many other system parameters. In structures, the "stress" is usually a function of the geometry, dimensions, boundary conditions, material properties, and loads, each of which may be random, and some may be correlated. The "strength" can also be a function of structural parameters (as will be demonstrated in some examples). Thus, the failure function is a surface in a multidimensional space, which will be referred to as hyper-surface. When the failure function is linear in all the parameters, this surface is a hyper-plane.

There are many textbooks in which evaluation of the basic concepts of probabilistic analysis of structures is described and demonstrated. The interested reader should consult these books (e.g., [15, 44–47]) to understand the basic concepts. In this book, only very basic analyses are presented and basic examples are demonstrated.

In Figure 5.1, the (R, S) space is shown. The Probability Density Function (PDF) of the R is given by $\phi_R(r)$, and the PDF of S is given by $\phi_S(s)$, both shown on the R and S axes, respectively. The joint PDFs of S and R are shown as ellipses, which represent a "hill" perpendicular to the (R, S) plane. The total volume of this "hill" is 1. The volume of the "hill" over the "fail" space represents the probability of failure of this example.

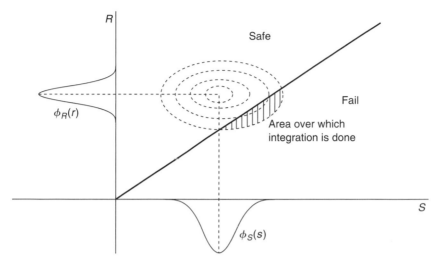

FIGURE 5.1 The $R - S$ space (stress-strength model).

In many published works (e.g., [16, 44]), the advantage of working in a trans-formed u-space was demonstrated. The "physical" variables R and S are transformed into a u-space. When the marginal distributions of R and S are mutually independent, and also normal, the following transformation is made:

$$u_R = \frac{R - \mu_R}{\sigma_R}$$
$$u_S = \frac{S - \mu_s}{\sigma_S}$$

(5.9)

where $\mu_{(.)}$ is the mean and $\sigma_{(.)}$ is the standard deviation of the relevant vari-able. The failure function $g(R, S) = 0$ is then transformed to $G(u_R, u_S) = 0$ in the u-space. In the u-space, the point on $G(u_R, u_S) = 0$ closest to the origin is called "the most probable point of failure," sometimes designated MPP. The distance of the MPP to the origin is called the reliability index β, and the probability of failure is obtained, using linear approximations, by

$$P_f = \Phi(-\beta)$$

(5.10)

where Φ is the standard normal CDF—a variable of zero mean with a standard deviation of 1. When nonlinear approximations are used, expressions similar to Eq. (5.10) exist, which include the curvature of the failure function. In some earlier works, the MPP is called the design point. It is somewhat incongruous to call the point of most probable failure such a misleading name.

When the random variables are not normally distributed, and/or are not mutually independent, there are other expressions for the transformation into the u-space, many of which are described in [16]. All the commercially available computer codes for probabilistic structural analysis contain internal algorithms that do the transformation into the u space and solve the problem in the u space. Then, an inverse transformation is done back to the physical random variables.

In some simple cases, there is a possibility to use closed form formulations for the computation of the reliability index β. One of these methods—the Lagrange multiplier method, described in [16]—was used in the beginning of this chapter, and is described in the next section. Other methods are also described there.

When both R and S are normally distributed and mutually independent, the reliability index can be written directly in the physical variables R and S as

$$\beta = \frac{\mu_R - \mu_S}{\sqrt{\sigma_R^2 + \sigma_S^2}} \tag{5.11}$$

and the probability of failure is then calculated using Eq. (5.10).

5.2 SOLUTIONS FOR THE PROBABILITY OF FAILURE

5.2.1 ANALYTICAL SOLUTION—THE LAGRANGE MULTIPLIER METHOD

When the failure function (of physical variables X_i, $i = 1, 2, \ldots, N$ is given by a closed form expression, it can be easily transformed to the u space by using

$$u_i = \frac{X_i - \mu_i}{\sigma_i}; \quad i = 1, 2, \ldots, N \tag{5.12}$$

where μ_i and σ_i are the mean and the standard deviation of the random variable X_i. Then, the failure function $g(X_1, X_2, \ldots, X_N) = 0$ is transformed to:

$$G(u_1, u_2, \ldots, u_N) = 0 \tag{5.13}$$

The reliability index β is the minimum distance between the origin of the u space and the MPP. This minimum is expressed as

$$\beta = \sqrt{u_1^{*2} + u_2^{*2} + \cdots + u_N^{*2}} \tag{5.14}$$

where u_n^* is the value of u_n at the MPP.

The following expression is constructed:

$$D = \beta^2 - \lambda \cdot G(u_1, u_2, \ldots, u_N) \tag{5.15}$$

where λ is a Lagrange multiplier. Eq. (5.15), plus the following N equations

$$\frac{\partial D}{\partial u_i} = 0; \quad i = 1, 2, \ldots, N \tag{5.16}$$

provide $N+1$ equations for n values of u_n, plus the value of λ. As $G(u_1, u_2, \ldots, u_N)$ is not necessarily a linear function, the $N+1$ algebraic equations obtained are not necessarily linear. A general closed form solution is not possible, but many numerical tools can be used to solve this set of equations, like the TK+ Solver and the MATLAB®. Once the numerical values of u_n^* are obtained, the reliability index is calculated with Eq. (5.14), and the probability of failure is obtained using Eq. (5.10). Then the values of the physical variables X_i^* at the MPP can be obtained using Eq. (5.12).

As an example, the problem described in Table 5.2 is solved using the Lagrange multiplier method. In this problem, the loading force P, the Young modulus E, and the thickness h are considered random variables. The allowed tip deflection δ_0 is also a random variable. The data given in Table 5.2 is adopted.

The failure function is

$$g = \delta_0 - \frac{4PL^3}{Ebh^3} = \delta_0 - \frac{4L^3}{b} \cdot \frac{P}{Eh^3} = \delta_0 - K \cdot \frac{P}{Eh^3} = 0 \tag{5.17}$$

Using Eq. (5.12), one can write

$$\delta_0 = \sigma_{\delta_0} \cdot u_1 + \mu_{\delta_0}$$

$$P = \sigma_P \cdot u_2 + \mu_P$$

$$E = \sigma_E \cdot u_3 + \mu_E$$

$$h = \sigma_h \cdot u_4 + \mu_h \tag{5.18}$$

where the μ's stand for means and σ's stand for standard deviations.

In the u space, the failure function is

$$G = \sigma_{\delta 0} \cdot u_1 + \mu_{\delta 0} - K \frac{(\sigma_P \cdot u_2 + \mu_P)}{(\sigma_E \cdot u_3 + \mu_E) \cdot (\sigma_h \cdot u_4 + \mu_h)^3} = 0 \qquad (5.19)$$

Thus

$$D = u_1^2 + u_2^2 + u_3^2 + u_4^2 - \lambda \left\{ \sigma_{\delta 0} \cdot u_1 + \mu_{\delta 0} - K \frac{(\sigma_P) \cdot u_2 + \mu_P}{(\sigma_E \cdot u_3 + \mu_E) \cdot (\sigma_h \cdot u_4 + \mu_h)^3} \right\}$$

$$(5.20)$$

Careful derivations according to Eq. (5.16) provide four nonlinear equations. These four plus Eq. (5.19) are solved to give

$$\begin{aligned} u_1^* &= -1.2224258 \\ u_2^* &= 0.785329 \\ u_3^* &= -0.58375338 \\ u_4^* &= -3.3501706 \end{aligned} \qquad (5.21)$$

From these values, the reliability index is

$$\beta = 3.69803716 \qquad (5.22)$$

Thus the probability of failure—i.e., the probability that the tip deflection is equal or greater than δ_0 (using Eq. (5.10))—is

$$P_f = 0.000108637 = 0.010864\% \qquad (5.23)$$

The physical combination of the structural parameters at the MPP is

$$\begin{aligned} \delta_0^* &= 6.59253 \, \text{cm} \\ P^* &= 10.2618 \, \text{kgf} \\ E^* &= 2.070812 \cdot 10^6 \, \text{kgf/cm}^2 \\ h^* &= 0.433 \, \text{cm} \end{aligned} \qquad (5.24)$$

This is the combination of physical random variables for which the probability of failure is maximal.

When performing these computations, the derivation phase (Eq. (5.16)) is very cumbersome and sensitive to algebraic errors as well as the solution of the $n + 1$ algebraic equation. This is the reason why this method should be applied carefully only for cases with a small number of random variables. Of course, when a probabilistic computer code is available, it can be done directly on the code.

5.2.2 THE MONTE CARLO SIMULATION

For many years, nondeterministic structures were tested using the Monte Carlo (MC) simulation. According to this method, the failure function is computed many times, each time with a set of the random variables selected randomly from the legitimate variables space. In some of the simulations, a safe state is obtained (failure function > 0) and in some of the simulations a failure is detected (failure function ≤ 0). The number of the failed simulations to the total number of simulations is approaching the true probability of failure, as the number of simulations increases. The MC method gives the exact solution when the number of simulation is infinite. The disadvantage of the MC method is the large number of simulations required. It was shown in the statistics literature that to estimate a given normal probability p within an error of $\pm D$ with confidence level of 95%, the required number of simulations is at least

$$N = \frac{2p(1-p)}{D^2} \tag{5.25}$$

As the required probability of failure of structures is very low, the required number of simulations is very large. For the example solved earlier, $p = 0.0001086$. Suppose an error of 3% is acceptable, $D = 0.0001086 \cdot 0.03 = 3.258 \cdot 10^{-6}$ and $N = 1.884 \cdot 10^7$. Thus, more than 18 million simulations are required. When one simulation run of a complex structure (using finite element program, for example) is in the order of several minutes to hours, the use of the MC method becomes prohibitive for practical industrial applications as a result of the large computation time that interferes with the schedule of any project. Naturally, the number of required simulations decreases 100 times when the expected probability of failure is increased 10 times.

Simulations other than the MC method were also developed. One of these is a directional simulation, in which the simulations are performed not on all the random variables space, but in a smaller region that is in the vicinity of the MPP. This decreases significantly the number of required simulations, but a good estimation of the MPP is required.

In the last two decades, numerical algorithms that use nonsimulative methods were developed (see, e.g., [43–46,51–53]). These resulted in several computer codes that solve the probabilistic structural analysis problem within a very reasonable and practical time frame, and therefore are suitable for industrial use. These programs use First Order Reliability Methods (FORM), Second Order

Reliability Methods (SORM), and other algorithms for the required solution. Such methods are described in the theoretical manuals of such programs.

5.2.3 SOLUTION WITH A PROBABILISTIC ANALYSIS PROGRAM

The problem solved by the Lagrange multiplier method is also solved using the ProFES code, a commercially available structural probabilistic program developed by ARA (Applied Research Associates). The preparation of the input file is done interactively, and so are the required internal functions, which are written using the program's GUI (Graphical User Interface) and compiled by a compiler included in the program. After defining the random variables, the failure function and the limit state functions, the user can select to solve the problem using many analysis options—FORM and Monte Carlo simulation are just two of these options. Performing a FORM solution using the same data used in the previous section, one obtains the following solution for the reliability index, the probability of failure, and the MPP:

$$\beta = 3.69815$$

$$P_f = 0.00010862$$

$$P^* = 10.2618 \, \text{kgf}$$ (5.26)

$$E^* = 2.07083 \cdot 10^6 \, \text{kgf/cm}^2$$

$$h^* = 0.432938 \, \text{cm}$$

$$\delta_0 = 6.59505 \, \text{cm}$$

which compares very well with the Lagrange multiplier results given in Eqs. (5.22) and (5.23).

The failure criterion for which the previous examples were solved is only one of the possible failure modes of the cantilever beam. Another failure criterion may be the existence of a bending stress at the clamped edge that is higher than a given stress, say the yield stress. For this, another failure function must be written. From slender beam theory (see Chapter 2), the stress at the clamped edge can be written [26] as

$$\sigma_b = \frac{6PL}{bh^2}$$ (5.27)

When the yield stress of the beam's material is σ_y, the failure function can be written as

$$g = \sigma_y - \sigma_b = \sigma_y - \frac{6PL}{bh^2} \le 0 \tag{5.28}$$

σ_y may also be a random variable. Suppose the yield stress has a normal distribution, with mean $\sigma_{y,mean} = 3000 \, kgf/cm^2$ and a standard deviation of $\sigma_{y,std} = 270 \, kgf/cm^2$, one can compute the probability of failure of this case by adding this random variable and the relevant failure function (Eq. (5.28)) to the ProFES program. The following solution is obtained:

$$\beta = 3.71155$$
$$P_f = 0.000103028$$

$$
\begin{aligned}
P^* &= 10.2604 \, kgf \\
E^* &= 2.1 \cdot 10^6 \, kgf/cm^2 \\
h^* &= 0.458134 \, cm \\
\delta_0^* &= 7 \, cm \\
\sigma_b^* &= 2199.81 \, kgf/cm^2
\end{aligned}
\tag{5.29}
$$

It can be seen that E and δ_0 are the nominal values of the problem, as these parameters do not influence the stresses at the clamped edge.

Listing and computing all the *possible failure modes* of the structure is the most important thing in the design process, even when a totally deterministic analysis is performed. The experience and skills of the designers can be "measured" by the way this "list of possible failures" is prepared. The more complete the list, the better and safer the design. In a separate section, structures with more than one failure modes are treated.

5.2.4 SOLUTIONS FOR CASES WHERE NO CLOSED-FORM EXPRESSIONS EXIST

In most practical cases treated in realistic design process, no closed-form expression is available for the failure surface. In many cases, this failure surface is known only by an algorithm; e.g., a finite element code.

It was suggested (e.g., [54]) that an approximated closed-form expression can be constructed by a finite number of deterministic solutions in the desired random variables space, and by best fitting a Taylor series based expression. This approximated closed-form expression can then be treated probabilistically with the methods used for this purpose. Demonstrations of Taylor series expansions of the failure surface and the use of these approximations in structural probabilistic analysis are shown in [55].

When there are n random variables and the failure surface is approximated by a second order Taylor series, the minimal number K of required deterministic solutions is

$$K = \left[\frac{1}{2}n(n+3)\right] + 1 \qquad (5.30)$$

In many practical cases, mixed terms (like $x_i \cdot x_j$) in the Taylor series can be neglected. The required number of deterministic solutions is then

$$K = 2n + 1 \qquad (5.31)$$

It is very important to properly select the evaluation point, around which the Taylor series is expanded. The best point to select is, of course, the MPP, but this point is not known in advance. Suggestions of best ways to select the evaluation point and a practical criterion for its validation are described in [16, Ch. 10].

In the last decade, the expansion of the approximated failure surface was automated in the large finite elements computer codes (e.g., NASTRAN and ANSYS). These programs now include a probabilistic module (called "probabilistic analysis" in ANSYS, "stochastic analysis" in NASTRAN). A finite number of deterministic runs of these programs (which include the variation of the random variables in their distribution range) are performed automatically, and the approximated closed-form expression for the "response surface" is automatically computed. Then, MC simulations are run using this approximated expression and the probability of failure is computed within these finite elements programs. The process is usually quite fast, although the accuracy depends, naturally, on the number of MC simulations used in the process.

In the commercially available structural probabilistic programs (e.g., Pro-FES, NESSUS), there is a possibility to introduce deterministic results for the "stress" term as computed separately using any other algorithm, and the programs automatically create the approximated failure surface that can then be solved using some of the algorithms used by these programs.

In addition, the developers of these programs now include an "interface module" between the structural probabilistic code and the finite element code of the interested user. The programs call the finite element code, perform solution iteration, change the required parameters, and then call again the finite element code, etc., until a required solution is obtained. This process is the most efficient, although it almost doubles the user's licensing cost, and sometimes may be "tricky" to an inexperienced user.

To demonstrate the Taylor series expansion method, the cantilever beam solved earlier with the Lagrange multiplier method was used again. Suppose the tip deflection of the beam (Eq. (5.1)) is not known explicitly, and the solutions for the tip deflection are obtained using a finite element program. The tip deflection is expressed by a second order Taylor expansion around an evaluation point E, neglecting mixed terms. There are three random variables—the tip force P, the Young modulus E, and the thickness h. The Taylor series expansion is

$$y_{tip} = \alpha_0 + \alpha_1(P - P_E) + \alpha_2(E - E_E) + \alpha_3(h - h_E)$$
$$+ \alpha_4(P - P_E)^2 + \alpha_5(E - E_E)^2 + \alpha_6(h - h_E)^2 \tag{5.32}$$

where the index E marks the evaluation point. The coefficients $\alpha_i, i = 0, 1, 2, 3, 4, 5, 6$ are determined by the procedure described below.

The evaluation point is selected at the following values:

$$P_E = 10\,\text{kgf}$$
$$E_E = 2000000\,\text{kgf/cm}^2 \tag{5.33}$$
$$h = 0.48\,\text{cm}$$

The reason for this selection will be explained shortly. Seven (2*3 + 1) deterministic solutions are created. The first is at the evaluation point. Then, two additional deterministic solutions are created for variation in each of the random variables, while the other variables are kept at their evaluation point. These are performed for

$$P = 10.166665\,\text{kgf};\ 10.33333\,\text{kgf}$$
$$E = 2050000\,\text{kgf/cm}^2;\ 2100000\,\text{kgf/cm}^2 \tag{5.34}$$
$$h = 0.46\,\text{cm};\ 0.44\,\text{cm}$$

In Table 5.3, the seven deterministic cases are listed, together with the finite element results for the tip deflection $\delta_{tip,FE}$. The last column in the table is the value computed later using the Taylor series, for comparison.

TABLE 5.3   Seven deterministic cases.

Case	P	E	h	$\delta_{\text{tip,FE}}$	$\delta_{\text{tip,Taylor}}$
1	10	2000000	0.48	4.88281250	4.88281250
2	10.166665	2000000	0.48	4.96419189	4.96419189
3	10.33333	2000000	0.48	5.04557129	5.04557129
4	10	2050000	0.48	4.76371951	4.76371951
5	10	2100000	0.48	4.65029762	4.65029762
6	10	2000000	0.46	5.54779321	5.54779321
7	10	2000000	0.44	6.33921863	6.33921863

The solution vector is given (according to Eq. (5.32)) by

$$\{\delta_{\text{tip,FE}}\} = \begin{Bmatrix} \delta_{\text{tip,FE,0}} \\ \delta_{\text{tip,FE,1}} \\ \delta_{\text{tip,FE,2}} \\ \delta_{\text{tip,FE,3}} \\ \delta_{\text{tip,FE,4}} \\ \delta_{\text{tip,FE,5}} \\ \delta_{\text{tip,FE,6}} \end{Bmatrix} = [Y] \cdot \begin{Bmatrix} \alpha_0 \\ \alpha_1 \\ \alpha_2 \\ \alpha_3 \\ \alpha_4 \\ \alpha_5 \\ \alpha_6 \end{Bmatrix} \qquad (5.35)$$

where $[Y]$ is the matrix of differences from the evaluation point. It has $m = 7$ columns and k rows, where k is the number of deterministic cases solved. In the demonstrated case the number of the deterministic solutions is equal to the number of coefficients in Eq. (5.32), and this is the minimum required. Sometimes, in order to increase accuracy, more than $2n + 1$ deterministic cases are performed.

The structure of the $[Y]$ matrix for this case is

$$[Y] = \begin{bmatrix} 1 & 0 & 0 & 0 & 0 & 0 & 0 \\ 1 & \Delta X_{1,1} & 0 & 0 & \Delta X_{1,1}^2 & 0 & 0 \\ 1 & \Delta X_{1,2} & 0 & 0 & \Delta X_{1,2}^2 & 0 & 0 \\ 1 & 0 & \Delta X_{2,1} & 0 & 0 & \Delta X_{2,1}^2 & 0 \\ 1 & 0 & \Delta X_{2,1} & 0 & 0 & \Delta X_{2,1}^2 & 0 \\ 1 & 0 & 0 & \Delta X_{3,1} & 0 & 0 & \Delta X_{3,1}^2 \\ 1 & 0 & 0 & \Delta X_{3,1} & 0 & 0 & \Delta X_{3,1}^2 \end{bmatrix} \qquad (5.36)$$

where $\Delta X_{i,1}$ and $\Delta X_{i,2}$ are the first and the second modification of the ith variable, respectively. For the given example, the numerical values of the $[Y]$ matrix are

$$[Y] = \begin{bmatrix} 1 & 0 & 0 & 0 & 0 & 0 & 0 \\ 1 & 0.166665 & 0 & 0 & 0.166665^2 & 0 & 0 \\ 1 & 0.33333 & 0 & 0 & 0.33333^2 & 0 & 0 \\ 1 & 0 & 50000 & 0 & 0 & 50000^2 & 0 \\ 1 & 0 & 100000 & 0 & 0 & 100000^2 & 0 \\ 1 & 0 & 0 & -0.02 & 0 & 0 & (-0.02)^2 \\ 1 & 0 & 0 & -0.04 & 0 & 0 & (-0.04)^2 \end{bmatrix}$$

$$(5.37)$$

The optimal set of coefficient $\{\alpha\}$ for Eq. (5.35) (based on minimum least-square error) is obtained by solving [56]:

$$[Y]^T[Y]\{\alpha\} = [Y]^T \{\delta_{tip, FE}\} \qquad (5.38)$$

which can easily be done using MATALB. The results obtained are

$$\{\alpha\} = \begin{Bmatrix} 4.8828125 \\ 0.48828119281 \\ -2.4385708 \cdot 10^{-6} \\ -30.08791775 \\ 1.80003596 \cdot 10^{-7} \\ 1.13422 \cdot 10^{-12} \\ 158.0558875 \end{Bmatrix} \qquad (5.39)$$

Substituting these values into Eq. (5.32) provides an approximate, close-form expression for the tip deflection of the cantilever beam. In the last column of Table 5.3, the tip deflections $\delta_{tip,Taylor}$ calculated with the Taylor series expression are shown. There is no difference between these values and the tip deflections calculated with the finite element program. Not always can such accuracy be obtained. The accuracy of the approximated closed-form solution depends on the complexity of the basic phenomena and on the selection of the evaluation points. When the response surfaces have complex shapes, with local extremis and highly curved shapes, their approximation by a Taylor surface can be very inaccurate. The selection of the evaluation point requires a qualitative analysis of the influence of each random variable. In the described case, it is clear that in the MPP, P must be higher that the nominal value of 10 kgf. The values selected for the approximation are the nominal values $+0.5\sigma$ and 1σ. It is also clear that at the MPP, E will be smaller than the nominal; thus, the evaluation point was selected at $E = 2000000\,\text{kgf/cm}^2$, with 1σ and 2σ above it. It is also clear that at the MPP, h is smaller than the nominal. As h influence tip deflections in a cubic manner, the evaluation point selected was lower (-1σ) from the nominal, with two more values at -2σ and -3σ from the nominal value. Experience may provide means to the selection of the evaluation point, together with some other criteria that can be found in [16].

Once an approximate expression is found for δ_{tip}, a closed-form expression can be written for the failure surface. This expression is called the response surface of the structure. In the present case, failure occurs when

$$g = (\delta_0 - \delta_{\text{tip,Taylor}}) \leq 0 \tag{5.40}$$

This expression was introduced to the NESSUS program (developed and provided by Southwest Research Institute) and a FORM analysis was performed. The results obtained are

$$\beta = 3.675495$$
$$P_{\text{f}} = 0.0001187258$$

$$\begin{aligned}
\delta_0^* &= 6.582\,\text{cm} \\
P^* &= 10.20\,\text{kgf} \\
E^* &= 2.079 \cdot 10^6\,\text{kgf/cm}^2 \\
h^* &= 0.4325\,\text{cm}
\end{aligned} \tag{5.41}$$

These results are quite similar to those in Eqs. (5.22), (5.23), (5.24), (5.26), and (5.29), but not identical. The accuracy can be improved if a second iteration is made, selecting the values obtained in Eq. (5.41) for P^*, E^*, and h^* as a new evaluation point, and selecting two more values for each P, E, and h, as plus or minus values around the evaluation points. Then a new set of $\{\alpha\}$ can be obtained, and a new probabilistic solution performed. This procedure is not repeated here, and is left to the interested reader.

It is clear that advanced knowledge of the MPP is of great benefit to the designer and the analyst. In [57], a method was suggested for the determination of the MPP using the optimization module of the ANSYS, without the need to apply any structural probabilistic program. This method is based on finding the maximum of a Modified Joint Probability Density Function (MJPDF). Use of this method also enables the computation of the FORM probability of failure using only the finite element code. The method is not described in this publication, and the interested reader can find a complete description in ([16], Ch. 11).

5.3 SOLUTION WITH A COMMERCIAL FINITE ELEMENT PROGRAM

In both most "popular" commercially available finite elements programs, NASTRAN and ANSYS, there is a "probabilistic analysis module" that enables a probabilistic analysis of finite elements based solutions. The solution is performed by either a direct or Latin Hypercube Monte Carlo simulation, or by generating a response surface and then doing Monte Carlo simulations using this response surface. An algorithm for optimizing the required number of sampling is used when the response surface method is used. The direct MC simulations are done by looping the case file many times, and then a statistical analysis of the results is performed. Generation of a response surface is done by a suitable sampling of the relevant structure, thus creating an approximate function for the desired random parameter (similar to the Taylor series expansion described in a previous section). This evaluation can be linear, pure quadratic, or quadratic with cross terms.

The looping file is prepared at the beginning of the analysis, and has to include all the random input variables as parameters. The analysis and the post-processing phases are to be included in the looping file, and then the required response random variables can be defined. One of these parameters must be the failure function, and the probability of failure is obtained by calculating the probability that this variable is less than or equal to zero. A looping file can also be prepared for a mathematical expression, thus using the ANSYS probabilistic module as a tool for computing the statistics of any mathematical failure function, with no finite elements computations at all.

The number of MC simulations may be very large for practical cases where the probability of failure is small, as mentioned previously. Performance of a very large number of simulations by a repeated solution of the looping file takes a lot of computer time. On the other hand, doing MC simulations of the fitted response surface usually takes a very short time, and therefore is more recommended.

ANSYS file (**probeam4.txt**) for a cantilever beam is provided in the Appendix. As the described problem has a very low probability of failure, millions of MC simulations are required. The results of this file run are not shown here, and are left for the reader to test. The file **probeam4.txt** in the Appendix is full of comments so the assumption is that there are no difficulties in understanding and running it. It should be emphasized again that a looping file is to be written first. This looping file is demonstrated in the listing of **probeam4.txt**. In addition, our experience showed that the most efficient procedure to apply the statistical results using the ANSYS probabilistic method is to use the GUI of the program. Instructions for such a procedure are also listed in the file. This module provides an extensive choice of statistical analysis of computation results. The reader can plot CDFs of variables, compute probabilities of failure, demonstrate the history of the sampling of each variable, and print a final report on the solved problem. Users of finite elements code should be encouraged to use this module.

In order to show that the ANSYS probabilistic module can also solve a mathematical "stress-strength" model, file **math.txt** was written, and listed in the Appendix. In this file, which is also self-explanatory due to the comments included, a solution of a mathematical problem is performed, without any reference to a specific structure and without any finite elements. Readers are encouraged to make use of this benefit.

The finite elements program probabilistic modules do not allow FORM and SORM analyses. Sensitivity and trend analyses are possible.

5.4 Probability of Failure of Dynamically Excited Structures

In a previous section, the probability of failure of structures with random variables (dimensions, material properties, loading) subjected to static load was described. In Chapter 3, the deterministic dynamic behavior of a deterministic structure subjected to random dynamic loads (for which a PSD function is given) was explored. In this chapter, the probability of failure of a nondeterministic structure subjected to random loads is discussed. The discussion is limited to a stationary Gaussian excitation process, in which the behavior of the system at a given time depends only on the time difference between the present and a previous time. A Gaussian process with zero mean is common to many practical structures, which vibrate as a result of external random loads. This case can be solved by explicit expressions so the phenomena can be demonstrated and, therefore, clearly explained. It can be shown that when the excitation is a Gaussian process, the response is also Gaussian. The statistical behavior of a Gaussian response is treated. A response of a structure can be a displacement in a selected location, a certain stress, a certain reaction force, certain acceleration, and many other structural parameters that may be of interest to the designer. Assume a certain response $S(t)$ that can be described by a Gaussian process with zero mean and a PSD function $G_S(\omega)$, where $0 \leq \omega \leq \infty$. The absolute value of the maximum of this process during a period τ is of interest to the designer, who can calculate the probability of the structure to have a response higher than a defined failure limit. The value of the response $S(t)$ is a function of several random variables $\{z\}$. Some basic literature results for the behavior of a Gaussian stochastic process are repeated here.

Assume s is a threshold value (which may also be random and, therefore, is a part of the vector $\{z\}$). It is shown that the conditional probability that the process is equal or higher than s after a time τ, for a given set of $\{z\}$ is given by ([16, 58])

$$P_f(z) = 1 - \left[1 - \exp\left(-\frac{s^2}{2\lambda_0}\right)\right] \cdot \exp[-v(z)\tau] \qquad (5.42)$$

In Eq. (5.42), $v(z)$ is the rate of crossing upward the value s:

$$v(z) = v_{+s} = \frac{1}{2\pi}\sqrt{\frac{\lambda_2}{\lambda_0}} \cdot \exp\left(-\frac{s^2}{2\lambda_0}\right) \qquad (5.43)$$

and

$$\lambda_m = \int_0^\infty \omega^m G_s(\omega) \cdot d\omega \quad m = 0, 1, 2 \qquad (5.44)$$

The values λ_m are called spectral moments of the process. It can be seen that λ_0 is the mean square value of the process, and λ_2 is the mean square value of the velocity of the process. The PSD of the response $G_s(\omega)$ is a function of the PSD of the excitation, and some other parameters of the structure that are included in the structure's transfer function $H(\omega)$.

In the problem described in Section 3.2 of Chapter 3, assume (without a loss of generality) that there are three random variables—the thickness h, the Young's modulus E, and the damping coefficient ζ—and all of them normally distributed, with means and standard deviations given in Table 5.4.

Using the nominal data of the other structural parameters, it can be shown that for this case the following relations exist for this system (in kgf, cm, radian, and seconds). Only the response of the first mode is considered, as it was shown that the contribution of the other modes is negligible.

$$\omega_1 = 0.099937527 \cdot h \cdot \sqrt{E}$$

$$\lambda_0 = 712.1539 \cdot \frac{\pi}{4\zeta_1 h^2 \omega_1^3}$$

$$\lambda_2 = 712.1539 \cdot \frac{\pi}{4\zeta_1 h^2 \omega_1} \qquad (5.45)$$

When these values are introduced into Eq. (5.42), the values of the conditional probability to get a tip amplitude equal or higher than a given threshold s at a given time t is shown in Figure 5.2.

TABLE 5.4 Random variables for the dynamic problem.

Variable	Mean	SD	Range $\pm 3\sigma$
H, Thickness (cm)	0.5	0.02	$0.44 \rightarrow 0.56$
E, Young (kgf/cm^2)	2100000	50000	$1.95 \cdot 10^6 \rightarrow 2.25 \cdot 10^6$
ζ, Damping coefficient	0.02	0.001	$0.017 \rightarrow 0.023$

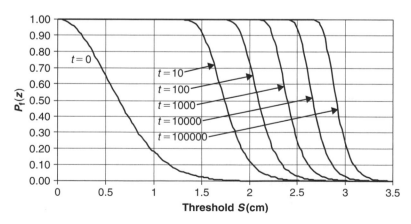

FIGURE 5.2 Conditional probability of threshold passage.

It can be seen that as time increases, the conditional probability to cross a given threshold is increased.

It was suggested ([59]) that the nonconditioned probability could be calculated using the following limit (failure) function:

$$g(z, u_{n+1}) = u_{n+1} - \Phi^{-1}[P_f(z)] \tag{5.46}$$

where u_{n+1} is a standard normal variable added to the structural random variables (in this case, h, E, and ζ) and Φ^{-1} is the inverse standard normal distribution function. With Eq. (5.46) the probability of crossing a threshold value s at time t can be computed for the dynamic problem. At this moment it is not possible to use most of the commercially available structural probabilistic computer codes, as most of them do not include an external function Φ^{-1}. At the author's suggestion, ARA (Applied Research Associates, developer and provider of ProFES [49]) introduced this function to a beta version of the code. To the user who does not have access to such a program, a nondirect method is used, and then compared to the ProFES beta version. The PDF function of $P_f(z)$ is computed separately and then approximated by a PDF function $f_{approx}(z)$. Then the inverse of this function, f_{inv} is computed with any mathematical solver, and the commercially available codes can then be used to compute the probability of failure (threshold crossing) of the limit function

$$g_{appr} = u_{n+1} - f_{inv} \tag{5.47}$$

In Figure 5.3, the PDF of $P_f(z)$ is shown for $t = 100$ sec, and for a threshold value of $s = 2.5$ cm. This PDF was calculated by 10,000 Monte Carlo simulations using MATLAB.

In Figure 5.4, the inverse transformation $\Phi^{-1}[P_f(z)]$ obtained from the simulations depicted in Figure 5.3 is shown. In the same figure, normal distributions with the same mean and standard deviation are also shown.

Although the obtained PDF is not exactly normally distributed (in fact it looks more like a lognormal distribution), normal distribution was used, for the case of simplicity. Then, the reliability index β was calculated using the Stress-Strength model described in Section 5.2, and the probability of threshold crossing $\Pr_{cross}$ was computed using

$$\Pr_{cross} = \Phi(-\beta) \tag{5.48}$$

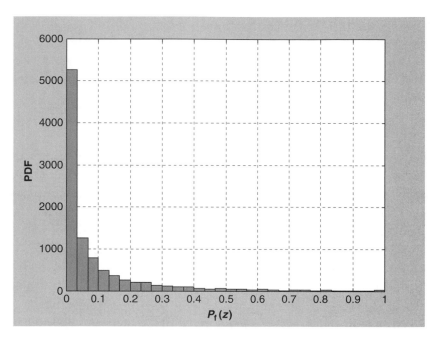

FIGURE 5.3 PDF of $P_f(z)$; Mean $= 0.10007$, Standard deviation $= 0.1685$.

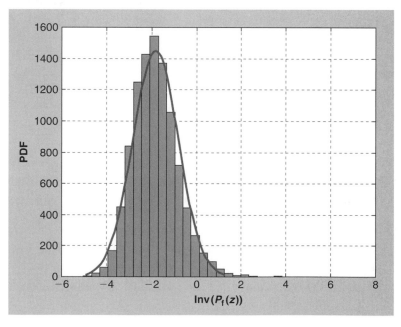

FIGURE 5.4 PDF of $\Phi^{-1}(P_f(z))$; Mean $= -1.8212, \sigma = 1.0168$.

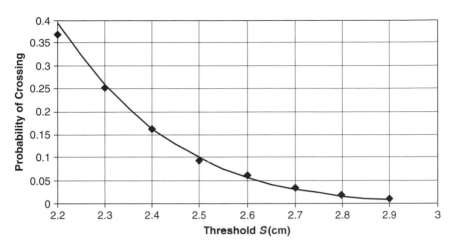

FIGURE 5.5 Probability of threshold crossing at $t = 100$ sec; solid line: indirect computation, symbols: ProFES program.

Results for different threshold levels, at time $t = 100$ sec, are shown in Figure 5.5.

The "automatic" solution of dynamic problems (i.e., the use of commercially available probabilistic computer codes) with many variables must wait until the formulation in Eq. (5.46) is incorporated in these programs. The interested reader should check whether this formulation has already entered these codes when this book is published.

As mentioned, the author was able to get access to a version of ProFES in which the inverse transformation $\Phi^{-1}(P)$ is included. The PDF function obtained running this version for $\Phi^{-1}[P_f(z)]$ is shown in Figure 5.6. The general shape, the mean, and the standard deviation of the results are quite similar to those shown in Figure 5.3. The probability of crossing a threshold $s = 2.5$ cm at $t = 100$ seconds can then obtained directly with the program, without the need to directly assume a probability for $\Phi^{-1}[P_f(z)]$. Results for this probability are shown in Table 5.5, and compared to the relevant value from the previous analysis (Figure 5.4).

Results of ProFES computations for other values of S are also shown in Figure 5.5. It can be seen that the indirect method, in which the distribution of $\Phi^{-1}[p_f(z)]$ was assumed normal, yields quite good results. Nevertheless, the use of a direct computation, which takes into account the true distribution, is much more recommended. Such a computation allows also computing the

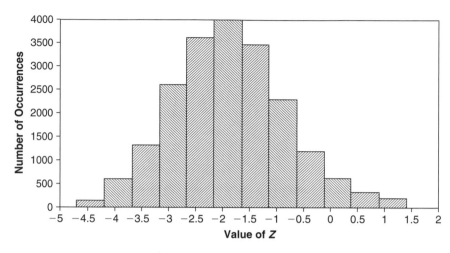

FIGURE 5.6 PDF of $\Phi^{-1}(P_f(z))$ obtained with ProFES; Mean $= -1.8167$, $\sigma = 1.0135$.

TABLE 5.5 Probability of crossing $s = 2.5$ cm at $t = 100$ seconds.

	MC (5000)	MC (20000)	FORM	Indirect Method
Pr_{cross}	0.1018	0.09501	0.09134	0.1008
β	-1.27153	-1.29495	-1.29345	-1.27701

distributions of the frequency, the statistical moments, and the upward rate of crossing. In addition, correlations between all the participating variables are obtained.

5.5 STRUCTURAL SYSTEMS

A structural element can fail in more than one possibility, according to the design criteria set on a specific problem. The cantilever beam described earlier fails, according to the design criteria set in the design specifications, when its tip deflection is larger than a specific threshold value. However, there is another possibility of failure that cannot be ignored—when the bending stress at the clamped edge is higher than a given value, say the yield stress. Suppose the beam has a hole in it, which gives rise to a stress concentration at its edge. Then, the designer can set yet another failure criterion, which is the demand that the stresses around the hole should be smaller than a specific value. Additional design criteria can be set to incorporate fatigue or crack propagation possibilities, etc.

The determination of the failure criteria of a design is the most important step in a successful design process. This is the stage where the skills and experience of the designer and the experience of the organization are mostly required in order to create a successful, safe design.

Sometimes, the design criteria are combined. It may happen that there is more than one failure mode that can cause failure to the structure. Then, combined criteria can be formulated. Such criteria may be: "the structure fails when *both* the tip deflection is larger than a given threshold *and* the stress at the clamp is higher than a prescribed value," or it may be: "the structure fails when *either* the tip deflection is larger than a given threshold *or* the stress at the clamp is higher than a prescribed value."

In order to set up the required design criteria, the fault tree analysis that is used extensively by safety and reliability experts can be adopted, using the tools that were established and used for many years in these important fields, and integrating these into the structural design process. Fault tree analysis is outside the scope of this book, and the interested reader can look at relevant textbooks of safety analysis. In order to understand the concepts, some background is described here, and examples based on the previous cases are demonstrated.

When building a fault tree, basic failure events can be listed as base event. Base event 1 is defined as "The structure fails when the tip deflection is larger than a given threshold." Base event 2 is defined as "The structure fails when the stress at the clamp is higher than a prescribed value." Two possibilities exist. The first possibility is the series combination, where system failure occurs when either of these events happens. The second is parallel failure, where the system fails only when both events occur. The fault trees for both cases, as well as the compatible "flow" diagrams, are shown in Figure 5.7.

A more complex structural fault tree is described in Figure 5.8. It depicts a structural system with three structural elements. The first structural element has three failure modes, and this element fails only when all of them occur, designated FS1. The second element has two failure modes, and it fails if

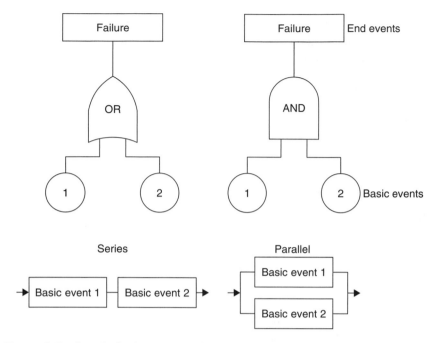

FIGURE 5.7 Simple fault trees and flow diagrams.

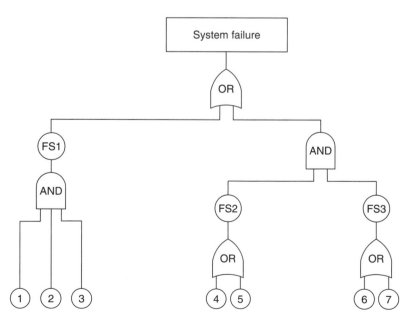

FIGURE 5.8 An example of a system with several structural failure modes.

either of one occurs, the failure designated FS2. The third element has two failure modes, and it fails if either of them occurs, designated FS3. If FS1 occurs or *both* FS2 and FS3 occur, the whole structural system fails.

It should be noticed that the base events depicted at the bottom of the fault tree depend on the structural parameters, which include geometry, dimensions, material properties, and loading conditions, all of which can be random variables. In addition, the base events may be correlated with each other, and are not necessarily independent.

In many cases, failure of one of the elements does not mean that the whole system fails. In these cases, after the failure of one element, the loads are redistributed, and a new system with the new conditions should be analyzed. It is clear that although the whole system does not fail, the probability of failure of the new system will be higher than the original system.

Both the ProFES [49] and the NESSUS [48] probabilistic codes can calculate the probability of structural systems. The systems can be modeled using the limit state functions of the base events, and combinations of *and/or* gates can be inserted.

In Section 5.2, the probability that the cantilever beam has a tip deflection equal to or higher than a given threshold was calculated using ProFES, with the following results (from Eq. (5.26)):

$$Pr_1 = 0.00010862; \quad \beta_1 = 3.69815 \tag{5.49}$$

The probability of having a bending stress equal or higher than the yield stress was also calculated (from Eq. (5.29)):

$$Pr_2 = 0.000103028; \quad \beta_2 = 3.71155 \tag{5.50}$$

The probability that both failures occur simultaneously (and condition) is calculated with ProFES and the result is

$$Pr(1\&2) = 1 \cdot 10^{-5}; \quad \beta(1\&2) = 4.26484 \tag{5.51}$$

and the probability that either of them occurs (or condition) is

$$Pr(1|2) = 0.0001342; \quad \beta(1|2) = 3.55375 \tag{5.52}$$

The two failure cases are interdependent through the random variables h and P that appear in both failure functions (Eqs. (5.17) and (5.28)).

It can be seen that the probability of occurrence of either of the events is higher than the probability of occurrence of both, a conclusion that may be easily explained intuitively.

In the following example, a structural system of three structural elements is described and solved. The structure is described in Figure 5.9. This is an undetermined hinged truss comprised of three members, each with a circular cross-section. The lengths of the members are L_1, L_2, and L_3, and the diameters are d_1, d_2, and d_3, respectively. L_1 and L_2 are of random values. The length L_3 is determined by the lengths of the first and the second members; therefore, it is assumed that the third member has some device that allows small changes in the length during installation of the system. All three diameters are considered random.

The location of the support hinges is deterministic as shown in Figure 5.9. The structure is loaded by a force F, which is also random, at the common hinge. The Young's modulus of all three members is E, a random value. The maximum allowable stress in the second member is $\sigma_{2,\max}$, and the maximum

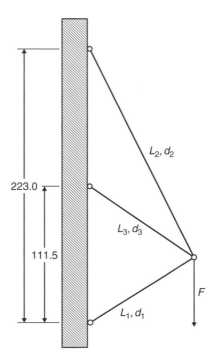

FIGURE 5.9 Three members truss.

allowable stress in the third member is $\sigma_{3,\max}$. The buckling load of the first member is given [60] by the following equation:

$$P_{cr,l} = P_{ratio} \cdot \frac{\pi^2 EI_1}{L_1^2}; \quad I_1 = \frac{\pi d_1^4}{64} \tag{5.53}$$

P_{ratio} is a parameter described in Section 5.5, which takes into account the fact that a simply supported compressed beam fails at a compressive force that is smaller than the Euler's calculated buckling load.

In Table 5.6, data for the random variables is given.

When loaded, the first member is under compression, while the other two are under tension. Therefore, failure of the first member is when it buckles ($P_{cr,l} \leq |F_1|$), and failure of the second and third members is when $\sigma_{2,\max} \leq \sigma_2, \sigma_{3,\max} \leq \sigma_3$, respectively. When either member fails, the whole system does not fail because the other two members can still support the load, although a redistribution of the forces takes place. Failure of at least two members is required for the system to fail.

In order to do a probabilistic analysis, expressions for the forces F_1, F_2, and F_3 in the three members as functions of the structural parameters are required. Although this problem can be solved analytically by applying equilibrium and displacement compatibility—the three members elongate to the same point—this is not always the case in practical problems. Therefore, this example was solved by formulating an approximate Taylor expansion expression for the

TABLE 5.6 Data for the random variables.

		Units	Distribution	Mean	Standard Deviation	Mean$^{\pm 3\sigma}$
1	L_1	cm	Normal	100	1	$100^{\pm 3}$
2	L_2	cm	Normal	200	1	$200^{\pm 3}$
3	d_1	cm	Normal	2	0.003333	$2^{\pm 0.01}$
4	d_2	cm	Normal	0.2	0.003333	$0.2^{\pm 0.01}$
5	d_3	cm	Normal	0.9	0.003333	$0.9^{\pm 0.01}$
6	E	kgf/cm^2	Normal	$2 \cdot 1 \cdot 10^6$	$5 \cdot 10^4$	$2 \cdot 1 \cdot 10^6 \pm 1.5 \cdot 10^5$
7	F	kgf	Normal	1200	33.333	$1200^{\pm 100}$
8	P_{ratio}	–	Weibull	0.8	0.03	$0.08^{\pm 0.09}$
9	$\sigma_{2,\max}$	kgf/cm^2	Normal	1500	120	$1500^{\pm 360}$
10	$\sigma_{3,\max}$	kgf/cm^2	Normal	2100	120	$2100^{\pm 360}$

TABLE 5.7 Deterministic cases solved with ANSYS.

	L_1	L_2	d_1	d_2	d_3	E	F	F_1	F_2	F_3
1	100	200	2	0.2	0.9	$2.1 \cdot 10^6$	1200	1056.866	38.73481	1163.098
2	97	200	2	0.2	0.9	$2.1 \cdot 10^6$	1200	1025.861	37.28883	1150.945
3	103	200	2	0.2	0.9	$2.1 \cdot 10^6$	1200	1087.947	40.24299	1175.392
4	100	197	2	0.2	0.9	$2.1 \cdot 10^6$	1200	1057.279	37.33972	1136.112
5	100	203	2	0.2	0.9	$2.1 \cdot 10^6$	1200	1056.467	40.12630	1189.825
6	100	200	1.99	0.2	0.9	$2.1 \cdot 10^6$	1200	1056.853	38.76050	1163.069
7	100	200	2.01	0.2	0.9	$2.1 \cdot 10^6$	1200	1056.878	38.70950	1163.126
8	100	200	2	0.19	0.9	$2.1 \cdot 10^6$	1200	1058.695	35.07713	1167.198
9	100	200	2	0.21	0.9	$2.1 \cdot 10^6$	1200	1054.956	42.55341	1158.817
10	100	200	2	0.2	0.89	$2.1 \cdot 10^6$	1200	1056.472	39.52147	1162.216
11	100	200	2	0.2	0.91	$2.1 \cdot 10^6$	1200	1057.247	37.9728	1163.952
12	100	200	2	0.2	0.9	$1.95 \cdot 10^6$	1200	1056.866	38.73481	1163.098
13	100	200	2	0.2	0.9	$2.25 \cdot 10^6$	1200	1056.866	38.73481	1163.098
14	100	200	2	0.2	0.9	$2.1 \cdot 10^6$	1100	968.7936	35.50691	1066.173
15	100	200	2	0.2	0.9	$2.1 \cdot 10^6$	1300	1144.938	41.96271	1260.022

forces in the system's members, with deterministic solutions obtained using the ANSYS finite elements program. The file for a deterministic ANSYS solution is **truss1.txt** (see Appendix). The parameters that influence the magnitude of the three forces in the systems are the first seven variables of Table 5.6. Therefore, at least $2 \cdot 7 + 1 = 15$ deterministic cases must be solved with the ANSYS in order to get a quadratic Taylor expansion series (without mixed terms, see Section 5.2). The deterministic cases solved with the resulting forces are shown in Table 5.7.

The following Taylor series are expressed for the forces F_1, F_2, and F_3:

$$
\begin{aligned}
F_1 = {} & \alpha_1 + \alpha_2(L_1 - L_{1,0}) + \alpha_3(L_2 - L_{2,0}) + \alpha_4(d_1 - d_{1,0}) \\
& + \alpha_5(d_2 - d_{2,0}) + \alpha_6(d_3 - d_{3,0}) + \alpha_7(E - E_0) + \alpha_8(F - F_0) \\
& + \alpha_9(L_1 - L_{1,0})^2 + \alpha_{10}(L_2 - L_{2,0})^2 + \alpha_{11}(d_1 - d_{1,0})^2 \\
& + \alpha_{12}(d_2 - d_{2,0})^2 + \alpha_{13}(d_3 - d_{3,0})^2 + \alpha_{14}(E - E_0)^2 + \alpha_{15}(F - F_0)^2
\end{aligned}
$$

$$
\begin{aligned}
F_2 = {} & \beta_1 + \beta_2(L_1 - L_{1,0}) + \beta_3(L_2 - L_{2,0}) + \beta_4(d_1 - d_{1,0}) \\
& + \beta_5(d_2 - d_{2,0}) + \beta_6(d_3 - d_{3,0}) + \beta_7(E - E_0) + \beta_8(F - F_0) \\
& + \beta_9(L_1 - L_{1,0})^2 + \beta_{10}(L_2 - L_{2,0})^2 + \beta_{11}(d_1 - d_{1,0})^2 \\
& + \beta_{12}(d_2 - d_{2,0})^2 + \beta_{13}(d_3 - d_{3,0})^2 + \beta_{14}(E - E_0)^2 + \beta_{15}(F - F_0)^2
\end{aligned}
$$

$$F_3 = \gamma_1 + \gamma_2(L_1 - L_{1,0}) + \gamma_3(L_2 - L_{2,0}) + \gamma_4(d_1 - d_{1,0})$$
$$+ \gamma_5(d_2 - d_{2,0}) + \gamma_6(d_3 - d_{3,0}) + \gamma_7(E - E_0) + \gamma_8(F - F_0)$$
$$+ \gamma_9(L_1 - L_{1,0})^2 + \gamma_{10}(L_2 - L_{2,0})^2 + \gamma_{11}(d_1 - d_{1,0})^2$$
$$+ \gamma_{12}(d_2 - d_{2,0})^2 + \gamma_{13}(d_3 - d_{3,0})^2 + \gamma_{14}(E - E_0)^2 + \gamma_{15}(F - F_0)^2$$

$$(5.54)$$

where $\{\alpha\}, \{\beta\}$, and $\{\gamma\}$ are vectors of the Taylor series coefficients. The values (parameter)$_{i,0}$ are the evaluation point, and the values of the first row of Table 5.7 (the nominal case) were selected for this point. Applying the procedures described in Section 5.2, these coefficients were solved using MATLAB.

$$
\{\alpha\} = \left\{ \begin{array}{c} 1056.866 \\ 10.3477 \\ -0.135333 \\ 1.25 \\ -186.95 \\ 38.75 \\ 0 \\ 0.880722 \\ 0.004222 \\ 0.0007778 \\ -5.0 \\ -405.0 \\ -65.0 \\ 0 \\ 0 \end{array} \right\} ; \quad
\{\beta\} = \left\{ \begin{array}{c} 38.73481 \\ 0.49236 \\ 0.46443 \\ -2.550 \\ 373.814 \\ -77.4335 \\ 0 \\ 0.032279 \\ 0.0034556 \\ -0.0002 \\ 1.90 \\ 804.60 \\ 123.25 \\ 0 \\ 0 \end{array} \right\} ; \quad
\{\gamma\} = \left\{ \begin{array}{c} 1163.098 \\ 4.0745 \\ 8.95217 \\ 2.85 \\ -419.05 \\ 86.80 \\ 0 \\ 0.969245 \\ 0.0078333 \\ -0.0143889 \\ -5.0 \\ -905.0 \\ -140.0 \\ 0 \\ 0 \end{array} \right\}
$$

$$(5.55)$$

It is interesting to note that the three forces do not depend on E and E^2 (because all the members have the same Young modulus) and not on the square of the applied force F (as the analysis is linear).

The approximate expressions for the forces (Eq. (5.54)) with the coefficients in Eq. (5.55), and the expression for the critical buckling load, Eq. (5.53), were introduced into the ProFES program, with the random variables described in Table 5.1. Stresses in the second and the third members were calculated using the cross-section area obtained by the given random diameters, as well as the cross-section moment of inertia I_1, also described in Eq. (5.53). Three limit state functions were formulated for the failure of a single member:

$$\text{Limit State 1:} \quad P_{cr} \leq F_1$$
$$\text{Limit State 2:} \quad \sigma_{2,\max} \leq \sigma_2 = F_2/\text{area}_2 \qquad (5.56)$$
$$\text{Limit State 3:} \quad \sigma_{3,\max} \leq \sigma_3 = F_3/\text{area}_3$$

In the first runs, the probability of failure of a single member was performed. This is not the probability of failure of the system, as failure of at least two members is required for a system failure. The computations were done by both the FORM method and 100,000 Monte Carlo simulations. Results are summarized in Table 5.8.

The reliability index obtained using the FORM analysis is quite close (2% and less) to those of the MC simulation. A larger error is observed for the probability of failure, because results are obtained in the tail of the distribution. Results obtained for Limit State 1 using FORM are less accurate, because the buckling load is proportional to the random diameter to the power of four and inversely proportional to the square value of the length; thus, they are more nonlinear than the other two limit states.

Although Monte Carlo simulations take more computation time, they have some extra benefits that allow better understanding of the behavior of the dependent random parameters of the problem. Histograms of the different variables (such as the forces in the members and the buckling load) can be obtained, and correlation coefficients between any couple of random variables are also computed. These coefficients may help the designer when a change in the design is required.

TABLE 5.8 Probability of Failure and Reliability Index for each member.

	Limit State 1	Limit State 2	Limit State 3
100000 MC	$p_f = 0.00198$	$p_f = 0.01792$	$p_f = 0.01335$
Simulations	$\beta = -2.88167$	$\beta = -2.11369$	$\beta = -2.21632$
Confidence	$P_{f,5\%} = 0.0018$	$P_{f,5\%} = 0.0167624$	$P_{f,5\%} = 0.0128855$
Level Range	$P_{f,95\%} = 0.00216$	$P_{f,95\%} = 0.0178176$	$P_{f,95\%} = 0.0138145$
FORM	$p_f = 0.0016346$	$p_f = 0.017699$	$p_f = 0.0128212$
	$\beta = -2.94124$	$\beta = -2.10377$	$\beta = -2.23159$

In some cases, it may be more convenient to solve the probability of failure of a single member using the probabilistic module, which exists in both ANSYS and NASTRAN. Thus, these probabilities are obtained without using any program except for the finite element code, and it is not necessary to write an approximate Taylor expansion of the forces in the members. The disadvantage, of course, is that system analysis is not possible with the existing modules.

When one of the members fails (with the probabilities shown in Table 5.8), the remaining system is a two-member truss. There are three new cases: (a) a truss with members 1 and 2; (b) a truss with members 2 and 3; and (c) a truss with members 1 and 3. In each case, failure of one member is a system failure. All the cases are of determined systems and can be easily formulated analytically. Each new case has to be analyzed separately.

In order to calculate the probability of failure of the system, three more system limit states were defined:

$$LS1 \ \& \ LS2 \equiv \text{Failure of member 1 and member 2}$$
$$LS1 \ \& \ LS3 \equiv \text{Failure of member 1 and member 3} \qquad (5.57)$$
$$LS2 \ \& \ LS3 \equiv \text{Failure of member 2 and member 3}$$

The case was solved with the help of the ProFES program using the FORM method. The results are shown in Table 5.9.

It can be seen that although the probability of failure of a single tension element is rather high (Table 5.8, 1.8% for member 2, 1.3% for member 3), the probability of two of the system failure modes (1&2, 1&3) is much smaller. This system (with this set of data—nominal and dispersion values) fails when the two tension members (2&3) fail together. The probability for this to happen is $0.00026 = 0.26\%$. The described analysis should also include the probability of failure of all the three members simultaneously, but this case was not examined here, as it was estimated qualitatively that this probability is very low.

TABLE 5.9 System probabilities of failure.

	LS1 & LS2	LS1 & LS3	LS2 & LS3
p_f	$7 \cdot 10^{-5}$	$2 \cdot 10^{-5}$	0.00026
β	-3.80825	-4.10748	-3.47042

5.6 MODEL UNCERTAINTIES

The importance of probabilistic analysis in the design process of structural elements and structural systems is well recognized today. Analytical methods and computational algorithms were developed, and are used in many design establishments and in R&D institutes. Randomness in structural geometric and dimensional parameters, material properties, allowable strength, and external loads can now be treated during the design process.

All of the methods described in this publication and in many others are based on a model that is built for the designed structure, either by a closed form expression (which may be analytical or approximated) or an algorithm, like a finite element computer code. A question may be asked about the validity of the model itself, which certainly has some uncertainties in it. The model used in the solution of a problem doesn't always truly describe the behavior of the observed system, and in many cases the discrepancy between the observed results and the model presents a random behavior. A well-known example is the buckling of a simply supported beam-column. The classical buckling load predicted by using the Euler model (based on the solution of an eigen-values equation (e.g., [60, 61]) is never met when experimental results are analyzed. We now know that this happens because of the initial imperfections of the original beam-column, which is never perfectly straight. The imperfect geometry is the reason for bending moments that are created in the beam. Consequently, bending stresses are created and induce a nonlinear behavior of the structure, causing the collapse of the beam-column when the compressive external force is (sometimes) much lower than the Euler buckling load. These imperfections, which are created during the manufacturing phase of the beam, may be random in their magnitudes. Results from experiments of many "identical" specimens show dispersion in the value of the buckling load. Similar phenomena are observed in experimental results of plates and shells buckling [60].

There is no way to avoid modeling in an intelligent design process. This is especially true for large projects in which many subsystems comprise the final product, where time to design and manufacture a prototype is long, and when the number of tests is limited. In the aerospace industry, products are frequently manufactured in small quantities (e.g., the space telescope, a satellite for a given mission, a small number of space shuttles, the Martian Lander, a special purpose aircraft). The problem of the verification of their reliability is not similar to the reliability verification of consumer goods, where a large number of tests can be conducted and statistical estimates can be verified. In some cases, complete ground tests are not possible at all, as is the case with spacecrafts. Therefore, in large projects of this kind, the importance of models is enhanced. Once a model is built and verified, simulations that use it can be conducted instead of real tests. The collection of a product's performance data can be replaced by these "model simulations," including extreme points in the required performance envelope that cannot usually be tested due to technical difficulties, time schedule reasons, and budget limitations.

The use of probabilistic models enables the determination of the probability of failure of structural elements and structural systems; thus, the reliability of the structure can be estimated and verified through the model, and incorporated into the reliability analysis of the system as a whole. It is clear that in such cases, incorporation of model uncertainties is extremely important. When building a model, many assumptions are made. Sometimes, the influence of some parameters is intentionally neglected, with the proper justification. In many cases there are parameters whose influence cannot be evaluated due to ignorance. Suppose that nobody ever thought about the initial imperfections and their ability to influence the buckling load of a beam-column. Then, the available model (e.g., the Euler buckling load) is unable to describe the real behavior of the structure, as observed from experiments. These experiments show dispersion in the buckling load, with a mean value smaller than the Euler load. When there is a discrepancy between carefully controlled experiments and a model, the chances are that something is wrong with the model, and not with the experiments. These discrepancies can originate from some (unknown?) parameters or physical phenomena that were not included in the model, and the designer is unaware of.

In many cases this uncertainty in the model can be formulated using an additional random variable or random process. Using this methodology, the model, which includes now a "device" that takes care of the model uncertainties, can predict probabilistic behavior of the structure that will be in agreement with the experimental result.

The proposed approach is first demonstrated on the Euler buckling model discussed previously. A beam-column of length L, width b, and thickness h made

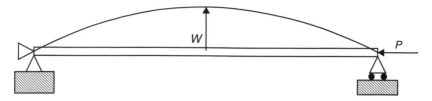

FIGURE 5.10   A simply supported beam-column under axial compression.

of a material with Young's modulus E is subjected to an axial compressive force P. The model is shown in Figure 5.10.

When the beam is perfectly straight, the buckling load is given by the Euler critical load (e.g., [60, 61]), which is a solution of the eigenvalues equation

$$P_{cr} = \frac{\pi^2 EI}{L^2} \tag{5.58}$$

where I is the cross-section moment of inertia, which in this case is

$$I = \frac{bh^3}{12} \tag{5.59}$$

According to this solution, the column will not have any lateral deflections when the force P is increased from zero to P_{cr}, and will collapse when the force reaches the critical value P_{cr}. The lateral deflection at this point will tend suddenly to infinity.

If the beam has an initial lateral deflection w_0, increase of the load creates lateral deflection w. Assuming the initial deflection is half sine wave with amplitude a_0:

$$w_0 = a_0 \sin \frac{\pi x}{L} \tag{5.60}$$

It can be shown that in this case, the lateral deflection w is given by

$$w = a_0 \sin \frac{\pi x}{L} \cdot \frac{\frac{P}{P_{cr}}}{1 - \frac{P}{P_{cr}}} \tag{5.61}$$

For any given x, Eq. (5.61) describes a nonlinear relation between the lateral deflection w and the applied force P. Due to the lateral deflection there is a bending moment in the beam that has a maximum at the beam center, $x = L/2$. This moment creates bending stresses. In addition to these are compression stresses due to the axial force itself. Suppose that the material fails when the total local stress reaches a value of σ_{max}. When σ_{max} is reached, the beam fails while the acting force P is still smaller than the Euler critical value P_{cr}.

In reality, initial imperfections may be a combination of many sine waves along the beam, but they can be described by a Fourier series of amplitudes (that may be random) for many wave lengths. In what follows, only one half sine wave is used, without a loss of generality, because expressions similar to Eq. (5.61) can be obtained for every combination of waves.

In Figure 5.11, the absolute maximum stress in the mid-length is described as a function of P for several values of a_0. The Euler solution is shown in the thicker L-shaped line. The numerical values used for the example are

$$L = 60\,\text{cm} = 0.6\,\text{m} = 23.62"$$
$$b = 8\,\text{cm} = 0.08\,\text{m} = 3.15"$$
$$h = 0.5\,\text{cm} = 0.005\,\text{m} = 0.197" \tag{5.62}$$
$$E = 2100000\,\text{kgf/cm}^2 = 2.058 \cdot 10^5\,\text{MPa} = 29842\,\text{ksi}$$
$$\sigma_{max} = 5000\,\text{kgf/cm}^2 = 490\,\text{Mpa} = 71\,\text{ksi}$$

The Euler buckling force in this case is

$$P_{cr} = 479.77\,\text{kgf} = 1056.76\,\text{lbs} \tag{5.63}$$

As we do not have experimental results for this case, such results will be "manufactured" artificially. It is assumed that the specimens in the "experiments" have

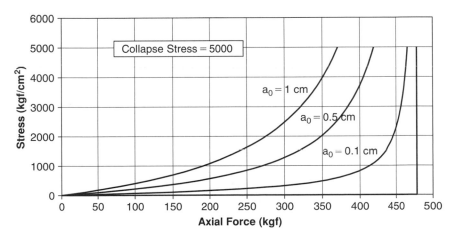

FIGURE 5.11 Stress in mid-beam for three values of imperfection; Euler load = 479.77 kgf.

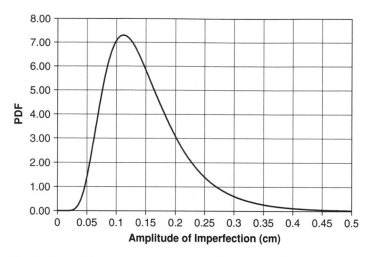

FIGURE 5.12 PDF of the amplitude of imperfection (log-normal); mean = 0.15, SD = 0.07.

a half sine wave initial imperfection, with amplitude a_0, which has a lognormal distribution with mean = 0.15 cm and standard deviation = 0.07 cm. The location and shape parameters (another pair of parameters sometimes used to define a log-normal distribution) for this case are $m = -1.9956338$ and $s = 0.443878028$. The distribution of the imperfection amplitudes is shown in Figure 5.12.

With this distribution, "virtual experimental data" can be created for a parameter P_{ratio}—the ratio of the collapse force (the force that creates a stress of 5000 kgf/cm^2 in the beam) to the Euler buckling force. The computations were done using 5000 Monte-Carlo simulations in a MATLAB program. The histogram of the results is shown in square symbols in Figure 5.13. The mean of these results is $\mu = 0.958$ and the standard deviation is $\sigma = 0.01803$. A Weibull distribution with the same mean and standard deviation was fitted to this data, and is shown by curved line in Figure 5.13. The parameters of the Weibull distribution obtained are

$$
\begin{aligned}
\mu &= \text{mean} = 0.958 \\
\sigma &= \text{standard deviation} = 0.01803 \\
\alpha\ &\text{parameter} = 67.42607 \\
\beta\ &\text{parameter} = 0.9660698
\end{aligned}
\tag{5.64}
$$

It can be seen that the fitted distribution agrees very well with the "experimental" generated results.

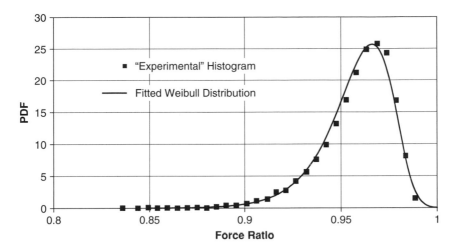

FIGURE 5.13 Distribution of "experimental" force ratio and fitted Weibull distribution.

A model for the collapse load based on the Euler buckling load and includes the model uncertainty can now be written as

$$P_{\text{collapse}} = \frac{\pi^2 EI}{L^2} \cdot P_{\text{ratio}} \tag{5.65}$$

P_{ratio} is an added random variable, which has a Weibull distribution shown in Figure 5.13, with the parameters given in Eq. (5.64).

Assuming that $E, L,$ and I are deterministic, and the only random variable is P_{ratio} (only model uncertainty exists), one can calculate the probability that the collapse load will be lower than a given value P_{lower}. Results computed using NESSUS [46] are shown in Figure 5.14. The same computation can be performed assuming that the geometry parameters $(L, b,$ and $h)$ and material property (E) are also random.

The next example treats the crack growth rate model, which is further discussed in more detail in Chapter 6. Some of it will be repeated here, for the completeness of the model uncertainties subject treated here.

Models for crack growth have been the subjects of thousands of papers published over the past 40 years. These range from simplified to more advanced models, most of which are based on experimental observations. The most

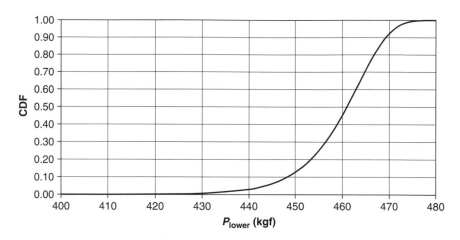

FIGURE 5.14 Probability that collapse load is lower than P_{lower}.

basic model is the one suggested by Paris and Erdogan [62], where the rate of crack growth is described by

$$\frac{da}{dN} = C \cdot (\Delta K_I)^n \tag{5.66}$$

where a is the crack length, N is the number of load cycles, ΔK_I is the stress intensity factor, and C and n are material properties extracted from tests. There are models that include the effects of the stress ratio R (i.e., Forman equation, which is used in the NASGRO® computer code [63]), with corrections for crack closure phenomena, and the "unified" approach model suggested by Vasudevan et al. (e.g., [64, 65]). All these models are deterministic.

When tests are performed on many "identical" specimens, typical results look like those in Figure 5.15 (e.g., [66, 67]).

These results are characterized by three major properties:

1. The behavior of crack length *is random*, even when very carefully controlled experiments are performed with "identical" specimens.

2. The crack length behavior is *nonlinear*.

3. The curves of different specimens *intermingle*.

When the growth rate *da/dN* is plotted against the stress intensity factor, experimental results yield the experimental circles depicted in Figure 5.16.

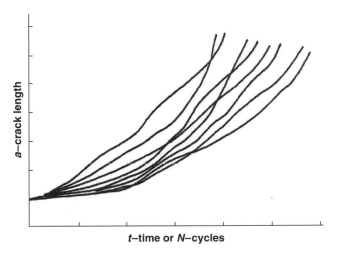

FIGURE 5.15 Generic curves for crack size as a function of time or cycle.

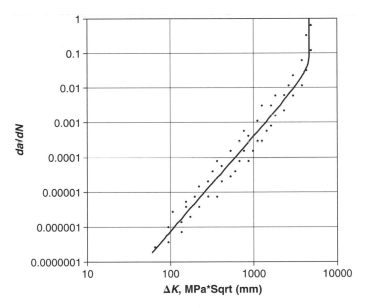

FIGURE 5.16 Crack growth rate vs. stress intensity factor.

On a log-log scale, the straight line shown is, in fact, the Paris Law. Other propagation models described in Chapter 6 model the rise toward infinity at the fracture toughness value. Thus, these are models that describe the mean behavior of the experimental data, but not the randomness expressed by the scatter in the experimental results, and therefore there is an uncertainty in the model.

There were many attempts to formulate a stochastic crack growth law. In some of them (e.g., [44, 68]), the material constants C and n were considered as random variables. This approach suffers from a serious physical interpretation—randomizing these constants also randomizes their units. In addition, the use of this approach does not describe the intermingling of the curves (Figure 5.15), which is observed experimentally.

Similar to the approach demonstrated for the Euler buckling model in the previous section, it was suggested (e.g., [69, 70]) to write the crack growth model in the following form:

$$\frac{da}{dN} = Q \cdot a^b \cdot X(t) \tag{5.67}$$

where Q and b are deterministic constants obtained by a best straight line fit (on a log-log scale) of the data as depicted in Figure 5.16, and $X(t)$ is *a stochastic process* that describes the dispersion of the experimental results around this straight line. As the crack growth is a local phenomenon, the stochastic process is correlated only within a certain distance near the crack tip. There are some suggestions (e.g., [69, 71, 72]) on how to select this correlation time. When the stochastic approach is applied, results obtained show both the *nonlinear behavior* of the crack's length and the *intermingling* phenomena (e.g., [16, 73]), and thus better describe the experimental results.

Once again it was demonstrated that, in this case, a stochastic process rather than a stochastic variable could be used to include a model's uncertainties in the evaluation of models. The parameters of either the stochastic variable or the stochastic process can be determined from experimental results; thus, they can represent properly, in a probabilistic way, the uncertainties in the model. This can be done even when the reasons for these uncertain outcomes are not completely understood, and the effects of uncertainties are introduced in an "integral" way.

Chapter 6 / Random Crack Propagation

6.1 CRACK PROPAGATION IN A STRUCTURAL ELEMENT

Failure of a structural element due to an initial crack, which propagates during loading until the element fails, is the major reason for catastrophic failure of aerospace and other structures. The description of this phenomenon is the subject of many textbooks, and is the main reason for the tremendous development of fracture mechanics in the last three decades.

Cracks are very small voids in the internal microscopic crystalline structure of the material, and form small discontinuities (flaws) inside the material. They are created either in the production process or due to environmental effects, such as corrosion and wear. When a cracked structure is loaded, stress concentrations are formed at the "tip" of the crack, and due to the material failure mechanism, they propagate through the structure. The phenomenon can occur due to an increasing static load. When the structure cannot carry more loads by the remaining uncracked material, it fails. Such crack propagation also occurs when the structure is loaded with a constant amplitude repeated load, when every additional load cycle causes the crack to "propagate." The crack propagation due to a repeated load is the explanation (given by fracture mechanics) of fatigue failure. In the literature, the terms "crack propagation" and "crack growth" are used synonymously.

For many years, failures due to fatigue were treated using the empirical S-N well-known curves, where the repeated stress to failure vs. the number of loading cycles required to cause this failure were mapped experimentally.

Handbooks for properties of materials include these *S-N* curves for many structural materials, especially metals [74].

In Figure 6.1, a generic experimental *S-N* curve is shown. In a log-log scale, these curves are described as straight lines. For some materials (but not all), there is a lower limit stress (endurance stress) below which no fatigue occurs. The experimentally determined points on this curve usually exhibit dispersion that, although seems to be small, is really large because the scales are logarithmic.

For many years, only experimental laws treated the fatigue of aerospace structural elements. Only when fracture mechanics began to develop some theoretical aspects were introduced, and it was possible to explain the fatigue failures by the fracture mechanics theories.

Models for crack growth (sometimes called crack propagation) have been the subjects of thousands of papers published over the last 40 years. These range from simplified models (e.g., [62]) to more advanced models (e.g., [75, 76]). In [77], an excellent list of 217 references is provided.

Today, theoretical and experimental terms of fracture mechanics are well covered and well explained in textbooks, papers, and conferences, and only

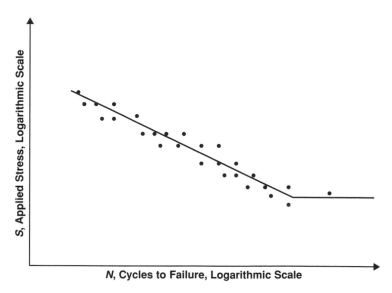

FIGURE 6.1 A generic *S-N* curve with random dispersion.

some of the basic concepts will be repeated here. Most of this chapter is concentrated on the random effects of crack growth. As the initial flaws and cracks in structural materials result from manufacturing and environmental effects, they are inherently random in nature, and using random analysis is called for even if the macro-structure and the loadings are deterministic. The cracks are propagating in and between material grains boundaries, which are highly random due to the randomness of the processes that initiate them. Although fracture mechanics today is well developed, most of the material properties entering the analysis must still be the results of extensive and expensive experimental work.

The failure of a structure due to the existence of a small crack starts when internal nucleation begins to form as a result of a very small and local yielding, and is expressed in a slip between the material's atom layers. This is a process whose laws are still not well understood. These nucleations form microcracks, which then start to interconnect to form macro-cracks. These macro-cracks grow until a structural failure occurs. The order of magnitude of cracks in each step is described in Figure 6.2.

A major important parameter in fracture mechanics is the stress intensity factor (SIF). This is a parameter that incorporates the applied stress, the dimensions of the crack, and the geometry of the structural specimen. There are SIFs in opening, shearing, and tearing of cracks, denoted K_{I}, K_{II}, and K_{III}, respectively. When a repeated stress ΔS is applied to a structural element the change in the opening SIF is

$$\Delta K_{\mathrm{I}} = Y(a)\Delta S\sqrt{\pi a} \tag{6.1}$$

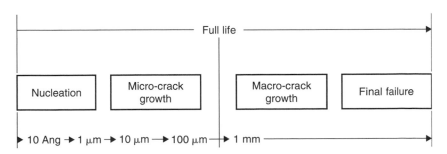

FIGURE 6.2 Nucleation, micro-crack, macro-crack, and crack growth to failure.

where $Y(a)$ is a geometry function and a is the crack length. Similar expressions can be written to the shearing and tearing SIFs. There are three major reasons for the indeterministic nature of the crack growth process:

1. The macro-properties of different structural specimens, which are the geometry, dimensions, and material properties, may differ slightly between specimens. Thus, the whole structural systems may be non-deterministic, as demonstrated in Chapter 5.

2. The external loadings in practical engineering cases are usually random, as demonstrated in Chapter 3.

3. The micro-properties of the structure or a specimen are random, which means that the microstructure is not homogenous, even for strictly controlled material production conditions. This random behavior is demonstrated in this chapter.

When tests are performed on many "identical" specimens, typical results look like those in Figure 6.3 (e.g., [66, 78, 79]).

These results are characterized by three major properties:

1. The behavior of crack length is random, even when very carefully controlled experiments are performed with "identical" specimens;

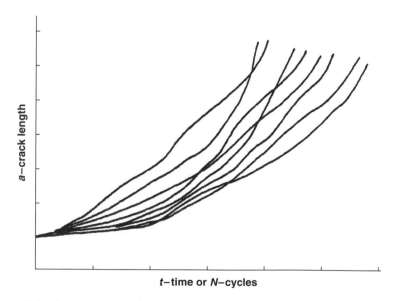

t−time or N−cycles

FIGURE 6.3 Generic curves for crack size as a function of time.

2. The crack length behavior is non-linear in time;

3. The curves of different specimens intermingle.

Four sets of basic experimental results are frequently quoted in the literature. In [80], two series of experiments for aircraft fasteners holes are described. In [66], a very well controlled test is described for a centrally cracked finite plate. More experimental results are shown in [67]. All results have characteristics similar to those shown in Figure 6.3.

When the growth rate da/dN is plotted against the stress intensity factor, experimental results yield the experimental circles depicted in Figure 6.4. On a log-log scale, the straight line shown is, in fact, the range of Paris Law [62]. The rise toward infinity at the fracture toughness value is modeled by Forman type laws. These laws are discussed later in this chapter. These are models that describe the mean behavior of the experimental data, but not the randomness expressed by the scatter in the experimental results, and therefore there is an uncertainty in the model, already discussed in Chapter 5.

In a large part of Figure 6.4, the data lies "around" a straight line. This observation is the origin of the Paris-Erdogan law, which states that

$$\frac{da}{dn} = C \cdot (\Delta K)^n \tag{6.2}$$

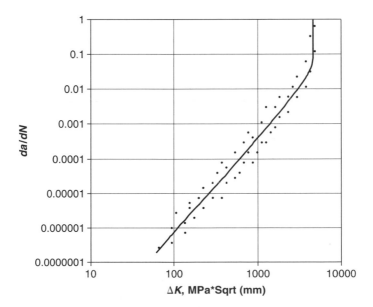

FIGURE 6.4 Crack growth rate vs. stress intensity factor.

C and n are material properties, obtained from data reduction of experimental curves as the one shown in Figure 6.4.

Stochastic models for crack growth were suggested in many publications. (see, e.g., [69, 78, 79, 81–83]). These models include evolutionary probabilistic models (Markov chain, Markov diffusion models), cumulative jump models (Poisson process, birth process, jump correlated models), and differential equation models. A comprehensive summary of the state of the art is given in [77]. From these, the models that can be easily used in practical engineering problems are the differential equation (DE) models.

Two major groups of models are used to investigate the random behavior of the crack growth:

1. Use of deterministic DE for the crack propagation, while assuming that different parameters in these equations are random variables (RV methods).

2. Use of a modified DE for the crack growth rate, where the stochastic nature of this rate is expressed by a random process (RP methods).

In the simplest case, the Paris-Erdogan equation (Eq. (6.2)), one assumes that C and n are random variables and therefore da/dn is also random. Use of such methods has two major disadvantages:

1. No intermingling of the crack lengths curves (as observed in experiments shown in Figure 6.3) is obtained, thus the results do not represent one of the experimentally observed phenomenon.

2. When C and n are assumed to be random variables, their units become also random. This conclusion has no physical meaning. It can be avoided only if a nondimensional equation is used, so that C and n are nondimensional parameters. In other words, the crack growth should be expressed as

$$\frac{da}{dn} = D \cdot \left(\frac{\Delta K}{\overline{K}} \right)^m \qquad (6.3)$$

where $\overline{K}$ is a normalizing factor (which can be random) with units of the SIF. In this case, reported experimental results should refer to such a normalizing factor. At present, all the reported experimental parameters C and n do not include such information.

In the RP method, the crack growth law is expressed by adding a random process to the deterministic law:

$$\frac{da}{dn} = C \cdot (\Delta K)^n \cdot X(t) \tag{6.4}$$

where $X(t)$ is the random process. According to this approach, C and n are deterministic material constants that are determined by regression of the experimental results, and the dispersion in the experimental results is introduced through the stochastic process. Such a description does reproduce the intermingling effects depicted in Figure 6.3. Experimental results were tested using a stochastic process with log-normal distributions and a correlation time and excellent results were obtained [69]. Suggestions for the determination of the correlation time can be found in [69, 71, 72]. Practical examples of this approach are described in [16].

6.2 EFFECTS OF A STATIC BIAS ON THE DYNAMIC CRACK GROWTH

In a practical design process, the structural element is subjected to both static and dynamic loads. The presence of a static load has an influence on the dynamic crack propagation.

The effects of static load together with a dynamic, harmonic loading is described in the literature using the load ratio R, which is defined as

$$R = \frac{\sigma_{\min}}{\sigma_{\max}} \tag{6.5}$$

where $\sigma_{\min}, \sigma_{\max}$ are the minimum and the maximum values, respectively, of the applied stress. Thus, when $R = 0$, the loading is harmonic between 0 and $\sigma_{\max}$, with a mean value of $0.5 \cdot \sigma_{\max}$. This is equivalent to a static load of $0.5 \cdot \sigma_{\max}$, plus a harmonic load with amplitude $\pm 0.5 \cdot \sigma_{\max}$. When $R = -1$, $\sigma_{\min} = -\sigma_{\max}$, and the loading is purely harmonic. When $R > 0$, the static "bias" is positive (tension) and the harmonic loading is positive.

In both fatigue and fracture mechanics analyses it was shown experimentally that the existence of a positive bias (tension) decreases the life of the structure, compared to a $R = 0$ loading, while the existence of compressive stresses during the dynamic loading increases its life. Thus, S-N curves for different values of R may be found in the fatigue literature, and da/dn vs. SIF curves (like the one depicted in Figure 6.4) also reflect the effects of R. The crack closure theory explains that during compression phase of the loading the crack does not propagate, as the faces of the crack are pressed one against the other and no stress concentrations occur. Crack closure theories are widely documented in the literature and references are quoted in Section 6.5. Another explanation for the effects of the static "bias" loads on crack growth are described by a well-documented literature (see references in the text of Section 6.5), in which it is argued that two driving forces are required in order to cause propagation of the crack. The interested reader is encouraged to check these references.

The simplest form of Paris-Erdogan propagation law, Eq. (6.2) does not include the effects of the load ratio R. Nevertheless, these effects are included in most of the crack growth computer codes. The most famous of these codes is NASGRO® [63] (see Chapter 7).

The general crack growth equation used in NASGRO is

$$\frac{da}{dN} = C\left[\frac{1-f}{1-R}\Delta K\right]^n \frac{\left(1 - \frac{\Delta K_{th}}{\Delta K}\right)^p}{\left(1 - \frac{K_{max}}{K_c}\right)^q} \tag{6.6}$$

ΔK —stress intensity factor (depends on stress, crack length, geometry factor)

f —crack opening function for plasticity induced crack closure, given in NASGRO [63]

ΔK_{th} —threshold stress intensity factors

K_c —critical stress intensity factor

$$K_{max} = \frac{\Delta K}{1-R}$$

p and q are constants of a given material. This formula contains the crack opening function in order to incorporate fatigue crack closure analysis. For materials that are less sensitive to crack closure, the crack growth equation can be written as

$$\frac{1}{f_s}\frac{da}{dt} = \frac{da}{dN} = \frac{C\Delta K^n\left(1 - \frac{\Delta K_{th}}{\Delta K}\right)^p}{\left(1 - \frac{K_{max}}{K_C}\right)^q} \qquad 0 \le R < 1$$

$$\frac{1}{f_s}\frac{da}{dt} = \frac{da}{dN} = \frac{CK_{max}^n \left(1 - \dfrac{\Delta K_{th}}{\Delta K}\right)^p}{\left(1 - \dfrac{K_{max}}{K_C}\right)^q} \qquad R < 0 \qquad (6.7)$$

$$K_{max} = \frac{\Delta K}{1 - R}$$

where $f_s \cdot dt$ replaces N. These equations are sometimes called Foreman equations. In order to demonstrate a numerical solution of Eq. (6.7), these equations were used for an Aluminum 2024 alloy. Because this material is sensitive to crack closure, Eq. (6.7) was used. These equations can be solved using a MATLAB® numerical procedure (described in the Appendix), while use of Eq. (6.6) requires the use of the much more expensive NASGRO program. A plate of width $b = 80$ mm has a central through initial crack whose length is 4 mm, thus the half initial crack length is $a_0 = 2$ mm. In NASGRO help files [63], it can be seen that for tension loading of ΔS, the SIF is given by the following expression, which includes a geometric correction for the result of an infinite plate:

$$\Delta K = \Delta S \sqrt{\text{secant}\left(\frac{\pi a}{b}\right)} \cdot \sqrt{\pi a} \qquad (6.8a)$$

When the plate is loaded in bending, the SIF is

$$\Delta K = \frac{1}{2}\Delta S \sqrt{\text{secant}\left(\frac{\pi a}{b}\right)} \cdot \sqrt{\pi a} \qquad (6.8b)$$

Assume that the stress (tension or bending at the crack location) is given by

$$S = \pm 200 \text{ MPa} \Rightarrow \Delta S = 400 \text{ MPa}$$

$$R = -1 \Rightarrow S_{max} = 200 \text{ MPa} \qquad (6.9)$$

Also assume that the frequency of the harmonic loading is $f_s = 10$ Hz, thus the dynamic load factor is 1 (DLF = 1). For Aluminum 2024 T3, the following data is obtained from the NASGRO database:

$$C = 0.2382 \cdot 10^{-11}$$

$$m = 3.2$$

$$p = 0.25 \qquad \text{(units are in MPa, mm)} \qquad (6.10)$$

$$q = 1$$

$$K_C = 2604$$

$$\Delta K_{th} = 200$$

Note the importance of declaring the system of units in the crack growth models. Results of any analysis may be distorted due to mistreating of the units!

Two MATLAB files are required for the solution: file **prop1.m** (see Appendix) is for the solution of the ordinary differential equation (ODE), and file **crack2.m** (also shown in the Appendix), which has the required data and calls file **prop1.m** during the solution phase.

In Figure 6.5, the crack length as a function of the bending load cycles as obtained using the MATLAB solution (Eq. (6.7)) is shown. In Figure 6.6, results for the tension loading are shown.

In Figure 6.7, the crack length as a function of the bending load cycles as obtained using the NASGRO demonstrator solution (Eq. (6.6)) is shown. In Figure 6.8, results for the tension loading obtained using the NASGRO demonstrator are shown.

The results obtained using the approximate formula (Eq. (6.7)) present shorter life (about 50% less) than those obtained using the NASGRO code. Remembering that the demonstrated case has $R = -1$, with a relatively high input stress (200 MPa, relative to yield of 365.4 MPa, UTS of 455.1 MPa for this material), this means that for the tension-compression case, half of the time is spent in high value compression and half of it in tension. The same goes for the bending case, where one side of the through crack is under

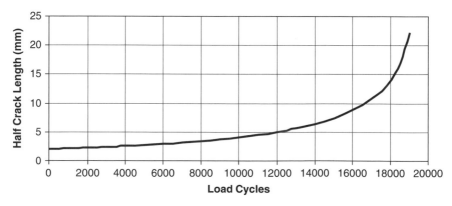

FIGURE 6.5 Crack length vs. bending load cycles, obtained using Eq. (6.7); failure—at 19000 bending load cycles.

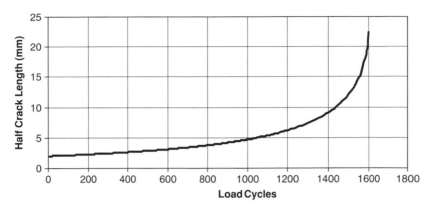

FIGURE 6.6 Crack length vs. tension load cycles obtained using Eq. (6.7); failure—at 1600 tensile load cycles.

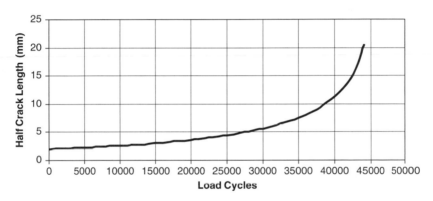

FIGURE 6.7 Crack length vs. bending load cycles, obtained using NASGRO (Eq. (6.6)); failure—at 44000 bending load cycles.

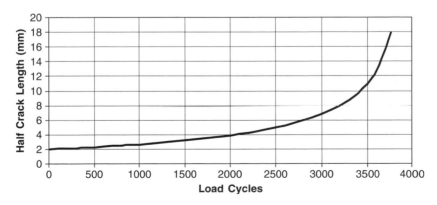

FIGURE 6.8 Crack length vs. tension load cycles obtained using NASGRO (Eq. (6.7)); failure—at 3770 tensile load cycles.

compression when the other half is in tension. This can explain the large error introduced when using the approximate equations. As the NASGRO contains much more effects and these effects are based on extensive experimental results, one can conclude that the use of NASGRO is preferred to the application of the approximate formulas. These effects include empirical equations for the threshold stress intensity factor and empirical expressions for $\frac{K_C}{K_{IC}}$. In any case, designers should be cautious when using any of the models, and include a suitable (large) factor of safety in this kind of problems.

6.3 STOCHASTIC CRACK GROWTH AND THE PROBABILITY OF FAILURE FOR HARMONIC EXCITATION

Stochastic behavior of crack length with time (or with applied load cycles) can be solved by using the (modified) Foreman equations, multiplied by a stochastic process $X(t)$, similar to what was written in Eq. (6.4) for the Paris-Erdogan law.

$$\frac{1}{f_s}\frac{da}{dt} = \frac{da}{dN} = \frac{C\Delta K^m}{\left(1 - \frac{\Delta K}{(1-R)K_{IC}}\right)^q} \cdot X(t) \quad 0 \leq R < 1$$

$$\frac{1}{f_s}\frac{da}{dt} = \frac{da}{dN} = \frac{C\left(\frac{\Delta K}{(1-R)}\right)^m}{\left(1 - \frac{\Delta K}{(1-R)K_{IC}}\right)^q} \quad R < 0 \tag{6.11}$$

K_{IC} is the material fracture toughness.

These equations were used for their simplicity, without a loss of generality. As the solution is performed using numerical solution of the differential equation for the crack length a, any equation can be used.

In order to include the geometry factor (which is shown in Eq. (6.8) for a finite plate) the SIF can be expressed as

$$\Delta K = \Delta S \cdot (\text{geom}) \cdot \sqrt{\pi a} \tag{6.12}$$

where (geom) is a geometry factor. Geometry factors can be found in [84].

Failure occurs when the crack growth rate tends to infinity, and it can be shown that the critical crack length (e.g., the length in which a violent crack growth occurs) is

$$a^* = \left[\frac{(1-R)K_{IC}}{(\text{geom})\, S_{\text{allow}}} \right]^2 \qquad (6.13)$$

S_{allow} is the maximum stress that the structural element can carry (its UTS), and K_{IC} is the material fracture toughness, which is a material property given, for instance, in the NASGRO materials database.

Eq. (6.11) was solved numerically using MATLAB, where at any time point the random process was created by a random numbers generator. Normal and lognormal stochastic processes were used in the demonstration.

Solving Eq. (6.11) many times is similar to performing many "virtual tests." Many curves of crack length a as a function of time are obtained. Then a statistical analysis of the results can be performed; for instance, the distribution of crack length for a given time point. In a "stress-strength" model, these results are the "stress" of the problem. The strength and its distribution can be obtained by finding the distribution of the critical crack length (Eq. (6.13)), assuming S_{allow} and K_{IC} may also be random variables. Using the "stress-strength" model, one can find the probability of failure at a given time point. As the variable a obtained from solving Eq. (6.11) and the variable a^* obtained from Eq. (6.13) both depend on K_{IC}, they are correlated, and this correlation has to be considered when the probability of failure is computed.

A diagram that describes the computation algorithm is shown in Figure 6.9. Two kinds of stochastic processes (normal and log-normal) are possible. Also there is a possibility to consider K_{IC} and S_{allow} as either deterministic or normally random distributions. The MATLAB required m-files can be found on the CD-ROM attached to this publication (see also the Appendix). Names of the relevant files are shown in Figure 6.9.

The calculated model is an infinite plate (thus geom $= \sqrt{\pi}$) with an initial half crack of length $a_0 = 0.1$ mm. The plate is loaded with a uniform stress with $S = 554.38$ MPa at a frequency of $f_s = 150$ Hz. The material is 4340 steel with a UTS value of 165 ksi. Data taken from NASGRO is summarized in Table 6.1.

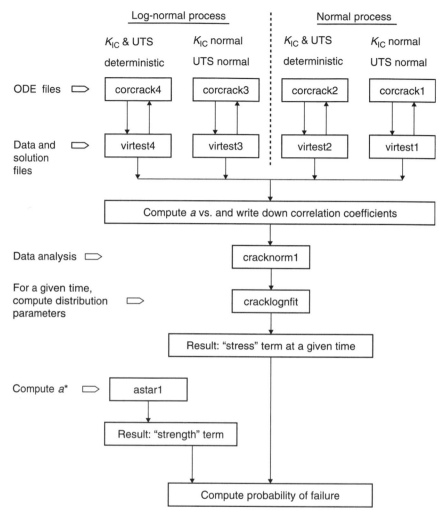

FIGURE 6.9 Flowchart for the computations (files are included on the CD-ROM and their names are written in the blocks of the flowchart).

When K_{IC} was assumed random, a normal distribution was assumed, with the mean value given in Table 6.1 and a standard deviation of 155 MPa · $\sqrt{mm}$. When S_{UTS} was assumed random, again a normal distribution was assumed with a mean value given in Table 6.1 and a standard deviation of 40 MPa.

First, a small number of 10 random cases were run, with $R = 0$, thus the loading is between $S = 0$ and $S = 554.38$ MPa, with a frequency of 150 Hz and assuming the stochastic process is normally distributed. In Figure 6.10, the

TABLE 6.1   Data for 4340 steel with UTS = 165 ksi.

Parameter	Value
S_{yield}—Yield stress	1069 MPa
S_{UTS}—Ultimate tensile strength	1172 MPa
K_{IC}—Fracture toughness	4691 MPa$\sqrt{mm}$
C—Material constant	$0.298 \cdot 10^{-11}$ (compatible units)
m—Material constant	2.7
q—Material constant	0.25

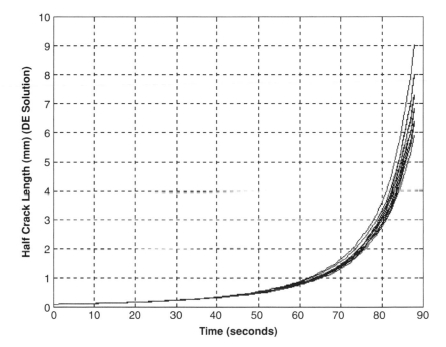

FIGURE 6.10   Half crack length vs. time, normal stochastic process, $R = 0$, 10 cases.

half crack length vs. time is shown for this case. In Figure 6.11, a zoom on part of Figure 6.10 is shown. The intermingling of the curves can be clearly observed. Then the same case was computed for 500 random cases (with the same data). Results of crack length vs. time are shown in Figure 6.12. The standard deviation of the half crack length at a given time is shown in

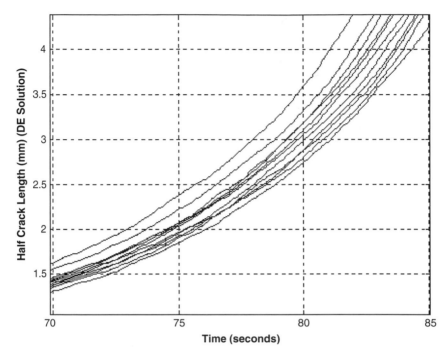

FIGURE 6.11 A zoom on part of Figure 6.10.

Figure 6.13. In Figure 6.14, the mean value of the half crack length plus and minus three standard deviations is described. In Figure 6.15, a histogram for the half crack length at $t = 84$ seconds is shown, and a lognormal distribution is fitted to this histogram.

The critical half crack length a^* was also computed for $R = 0$ and $R = 0.25$, by running 20,000 Monte Carlo simulations. The results are

For $R = 0$:

$$\mu_{a^*} = \text{mean} = 5.119212 \text{ mm}$$

$$\sigma_{a^*} = \text{standard deviation} = 0.49068 \text{ mm}$$

$$\text{parameters for the log} - \text{normal distribution:} \qquad (6.14)$$

$$\xi_{a^*} = 1.62843$$

$$\varepsilon_{a^*} = 0.095632$$

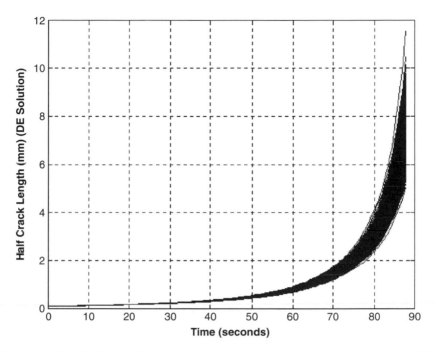

FIGURE 6.12 Half crack length vs. time, normal stochastic process, $R = 0$, 500 cases.

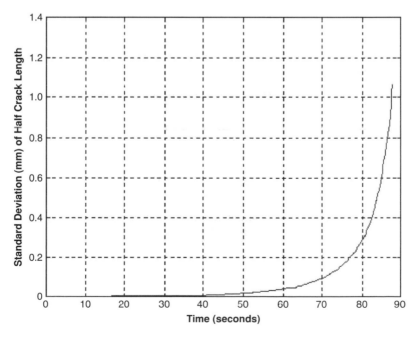

FIGURE 6.13 Standard deviation of half crack length as a function of loading time.

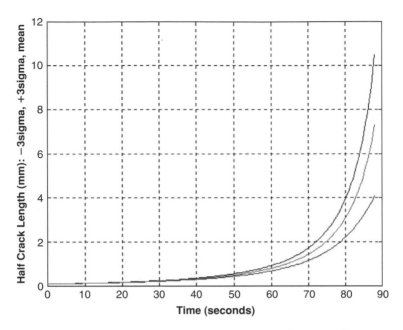

FIGURE 6.14 Mean value ±3 standard deviations, as a function of time.

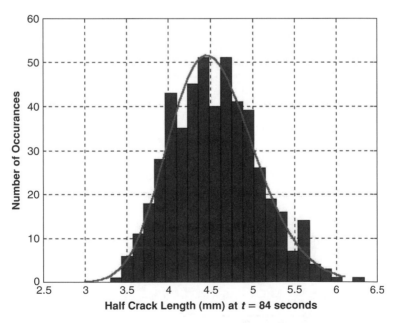

FIGURE 6.15 Histogram of half crack length at $t = 84$ sec. and a fitted log-normal distribution.

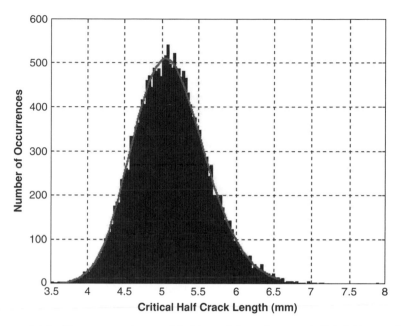

FIGURE 6.16 Histogram of critical half crack length, $R = 0$ (20,000 samples) and a fitted log-normal distribution.

Similar results can be obtained for $R = 0.25$. In Figure 6.16, a histogram of the results for $R = 0$ is depicted, and a fitted log-normal distribution is also shown.

Note that in both Figures 6.15 and 6.16, the fitted lognormal distributions can easily be replaced by normal distribution.

The distributions shown in Figures 6.15 and 6.16 were introduced into a "stress-strength" model, and the probability of failure was calculated using ProFES®. In Figure 6.17, this probability is shown for both values of R as a function of time.

The same computations were performed for a log-normal stochastic process. In Figure 6.18, the probability of failure for the two different stochastic processes is shown for $R = 0$.

The demonstrated crack length behavior with time (or number of loads) includes a correlation between the crack length and the critical length, due to the dependence of both lengths on the random K_{IC}. It does not include a correlation, which may exist in the stochastic process used for the

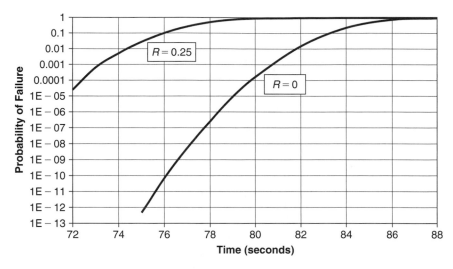

FIGURE 6.17 Probability of failure as a function of time, normal stochastic process.

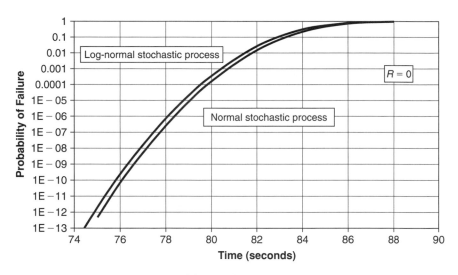

FIGURE 6.18 The probability of failure for two stochastic processes, $R = 0$.

demonstration. Correlation times for the stochastic process are discussed and described in [69, 71], and can be easily added to the solution process. The randomness of K_{IC} and S_{UTS} was described by both normal and lognormal distributions. There is no difficulty in applying other distributions to these parameters. The distribution is treated in the MATLAB files through the relevant random number generator and thus can be modified. In addition, other geometric factors can be used to solve other structural geometries. Such factors can be found in [84].

6.4 INITIAL CRACKS AND FLAWS

The propagation of cracks depends on the size of the initial cracks that are contained in the structural element. Several reasons are responsible for the existence of such initial cracks. All of the initial cracks formations are random in nature. Different values of small initial cracks may cause different propagation of these cracks. These small cracks are typically caused and affected by material and treatment factors (particles and inclusion), manufacturing processes (scratches and dents), working and loading conditions (corrosion), and geometrical factors (holes and corners). This is the reason why much effort is invested in the definition and the mapping of the statistical distributions of the initial cracks. Some of the research efforts are described in [85–87].

One of the most applicable methods to define the initial cracks distribution is the Equivalent Initial Flaw Size (EIFS), where cracks are measured in a certain time in the history of the structure, and a backward process is performed in order to estimate the initial flaws' sizes that existed at the beginning of the propagation process. In the last two decades, a significant amount of preliminary experimental work has begun to be carried out to build raw databases for accurate determination of EIFS values.

In [80], experimental results obtained at Wright Patterson Air Force Base were published, for two sets of fastener holes. The two tested configurations are called WPF series (33 specimens) for single fastener hole (no load transfer), and XWPF series (38 specimens) for double fastener hole (15% load transfer).

In [88, 89], the Time To Crack Initiation (TTCI) of the cracks described in [80] was statistically evaluated. It was found that the statistical distribution of the TTCI results is a Weibull distribution.

The classical expression for the cumulative probability function (CDF) of a Weibull distribution is

$$F_T(t) = P[T \leq t] = 1 - \exp\left\{-\left(\frac{t-\varepsilon}{\beta}\right)^\alpha\right\}; \quad t \geq \varepsilon \qquad (6.15)$$

α and β are the distribution parameters. In some cases, the Weibull distribution is described in another form:

$$F_T(t) = 1 - \exp(-a_E t^{b_E}); \quad \text{(for } \varepsilon = 0) \tag{6.16}$$

This is the form used, for instance, by MATLAB. Therefore, depending on the tool that is applied to create random numbers, one should be careful as to what distribution parameters are used. It is easily shown that

$$a_E = \left(\frac{1}{\beta}\right)^{\alpha}$$

$$b_E = \alpha \tag{6.17}$$

The Weibull parameters found for the TTCI distribution are listed in Table 6.2.

Also, the probability density function (PDF) can be computed using the equation (in the "MATLAB" form)

$$f_T(t) = a_E b_E t^{b_E-1} \exp(-a_E t^{b_E}) \tag{6.18}$$

Results for the TTCI values distributions are shown in Figure 6.19 (CDF) and Figure 6.20 (PDF) for the two series of experimental data. Calculations were done using the MATLAB m-file **weib1.m**, included in the Appendix.

The CDF of the EIFS for both cases are also obtained by [88, 89]. A value of $a_0 = 0.03''$ (initial crack size) was assumed by Yang. These results were plotted using a MATLAB m-file **weib2.m**, listed in the Appendix. The CDF functions were plotted on a semi-logarithmic paper (this is how they are shown in [88, 89]) in Figures 6.21 and 6.22 for both series of experiments.

TABLE 6.2 Weibull parameters for TTCI distributions of WPF and XWPF series.

	WPF Series	XWPF Series
α	4.9174	5.499
β	15936	11193
ε	0	0

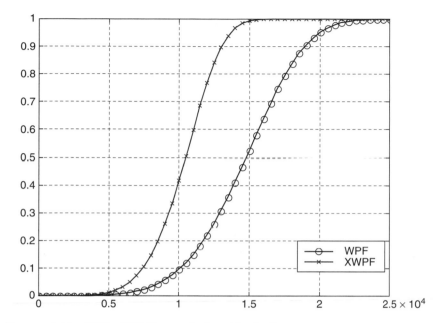

FIGURE 6.19 CDF of TTCI, as calculated by [88]; time axis in hours.

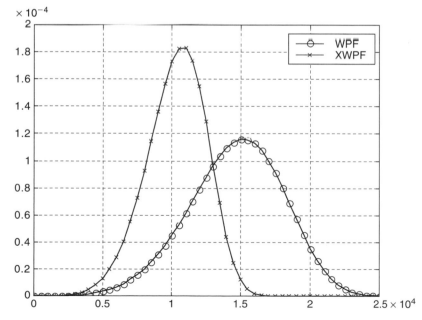

FIGURE 6.20 PDF of TTCI, as calculated by [88]; time axis in hours.

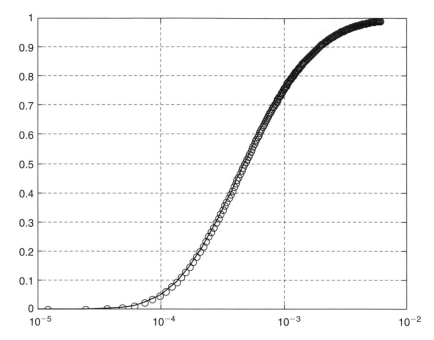

FIGURE 6.21 CDF of EIFS, WPF series; crack lengths are in inches.

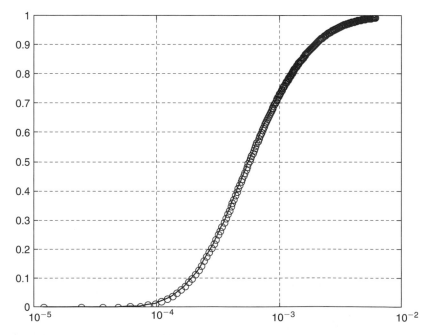

FIGURE 6.22 CDF of EIFS, XWPF series; crack lengths are in inches.

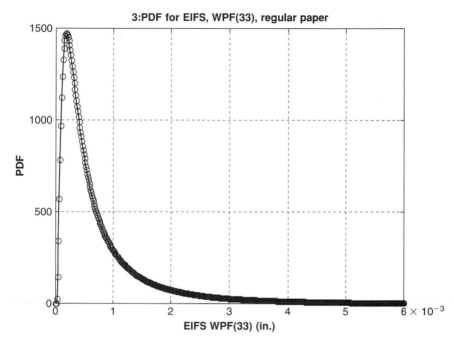

FIGURE 6.23 PDF of EIFS, WPF series, regular paper; crack lengths are in inches.

Then, the PDF functions were obtained by numerical differentiation of the CDF functions, and are shown in Figures 6.23 and 6.24.

The distributions obtained are not exactly Weibull, but Weibull distributions were approximated by a trial and error procedure. The parameters of these approximated distributions are listed in Table 6.3. The values of the Weibull distribution parameters depend on the units used for the crack length. The table refers to length given in inches. The approximated distributions with the original ones are described in Figures 6.25 and 6.26 for the WPF and the XWPF cases, respectively.

The preceding fitted Weibull distributions can be used in the probabilistic analysis of these kinds of structures, as these distributions provide an analytical expression.

Another source of data for initial flaws distribution can be found in [90]. In this paper, the initial flaws depth and width for Aluminum 2024-T3 is described, as shown in Figures 6.27 and 6.28, respectively. These distributions can be easily approximated by log-normal distributions. Results for Aluminum 7075-T6 and 2524-T3 are also available.

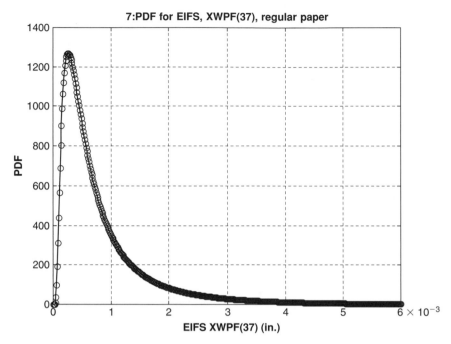

FIGURE 6.24 PDF of EIFS, XWPF(37), regular paper; crack lengths are in inches.

TABLE 6.3 Parameters for Weibull approximation (for crack length in inches).

	WPF(33)	*XWPF(37)*
α	0.918	1.0
β	$7.320582\,\mathrm{g}10^{-4}$	$8.38062\,\mathrm{g}10^{-4}$
a_E	755.696012	1193.22914
b_E	0.918	1.0

In [91], EIFS distributions for corroded and noncorroded aircraft fuselage splices are described. The distribution can be approximated by a Weibull distribution, as shown in Figure 6.29.

In [86], EIFS of inclusions in aluminum 7050-T7451, which can be approximated by a log-normal distribution, is shown in Figure 6.30.

It can be seen that the magnitudes and the distributions of initial flaws are not always the same. In some results, a Weibull distribution is a good fit to the experimental result. In other cases, a lognormal distribution is a better

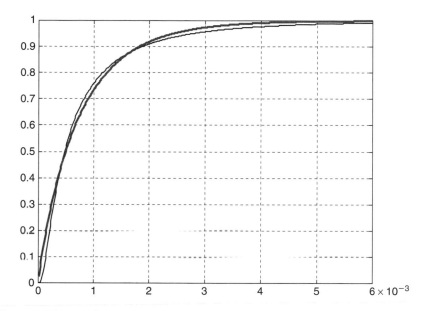

Figure 6.25 A fitted Weibull distribution (bolder line) to the EIFS CDF of Figure 6.21; crack lengths are in inches.

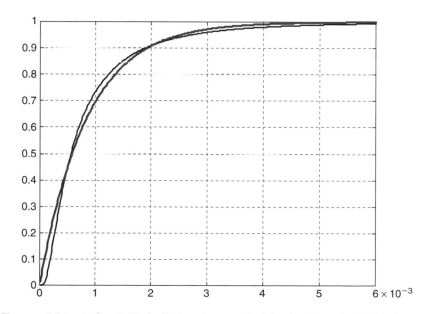

Figure 6.26 A fitted Weibull distribution (bolder line) to the EIFS CDF of Figure 6.22; crack lengths are in inches.

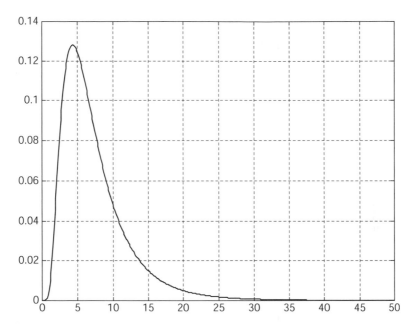

FIGURE 6.27   PDF of EIFS depth for Aluminum 2024-T3; depth is in microns.

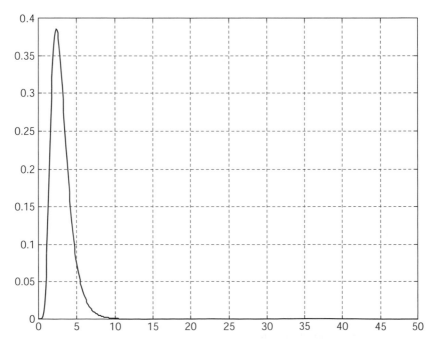

FIGURE 6.28   PDF of EIFS width for Aluminum 2024-T3; width is in microns.

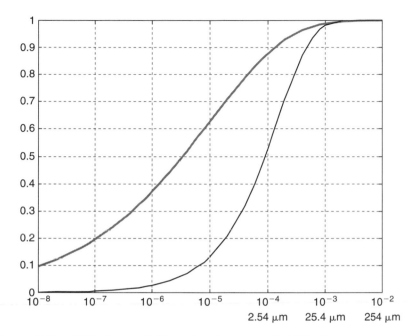

FIGURE 6.29 CDF of corroded (grey) and noncorroded (black) aluminum; crack length is in inches, values in microns are also shown.

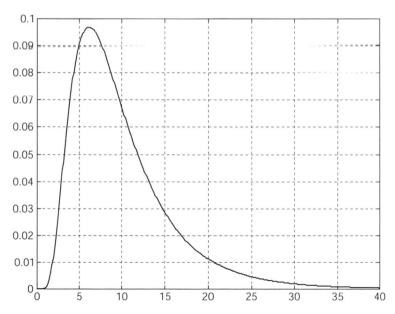

FIGURE 6.30 CDF of inclusions in 7050-T7451 aluminum, approximated by log-normal distribution; crack length is in microns.

approximation. It seems that every case should be treated separately, with a good inspection of the experimental results. Anyway, experimental results are essential to the characterization of the initial cracks in a given structure. Such experiments are very expensive and time consuming. The best thing to do is to initiate a joint research effort through the academy and the industry, in order to create a common database for the probabilistic analysis of certain types of structures.

6.5 PROBABILISTIC CRACK GROWTH USING THE "UNIFIED" APPROACH [92, 93]

The problem of crack growth is a major issue in the prediction and maintenance of aerospace structures, as well as other structural elements in mechanical and civil engineering projects. Prediction of expected life of a structural element due to constant (static) and alternating loading (fatigue) is of major concern to the designers. Prediction of remaining life of the structural elements influences the decisions of maintenance engineers (checking intervals, corrections, replacements).

For the last three decades, fracture mechanics have been the main tool with which such problems have been treated. During the last three decades, fracture mechanics scientists and engineers have made tremendous advances, from the basic practical approach dominated by Paris-Erdogan law to increasingly sophisticated crack growth models. Mathematical and metallurgical models (both deterministic and probabilistic), and experimental analysis of simple models and testing of complex structures have resulted in thousands of publications, dozens of models for crack growth and life prediction, maintenance decision-making processes, and numerous computer codes for crack growth analysis.

A major concern of fracture mechanics is the influence of the load ratio on the behavior of cracks. In "real life," both static and alternating (i.e., vibrations) loadings exist simultaneously. This is expressed in the load ratio R, which is classically defined as

$$R = \frac{S_{min}}{S_{max}} \tag{6.19}$$

S_{min} and S_{max} are the minimal and maximal applied stresses, respectively, in the far field. Thus, $R = -1$ refers to "pure" vibrations (mean value of the load is zero), and $R = 0$ is a loading between zero level ($S_{min} = 0$) and a maximum value (S_{max}).

Experiments have shown that the value of R influences the crack growth rate (da/dn, where a is the crack length and n is the number of load cycles). It was argued that the reason for this influence is the crack closure effect. Basically, this approach claims that the crack is not propagating while the crack's faces are compressed to each other. Crack closure was first introduced by Elber [94, 95]. Since then, many publications treated the phenomena, suggesting models for the closure phenomena (e.g., [96]) and methods to measure the driving force due to crack closure. The ASTM introduced a standard method (ASTM standard E647) for its measurement, and a number of numerical codes for its computation were developed (e.g., [97, 98]). The crack closure is also claimed to be the reason for the different (and sometimes contradicting) behavior of very short cracks and micro-structural cracks. In the last decade, some experimental results have shown that the role of closure in the crack growth process might have been exaggerated. Measurements of closure stresses were done in 10 separate labs (i.e., [99, 100]), with completely differing results. Experiments done in vacuum did not show the R effects on the results (i.e., [64, 101, 102]). Some approaches, which are considered controversial, have been published, claiming that the crack growth driving forces depend not only on the change in the stress intensity factor (SIF) ΔK, but on additional (mainly local internal) stresses.

One of these approaches, the so-called "unified" approach, is described in many papers by the NRL research group (i.e., [65, 103–107] and cited references). According to these works, the growth of the crack depends on both ΔK and K_{max}, and in order for a crack to grow, two thresholds values must be met. According to this approach, the local driving force of short cracks is comprised of two stresses—one originated from the far field stress intensity factor and one from local internal stresses close to the crack tip. These local stresses "create" local R values near the crack tip, which are different from the far field R value. The internal stresses were measured [108] and compared to finite element computations. By using this approach, the behavior of both short cracks and long cracks can be treated using one "unified" growth law. This approach is supported by data analysis of many previous experimental results for many materials described by numerous authors and laboratories.

Another approach, which somehow emerges from the previous one, is described in many publications of Kujawski (i.e., [109–111]). According to this approach, the driving force for crack growth is not the stress intensity factor alone, but a combined parameter $\left(\Delta K^+ \cdot K_{\max}\right)^{0.5}$, where ΔK^+ is the positive part of the applied stress intensity factor. When plotting previously published experimental results for aluminum, steel, and titanium alloys, the curves of da/dn vs. this combined parameter were collapsed into an almost single curve (for a given material), showing almost no effects of R.

In this section, these two approaches are used in order to write two empirical crack growth models for an Aluminum 2024-T351 (but can be performed for any other material for which experimental results are available), which are different from those used in industry today. The main tool used today for crack propagation analysis in the industry is the NASGRO 4 code, which was developed by NASA and is maintained presently by Southwest Research Institute in San Antonio, TX. After these empirical rules (which best fit the experimental results) are written, they are used to demonstrate the prediction of the crack length as a function of the load cycles, without the need to use crack closure based methods.

The experimental raw data for long and short cracks in Al 2024-T351 is taken from [110], quoted from [112] and [113]. The data presented here were taken from [110] for 8 values of da/dn and are shown in Figures 6.31 and 6.32. For clarity, the sampled data points are connected by straight lines in a log-log scale, which do not represent all of the intermediate results. Crack length histories for two pairs of $(\Delta S; R)$ were taken directly from [112], and are shown and discussed later in this section.

The data from the regular tests (long cracks) were plotted on a $\Delta K - K_{\max}$ plane in Figure 6.33. This is a planar projection (on the $\Delta K - K_{\max}$ plane) of a 3-D plot, where the Z-axis is da/dn. On this plot, points of equal R lie on straight lines, whose slope is $(1 - R)$. Three R lines are also shown. A family of experimental data points for the L-shaped curves is thus obtained. Such models were successfully depicted in the past for many dozens of materials in the works described in [104–108].

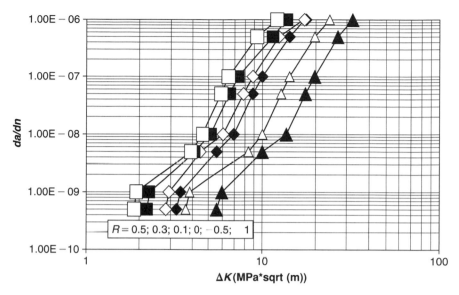

FIGURE 6.31 Sampling of the experimental data, long cracks, five values of R (see [110]).

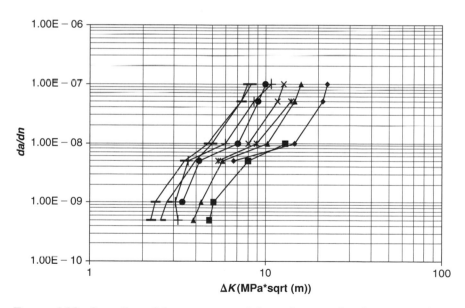

FIGURE 6.32 Sampling of the experimental data, short cracks, five values of R, for several values of initial crack (see [110]).

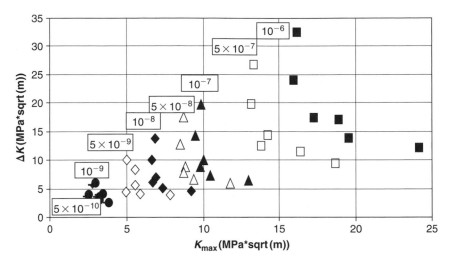

FIGURE 6.33 The experimental data of Figure 6.31 in the $K_{max} - \Delta K$ plane (values of da/dn for each family of symbols are shown in boxes).

In Figure 6.34, the experimental results of Figure 6.33 were smoothed. The L-shaped curves were drawn for the smoothed data, and "virtual data points" from the smoothed curves were selected. These are shown in Figure 6.34. A straight line was made through the corners of the L-shaped curves, whose slope is 0.73. Thus $R^* = 0.27$. The asymptotic values of ΔK_{th}^* were plotted against $K_{max_{th}}^*$ (where the subscript "th" stands for "threshold"), and are shown in Figure 6.35. In the same figure, the straight line

$$\Delta K_{th}^* = 0.73 \cdot K_{max_{th}}^* \tag{6.20}$$

is also plotted. The single straight line obtained suggests that the crack growth physical mechanism is not changed (for this material) during the propagation process. For each level of da/dn there are asymptotes for $K_{max_{th}}^*$ and ΔK_{th}^*, the values of which are shown in Figure 6.35. Also shown are best fits to a second order function of $\log(da/dn)$:

$$K_{max_{th}}^* = 84.914 + 16.503 \cdot \log\left(\frac{da}{dn}\right) + 0.8252 \cdot \left[\log\left(\frac{da}{dn}\right)\right]^2$$

$$\Delta K_{th}^* = 61.793 + 12.005 \cdot \log\left(\frac{da}{dn}\right) + 0.6002 \cdot \left[\log\left(\frac{da}{dn}\right)\right]^2 \tag{6.21}$$

This is in agreement with Eq. (6.20).

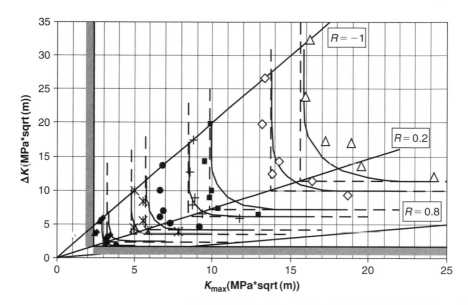

FIGURE 6.34 Experimental data (symbols) and smoothed data (lines), $K_{max} - \Delta K$ plane (da/dn values omitted for clarity).

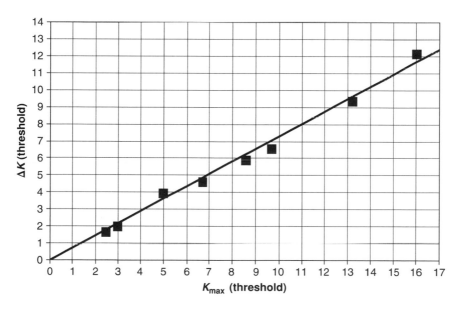

FIGURE 6.35 ΔK at threshold vs. K_{max} at threshold for different values of da/dn (symbols are from the smoothed experimental results; line is Eq. (6.20)).

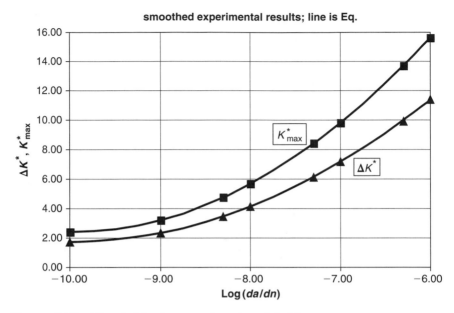

FIGURE 6.36 Threshold values as a function of da/dn.

The thresholds below which a crack arrest is obtained were defined arbitrarily when $\frac{da}{dn} = 1 \cdot 10^{-10}$, and the following values were obtained:

$$\Delta K_{th}^*(\text{lowest}) = 1.75 \,\text{MPa} \,\sqrt{m}$$
$$K_{max_{th}}^*(\text{lowest}) = 2.4 \,\text{MPa} \,\sqrt{m} \tag{6.22}$$

These threshold values are also shown as limits in Figure 6.34. An optimal surface was fitted to the smoothed results described in

$$\frac{da}{dn} = A \cdot \left(K_{max} - K_{max_{th}}^*\right)^n \cdot \left(\Delta K - \Delta K_{th}^*\right)^m$$

Figure 6.34. This is given by

$$A = 1.08033356 \cdot 10^{-7}; \quad n = 1.968271; \quad m = 1.680724;$$

$$K_{max} \text{ and } \Delta K \text{ are in MPa} \,\sqrt{m}, \quad \frac{da}{dn} \text{ is in } \frac{m}{\text{cycle}} \tag{6.23}$$

Eq. (6.23) is the differential equation for the "unified" approach model. Note that $K_{max_{th}}^*$ and ΔK_{th}^* depends on da/dn, according to Eq. (6.21), and ΔK_{th}^* can be expressed as a function of $K_{max_{th}}^*$ with Eq. (6.20).

The experimental raw data of Figure 6.31 was processed for a da/dn vs. $(\Delta K^+ \cdot K_{max})^{0.5}$ presentation, using the following rules:

$$K_{max} = \frac{\Delta K}{1 - R};$$
$$\Delta K^+ = \Delta K \qquad \text{for } R \geq 0;$$
$$\Delta K^+ = K_{max} \qquad \text{for } R < 0$$

(6.24)

Results are shown in Figure 6.37. The previously smoothed results were also processed, and are shown in Figure 6.38. The results of both Figures 6.37 and 6.38 were approximated by the following equation:

$$\frac{da}{dn} = C \cdot (PK)^m$$
$$PK = (\Delta K^+ \cdot K_{max})^{0.5}$$
$$C = 7.648179 \cdot 10^{-12}$$
$$m = 4.05$$

(6.25)

PK units are MPa $\sqrt{m}$, $\dfrac{da}{dn}$ are $\dfrac{m}{cycle}$

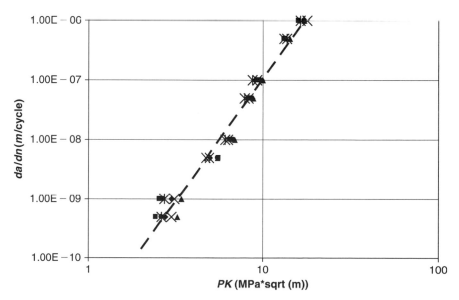

FIGURE 6.37 Collapsed experimental data for different values of R.

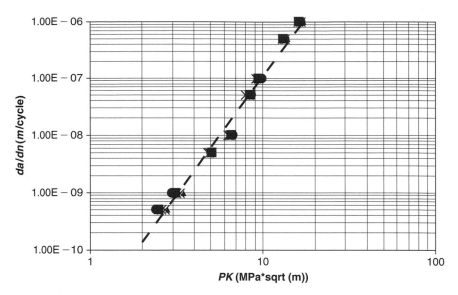

FIGURE 6.38 Collapsed smoothed experimental results for different values of R.

In log-log scale, this is a straight line, which is also plotted in the figures. Eq. (6.25) is the differential equation for the $\left(\Delta K^+ \cdot K_{max}\right)^{0.5}$ approach model.

In describing the NASGRO model, some of the equations already written are repeated for clarity.

The general crack growth equation used in NASGRO is

$$\frac{da}{dN} = C\left[\frac{1-f}{1-R}\Delta K\right]^n \frac{\left(1 - \dfrac{\Delta K_{th}}{\Delta K}\right)^p}{\left(1 - \dfrac{K_{max}}{K_c}\right)^q} \tag{6.26}$$

where $C, n, p,$ and q are material properties and

ΔK —stress intensity factor (depends on stress, crack length, geometry factor)

f —crack opening function for plasticity induced crack closure

ΔK_{th} —threshold stress intensity factors

K_c —critical stress intensity factor

$$K_{max} = \frac{\Delta K}{1-R}$$

p and q are material constants. A more simplified model, which does not include crack closure effects, is given by the following Forman equation:

$$\frac{da}{dn} = \frac{C\,(\Delta K)^n}{\left(1 - \dfrac{\Delta K}{(1-R)\,K_c}\right)^q} \quad \text{for} \quad 0 \le R < 1 \tag{6.27}$$

For Aluminum 2024-T351, the following material data are given in the NASGRO database:

$$C = 6.054 \cdot 10^{-12}$$

$$n = 3$$

$$q = 1 \tag{6.28}$$

$$K_c = 1181\,\text{MPa}\sqrt{\text{mm}}$$

$$\frac{da}{dn} \text{ is in mm/cycle}$$

Eq. (6.27) can also be solved numerically, using the TK Solver™ program (see Chapter 7) to produce curves of the crack length a as a function of the load cycles n. These computations are not presented here, as the procedure was demonstrated in [73]. Nevertheless, in the NASGRO database, the slope of the straight line fitted to the experimental data results of da/dn vs. ΔK is smaller than the slope of the experimental results depicted in Figures 6.34 and 6.35, and suggests that the number of cycles required reaching a critical crack length is higher, and therefore NASGRO predicts longer life to the cracked structure.

The experimental results on which the NASGRO material data is based (i.e., [114, 115] and internal industry reports) are different from those described in [112, 113]. The experimental points lie in the same region described in Figure 6.31, but the slope is smaller. This will yield a slower growth of the crack, and consequently a higher value for the life of the structural element is obtained. The fact that different experimental results presented in the literature are sometimes inconsistent was already pointed out in the past [116], and should be further investigated in the future.

Eq. (6.23) describes the "unified" crack growth model. It should be noted that both $K^*_{\max}$ and ΔK^*_{th} of this equation are functions of da/dn, as described in Eq. (6.21). Therefore, a numerical solution was applied in order to compute the crack length as a function of the load cycles. Without loss of generality, a 2-D crack in an infinite plate was numerically solved. Therefore, a and a^* refer to half crack lengths. The numerical solution was performed using the TK Solver program [41]. Eq. (6.25) describes the $\left(\Delta K^+ \cdot K_{max}\right)^{0.5}$ driving force approach, and was also numerically solved using the TK Solver program. Computations were performed until a half crack length 0.005 m (5 mm) was obtained.

In Figure 6.39, these two methods are shown for $\Delta S = 200$ MPa, $R = 0$. In Figure 6.40, a smaller stress is applied, $\Delta S = 75$ MPa, for $R = 0.5$. Both methods agree very well. Nevertheless, it should be noted that the results of the $\left(\Delta K^+ \cdot K_{max}\right)^{0.5}$ method are very sensitive to the "best" slope calculated from the data points described in Figures 6.37 and 6.38, and therefore should be examined with extreme care.

The model described in Eq. (6.23) was compared to experimental results described in [112]. In the latter, crack length vs. load cycles curves for single edge notched specimens under bending are shown. The model predictions and experimental results for the crack growth (length above the initial crack length) are shown in Figure 6.41, for two pairs of $\Delta S; R$. Good agreement is demonstrated for the first half lifetime of the specimens,

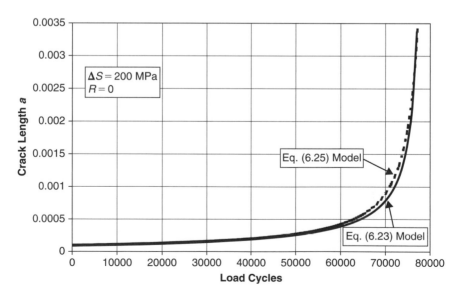

FIGURE 6.39 Crack length propagation for $\Delta S = 200$ MPa, $R = 0$, calculated with two methods.

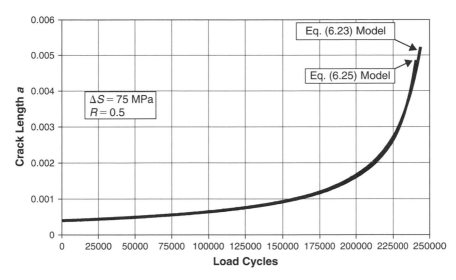

FIGURE 6.40 Crack length propagation for $\Delta S = 75\,\text{MPa}$, $R = 0.5$, calculated with two methods.

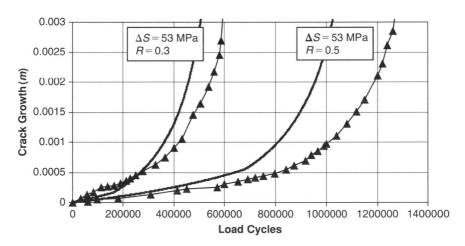

FIGURE 6.41 Model predictions (lines, Eq. (6.23)) and experimental results [112] (triangles) for two pairs.

while somewhat less agreement exists for the second half lifetime (toward failure).

The "unified" approach states that unless $\left(K_{\max} - K^*_{\max_{\text{th}}}\right) > 0$ and $\left(\Delta K - \Delta K^*_{\text{th}}\right) > 0$, there will be no growth in the crack length (as can also be seen from Eq. (6.23)). A value of a critical initial crack length a^* below which no crack growth can occur can be computed by using the numerical algorithm developed for the computation of the crack growth (based on Eq. (6.23)). Such values are shown in Figure 6.42, for different values of applied stress ΔS and stress ratio R. In Figure 6.43, a^* values for the right end of Figure 6.42 are shown, in microns (1 micron $= 1$ micrometer $= 1 \times 10^{-6}$ meter).

The threshold values of a^* shown in Figures 6.42 and 6.43 can be used to check, in future experiments, the validity of the controversial "unified" approach. According to this approach, initial cracks with lengths smaller than a^* should not propagate during a loading with the compatible ΔS and R values.

The use of Eq. (6.28) yields a longer lifetime for the cracked structural element. Typically, about twice the load cycles were required using the NASGRO model (to get a given crack length) than those required by the models described by Eq. (6.23) or Eq. (6.25). Thus, this model seems to be less conservative in a design process.

To demonstrate the probabilistic analysis, one has to know or to estimate the initial values of crack lengths. In the previous section, the EIFS method was described, and results based on [88] and [89] were depicted. Care must be taken as to what units are used in the analysis. In Section 6.4, crack lengths

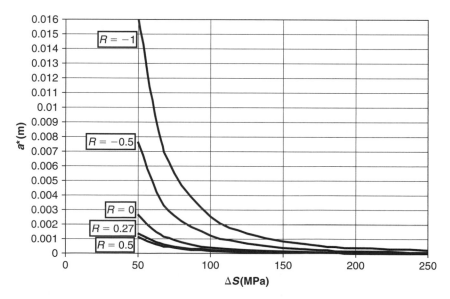

FIGURE 6.42 Critical crack length a^* (below which crack will not propagate).

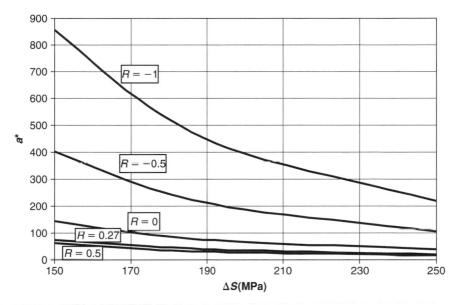

FIGURE 6.43 Enlargement of part of Figure 6.42; note that a^* is in microns.

were shown in inches. In this section, most of the length units are in meters and microns (10^{-6} meter). Therefore, for the numeric example some conversion of the EIFS must be performed.

In Figure 6.44, the EIFS of the WPF series experiments (see Section 6.4) is plotted on a regular paper. The values in this figure are identical to those in Figure 6.22 (EIFS), which was on semi-log paper. Then, values are changed into meters. In Figure 6.45, the results for the CDF and PDF of crack length in meters are described.

The Weibull distribution for the EIFS (in meters) is given by

$$F_{A_0}(a_0) = 1 - \exp\left(-A_E \cdot a_0^{B_E}\right)$$

$$A_E = 22014.4; \quad B_E = 0.918$$

or

$$F_{A_0}(a_0) = 1 - \exp\left[-\left(\frac{a_0}{\beta}\right)^{\alpha}\right]$$

$$\alpha = 0.918; \quad \beta = 0.000018594$$

(6.29)

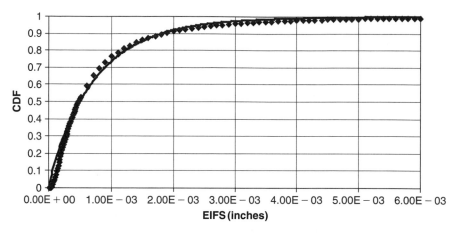

FIGURE 6.44 Calculated CDF of Yang's results (diamonds) and approximated Weibull distribution (line); regular scales.

where a_0 is the initial crack length (in m), and A_E and B_E (or α and β) are the Weibull distribution parameters.

The mean value μ_{a_0} and the standard deviation σ_{a_0} of the EIFS value of a_0 are given by

$$\mu_{a_0} = \beta \cdot \Gamma\left(\frac{1}{\alpha} + 1\right) = 0.00001936 \text{ m} = 19.36 \text{ micron}$$

$$\sigma_{a_0} = \beta^2 \left[\Gamma\left(\frac{2}{\alpha} + 1\right) - \Gamma^2\left(\frac{1}{\alpha} + 1\right)\right] = 0.0000211 \text{ m} = 21.1 \text{ micron}$$

Coefficient of variation $= 1.09$ \hfill (6.30)

The crack growth model (Eq. (6.23) and Eq. (6.21)) was numerically solved using the TK Solver program by UTS [41]. Solution was done for the following 2-D case: an infinite plate with a hole of radius $R = 0.005$ m (5 mm), and symmetric double through cracks of lengths a, under tension (see Figure 6.46). The stress intensity factor for this case is [33]:

$$\Delta K = \Delta S \cdot \sqrt{a} \cdot \sqrt{\pi} \cdot \left[1 + 2.365\left(\frac{R}{R+a}\right)^{2.4}\right] \hfill (6.31)$$

Other cases can be solved by introducing a suitable expression for the stress intensity factor. Stress intensity factors for numerous cases can be found in Murakami et al. [84].

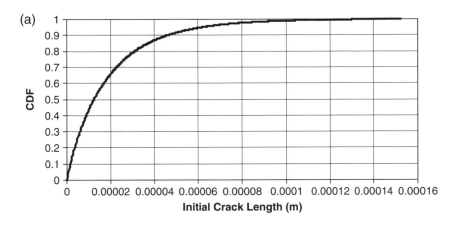

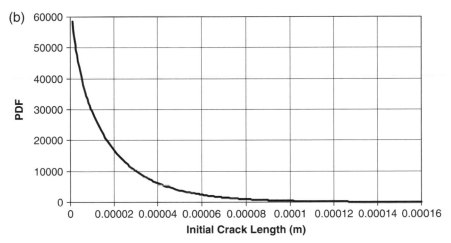

Figure 6.45 Approximated Weibull distribution (a) CDF, (b) PDF.

The EIFS CDF is approximated by Eq. (6.29). These values were obtained for a three-dimensional case. Nevertheless, for demonstration purposes, these EIFS values were used in the numerical computations described here, as the 2-D case is very similar to the 3-D case.

A reciprocating tensile load added to a static tension was used for the numerical example described. A value of $\Delta S = 75\,\mathrm{MPa}$ and a stress ratio $R = 0.5$ were selected. Deterministic analysis of such a case (according to [92]) showed that the deterministic initial crack threshold (the value below which the crack will not propagate) is $a^* = 0.0000168\,\mathrm{m} = 16.8\,\mathrm{microns}$.

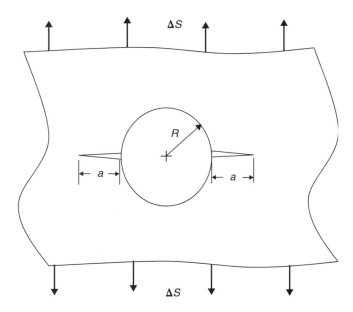

FIGURE 6.46 Geometry of the numerically solved case.

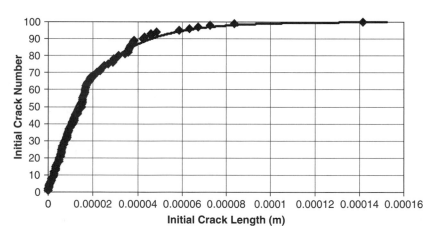

FIGURE 6.47 100 sampled initial crack lengths (diamonds) and the compatible trimmed Weibull CDF.

A sample of 100 initial crack values with the distribution given by Eq. (6.29) was created during the program run. In Figure 6.47, these initial values are shown in ascending order as diamond symbols. On the same figure, the corresponding Weibull distribution (Eq. (6.29), approximated from [88]), is also shown in full line.

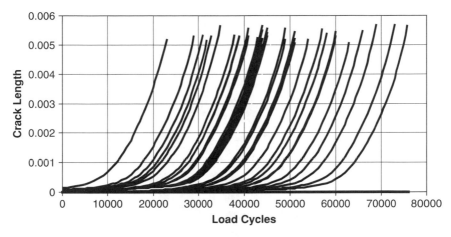

FIGURE 6.48 Crack length growth for 100 samples (35 of which propagate).

In Figure 6.48, the crack lengths vs. load cycles curves for these 100 cases are plotted as a function of the load cycles. Out of the 100 initial cracks, only 35 cracks were propagating, while 65 cracks (65%) were below the threshold value and did not show any growth. It should be noted that due to the random aspects of the solution, every computation provides different results, although mean and variance should be similar, converging to real values as the number of samples increases. A large dispersion of the crack growth behavior can be seen in Figure 6.48. In order to analyze the dispersion in the 35 propagating cracks, the number of load cycles required to reach a length of $a = 0.005$ m = 5 mm was extracted from the computations. The distribution of the load cycles required to produce this length of crack was analyzed statistically. Although a large dispersion is shown in the figure, the dispersion in the load cycles required to get a given crack length is smaller than the dispersion in the initial crack lengths. The mean number of required load cycles is 47,215 cycles, with a standard deviation of 12,363 cycles. The COV is 0.262, while the COV in Eq. (6.30) is 1.09, thus the dispersion in load cycles (to obtain a given crack length) is smaller than the dispersion of the initial EIFS values. Nevertheless, it should be noted that a factor of more than 3.3 exists between the maximum number of load cycles shown (75,100 cycles) and the minimum value (22,500 cycles). Such dispersion in "life cycles" should certainly be taken into account during any design process.

The number of nonpropagating cracks in the computed sample should tend, for an infinite number of samples, to the CDF of the Weibull distribution for

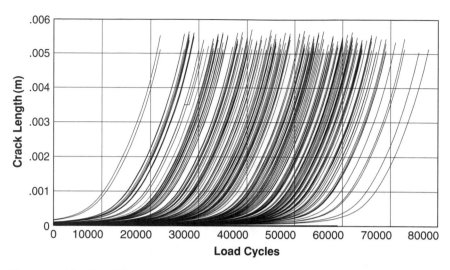

FIGURE 6.49 Crack length growth for 500 samples (182 of which propagate).

the threshold value $a^* = 0.0000168\,\text{m} = 16.8\,\text{microns}$, which is 0.598. Thus, for the EIFS distribution described by [88], about 60% of the flaws will not propagate. This was also checked by running a sample of 500 specimens. In Figure 6.49, the propagation of 182 cracks of this computation is shown. The number of nonpropagating cracks is 318, which is 64% of the total number of specimens. This percentage should approach 60% for an infinite number of specimens. For this case, the number of load cycles required to obtain a crack of length $0.005\,\text{m} = 5\,\text{mm}$ for the 182 propagating cracks has a mean value of 49,123 cycles, and the standard deviation is 12,671 cycles. The COV is 0.258, very similar to the results of 100 specimens. The ratio between the highest value to the smallest one is 3.65.

Statistical analysis of the results shown in Figure 6.49 was described in [93]. Suppose that the structure fails when the cracks reach a value of $0.005\,\text{m}$ (5 mm). Then the time in which the curves intersect this length is the time to failure. But only X% of the cracks propagate, which means that $(100 - X)\%$ do not propagate; i.e., their life expectancy is infinite. This concept is depicted in Figure 6.50. In this sketch, a distribution is shown for the time to failure, whose integral is not 1, but X%. The other part of the distribution is at time $= \infty$, and is very similar to the known delta function. The area under the spike is $(100 - X)\%$, thus the two parts together add to a 100% probability.

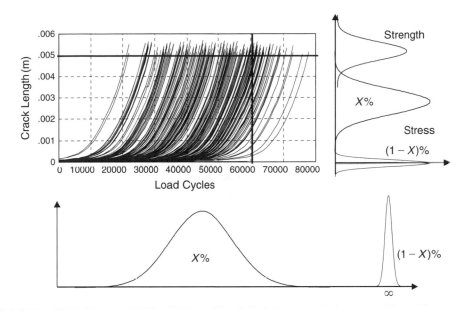

FIGURE 6.50 Distributions of cycles to failure and crack length (at given time).

Another cut can be made vertically; i.e., the crack length distribution at a given number of cycles, or a given time. This distribution has a probability of X%, while all the cracks that do not propagate are in a spike at a value of crack length that is the initial crack length (practically zero compared to the lengths of the propagating cracks). On this cut, the critical value of the crack (length of crack in which the structure fails) can also be plotted, because this quantity is also a statistical parameter. Then, a "Stress-Strength" model is created, from which probability of failure can be calculated.

The demonstrated numerical example is for a specific configuration. Nevertheless, any other 2-D configurations may be computed using the same approach. This can be done by replacing the expression for the SIF (Eq. (6.27)) in the numerical solution program with another expression. Also, the EIFS values were approximated by a Weibull CDF. When experimental results show clearly that the distribution is of another kind, the numerical procedure can be updated very easily.

No attempt was made in this section to include the stochastic behavior of the crack's growth except the random behavior of the EIFS values. In order to include such effects, a stochastic process based on the experimental results

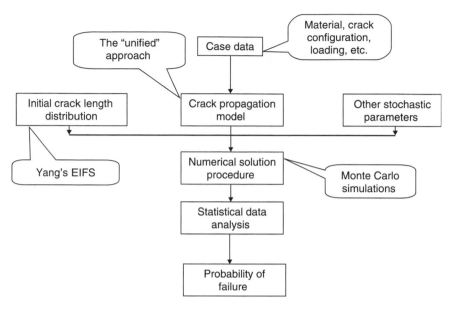

FIGURE 6.51 Flowchart of the demonstrated computation process.

can also be incorporated into the computation tool by multiplying the crack model (Eq. (6.23)) with a stochastic process of given distribution and statistical moments.

In Figure 6.51, a schematic flowchart of the demonstrated procedure is described. In this chart, the rectangular elements represent the basic procedure, while the "balloons" represent the specific methods used in the demonstrated procedure.

6.6 STOCHASTIC CRACK GROWTH AND THE PROBABILITY OF FAILURE FOR RANDOM EXCITATION [118]

Practically all the dynamic excitation of aerospace structure is random, and not harmonic. Harmonic excitation can be found only in a small number of cases, such as rotating machinery (e.g., propeller and jet engines, helicopter blades rotation), but the main excitation originates from the turbulent flow around the structure. Therefore, the use of harmonic alternating loads of constant amplitude in both fatigue and fracture mechanics may be impractical.

Even the case of several harmonic loadings, each having different amplitude and acting for a different number of load cycles, may present difficulties to the analyst.

Suppose that a structure is loaded by a stress S_1 for n_1 cycles, S_2 for n_2 cycles, etc. From the S-N curves of the relevant material one can find N_1, the number of cycles to failure in stress S_1, N_1 the number of cycles to failure in stress S_2, etc. A well-known law, the Miner equation, was set up to define the accumulated damage in the structure. Miner's law states that

$$D = \sum_1^m \frac{n_i}{N_i} \leq 1 \qquad (6.32)$$

This means that as long as the sum D is smaller than 1, the structure does not fail.

There are many limitations to Miner's law. It treats several harmonic loadings (and not random loads), it does not take into account the order of the loading, and experimental results showed that sometimes structures fail when D does not reach the value of 1, and sometimes they fail with values of D larger than 1. Some design codes state, to be on the safe side, that the structure is considered to fail when $D = 0.25$. With all the advances in fatigue and fracture mechanics analyses, there is still no practical law better than Miner's law, and thus it is used extensively in industry.

When trying to establish some rules to the failure of structures under random loading, a direct use of Miner's law is impractical, mainly because there are no well defined expressions that connect the applied stress to the number of cycles, which does not exist for a random load. There are several methods to express an equivalent number of load cycles; the best known is the "rain fall" analysis, which is frequently used in practical structures analysis.

The method described in this chapter, based on suggestions of [77], may give some reasonable results for the random loading cases. It is based on the method described for harmonic excitation (described earlier), where harmonic loads are replaced with some equivalent load, the stress ratio R is replaced by another expression, and analysis is made for time to failure rather than cycles to failure. The computation of the equivalent parameters is based on the analysis of stochastic processes; therefore, some basic results are presented here.

A stochastic loading cycle may be wide band or narrow band loadings, with power spectral density function (PSD) similar to those shown in Chapter 3. A wide band excitation presents randomness in both the local peaks of the stress and the frequency (a wide range of frequencies exists, see Figure 3.3), and the rate of crossing the time axis (also called zero crossing) is random. A narrow band process is random in its local peaks, but there is one major frequency, and the rate of zero crossings is almost constant (see Figure 3.4).

One should remember that a structure behaves like a filter to random external excitation. It responds to wide band external excitation with narrow band responses, centered on the resonance frequencies of the structure. Also, the stress response of a practical structure is usually centered on the first resonance frequency. Therefore, in many practical cases, narrow band stresses are the reason for the structure stress response, even when the external loading is wide-banded. Spectral moments λ_i were described in Chapter 5, Eq. (5.44). A stochastic signal can be characterized by a spectral width parameter ε

$$\varepsilon = \sqrt{1 - \alpha^2} \tag{6.33}$$

α is the regularity parameter, defined with the spectral moments:

$$\alpha = \frac{\lambda_2}{\sqrt{\lambda_0 \lambda_4}} \tag{6.34}$$

For a narrow band process, $\varepsilon \to 0$ and $\alpha \to 1$.

It can be shown that the mean value of the number of local maximum points per unit of time, which are above a given value u, is given by [77]:

$$\frac{\langle M_u(0,T) \rangle}{T} = v_2 \left[1 - \Phi\left(\frac{u}{\varepsilon \sqrt{\lambda_0}} \right) \right] + v_0 \cdot \exp\left(-\frac{u^2}{2\lambda_0} \right) \cdot \Phi\left(\frac{u}{\varepsilon \sqrt{\lambda_0}} \cdot \frac{v_0}{v_2} \right) \tag{6.35}$$

In this equation, Φ is the cumulative distribution function (CDF) of the standard normal distribution and

$$v_0 = \frac{1}{2\pi} \cdot \left(\frac{\lambda_2}{\lambda_0} \right)^{1/2} ; \quad v_2 = \frac{1}{2\pi} \cdot \left(\frac{\lambda_4}{\lambda_2} \right)^{1/2} \tag{6.36}$$

v_0 is the zero upward crossing rate of the process.

For a narrow band process ($\varepsilon \to 0$), Eq. (6.35) becomes

$$\frac{\langle M_u(0,T) \rangle}{T} = \frac{1}{2\pi} \sqrt{\frac{\lambda_2}{\lambda_0}} \cdot \exp\left(-\frac{u^2}{2\lambda_0} \right) \tag{6.37}$$

For a stationary Gaussian (normal) process with zero mean and a variance λ_0, the probability density function of the peaks' heights is given by

$$f_{max}(z) = \frac{\varepsilon}{\sqrt{\lambda_0}}\phi\left(\frac{z}{\varepsilon\sqrt{\lambda_0}}\right) + \sqrt{1-\varepsilon^2}\cdot\frac{z}{\lambda_0}\cdot\exp\left(-\frac{z^2}{2\lambda_0}\right)\cdot\Phi\left(\frac{z\left(1-\varepsilon^2\right)^{1/2}}{\varepsilon\sqrt{\lambda_0}}\right)$$

(6.38)

where ϕ is the probability density function (PDF) of the standard normal process. For a narrow band process ($\varepsilon \to 0$), Eq. (6.38) becomes

$$f_{max}(z) = \frac{z}{\sigma_s^2}\cdot\exp\left(-\frac{z^2}{2\sigma_s^2}\right), \quad z \geq 0$$

(6.39)

This function describes a Rayleigh distribution for the peaks' heights. If α is small ($\varepsilon \to 1$); i.e., a wide band process the PDF of the peaks' heights is

$$f_{max}(z) = \frac{1}{\sigma_s\sqrt{2\pi}}\cdot\exp\left(-\frac{z^2}{2\sigma_s^2}\right)$$

(6.40)

which describes a normal distribution of the peaks' height. For a narrow band process, the peaks and valleys are in the range between $S_{max} = m_s + Z$ and $S_{min} = m_s - Z$, where Z is the peak's height, distributed according to Eq. (6.38). The range $\Delta S = H = S_{max} - S_{min}$ is equal to $2Z$.

To demonstrate the use of the preceding equations, two random stress processes are defined. Both processes have a RMS value of $S_{rms} = 2000\,\text{kgf}/\text{cm}^2 = 196.13\,\text{MPa}$. One is a wide-band process, with constant PSD between 20 Hz and 2000 Hz, and the other is a narrow-band process, between 145 Hz and 155 Hz. In Figure 6.52, the PSD functions for both processes are shown.

In Table 6.4, some statistical parameters for both examples are described. The average number of peaks per unit time is obtained using Eqs. (6.39) and (6.40). Results are shown in Figure 6.53.

In Figure 6.54, the PDF of the peaks' heights is shown for the narrow band (Rayleigh) process, and the wide-band process. For comparison, PDF is shown also for a white noise process (white noise has a constant PDF on the frequency range $-\infty < \omega < +\infty$).

In [77] it was suggested to use the tools and expressions developed for a harmonic excitation for the computation of crack propagation in a structure

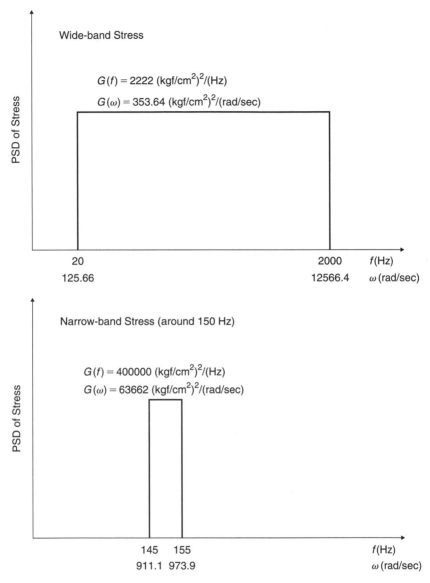

FIGURE 6.52 Two PSD functions for numerical examples.

subjected to random excitation, but to change the stress difference as follows:

$$S_{mr} = \langle \Delta S \rangle = \langle S_{max} \rangle - \langle S_{min} \rangle$$

$$S_{max} = m_s + Z \quad\quad\quad (6.41)$$

$$S_{min} = m_s - Z$$

TABLE 6.4 Statistical parameters for the examples (units are in kgf, cm, sec, radians).

Parameter	Narrow-Band Process	Wide-Band Process
λ_0	4000000	4000000
λ_2	$3.5545 \cdot 10^{12}$	$2.1268 \cdot 10^{14}$
λ_4	$3.1634 \cdot 10^{18}$	$2.0151 \cdot 10^{22}$
ν_0	150.03	1160.52
ν_2	150.14	1549.19
α	0.99924	0.7491
ε	0.038882	0.6624

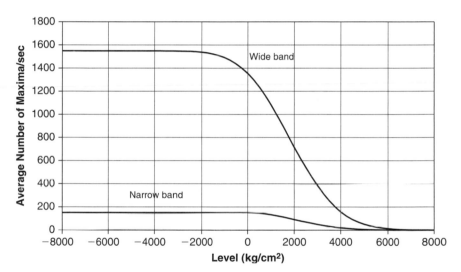

FIGURE 6.53 Mean number of peaks above a given level, per unit time.

S_{mr} is the mean value of the applied stress difference, where Z is the random heights of the peaks as described in Eq. (6.38). Also $S_{mr} = 2\langle Z \rangle$, and by integration of Eq. (6.38) the following expression is obtained:

$$S_{mr} = 2 \cdot S_{rms} \cdot \sqrt{\frac{\pi}{2} \cdot (1 - \varepsilon^2)} \qquad (6.42)$$

The frequency f_s in the equations is the number of the maximums per unit time above all the values of the stochastic process, which means the value of the plateau in Figure 6.53. The stress ratio R in the harmonic loading crack propagation equations is replaced by

$$Q = \frac{\langle S_{min}\rangle}{\langle S_{max}\rangle} = \frac{m_s - \langle Z\rangle}{m_s + \langle Z\rangle} \tag{6.43}$$

The following example demonstrates the application of the described procedure. A plate of infinite width with a centered initial crack of width $2a_0$ is loaded perpendicular to the crack direction. The plate is made of 4340 steel, unnotched with UTS $= 158$ ksi. The S-N curve for this material is shown in Figure 6.55. The crack propagation parameters are described in Table 6.1. The plate is loaded by a narrow band according the PSD described in Figure 6.52. An initial half crack length $a_0 = 0.1$ mm is assumed. Results for different values of R are described in Table 6.5.

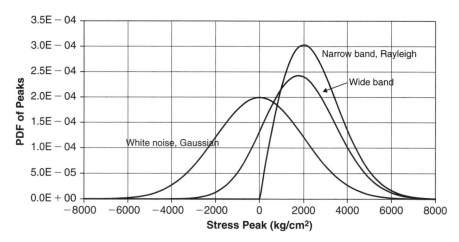

FIGURE 6.54 PDF of peaks' heights for the two examples, and for a white noise process.

TABLE 6.5 Results for narrow-band excitation.

0.5	0.3	0	−0.1	−0.2	−1	R
490.9	490.9	490.9	490.9	490.9	490.9	S_{mr}
245.5	245.5	245.5	245.5	245.5	245.5	$\langle Z\rangle$
736.4	455.9	245.5	200.8	163.6	0	S_{mean}
981.9	701.4	490.9	446.3	409.1	245.5	S_{max}
1.27	2.5	5.1	6.2	7.3	20.4	a^*
93	108	120	159	204	887	Time to Failure (sec)
13500	16200	18000	23850	30600	133050	Cycles to Failure

The values of the cycles to failure for the different R values are depicted in Figure 6.55, together with the S-N curve of the material for $R = 0$. A good agreement can be seen for $R = 0$.

For a wide-band excitation with the same parameters, results are described in Table 6.6.

Note that in Table 6.6, the peaks to failure are written, as there is no meaning to cycles to failure. The time to failure is smaller than the one found for narrow-band excitation. Although the loading is generally smaller, more peaks exist, and these may be equivalent to more loading cycles.

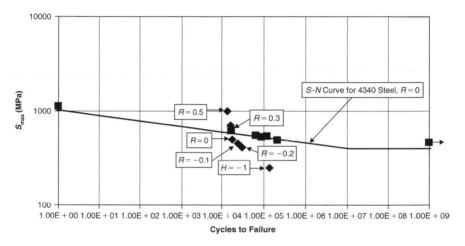

FIGURE 6.55 S-N curve for 4340 steel (squares) and results for random loading (diamonds).

TABLE 6.6 Results for wide-band excitation

0.5	0.3	0	−0.1	−0.2	−1	R
368.1	368.1	368.1	368.1	368.1	368.1	S_{mr}
184.0	184.0	184.0	184.0	184.0	184.0	$\langle Z \rangle$
552.1	341.8	184.0	150.6	122.7	0	S_{mean}
736.1	525.8	368.1	334.6	306.7	184.0	S_{max}
1.27	2.5	5.1	6.2	7.3	20.4	a^*
19.95	22.75	25.35	33.53	43.25	187.8	Time to Failure (sec)
30903	35240	39267	51938	66994	290902	Peaks to Failure

In Figure 6.56, a flowchart for the computation process is described.

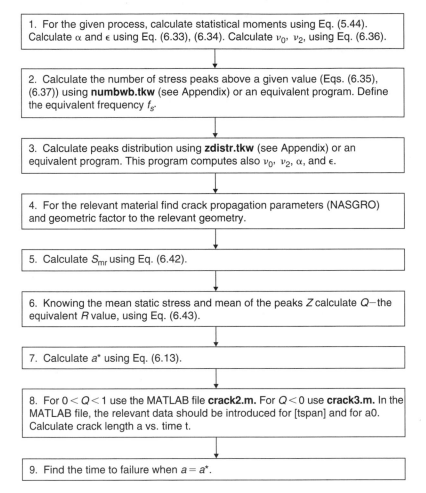

1. For the given process, calculate statistical moments using Eq. (5.44). Calculate α and ϵ using Eq. (6.33), (6.34). Calculate ν_0, ν_2, using Eq. (6.36).

2. Calculate the number of stress peaks above a given value (Eqs. (6.35), (6.37)) using **numbwb.tkw** (see Appendix) or an equivalent program. Define the equivalent frequency f_s.

3. Calculate peaks distribution using **zdistr.tkw** (see Appendix) or an equivalent program. This program computes also ν_0, ν_2, α, and ϵ.

4. For the relevant material find crack propagation parameters (NASGRO) and geometric factor to the relevant geometry.

5. Calculate S_{mr} using Eq. (6.42).

6. Knowing the mean static stress and mean of the peaks Z calculate Q—the equivalent R value, using Eq. (6.43).

7. Calculate a^* using Eq. (6.13).

8. For $0 < Q < 1$ use the MATLAB file **crack2.m.** For $Q < 0$ use **crack3.m.** In the MATLAB file, the relevant data should be introduced for [tspan] and for a0. Calculate crack length a vs. time t.

9. Find the time to failure when $a = a^*$.

FIGURE 6.56 Flowchart for the computation of time to failure under random stress.

Chapter 7 / Design Criteria

7.1 Dynamic Design Criteria

Real practical structures are seldom subjected to static-only or dynamic-only excitation. Usually, both kinds of excitation exist simultaneously. An airplane structure is subjected to quasi-static loads during maneuvering, together with random vibration excitation of boundary layer noise. A bridge or a building is subjected to static loads due to structural weight and to dynamic winds (and other dynamic excitations, like base movements) loads. An automobile is subjected to static load due to weight and passengers' loads, and to a dynamic (vibration) excitation of rough roads. The structural design has to answer both kinds of loads. A good, optimal design is the one where the acting stresses are lower than the allowable stresses with a pre-determined factor of safety. When this factor of safety is higher than the one required, the design is too conservative, not optimal, and may be too expensive. Of course, when the stresses are higher than the allowables, the structure fails and this design is bad for all purposes.

Usually, structures are designed according to input specifications. These specifications may be formal (like military standards or civil engineering design codes). They also may be based on past experience of the designers (both personal and institutional experience), or based on data collected on previous models' designs. It is quite advisable that a new model design of an existing structural system will be based as much as possible on knowledge collected in the previous design, as well as feedback from field experience and tests of this previous design.

It is essential that the design criteria set at the beginning of a project will be the best and clear as possible. Erroneous criteria lead to a bad designs. Bad design is not only the one that fails, but also the one that is too conservative—too heavy, too expensive.

Many young engineers and many design firms use the following criterion for the combination of static and dynamic loadings:

"Design the structure to an equivalent *static load* which is equal to the existing required static load + 3 standard deviations of the dynamic loads." This, they think, will take care of the static-dynamic combination and will include at least 99.73% (3σ) of the dynamic excitation, while doing only a static analysis. It is done mainly because these people are afraid of the structural dynamics analysis! This approach will be called "the equivalent static input design criteria," and is well criticized here. This design criterion may lead to an erroneous design. The following two examples demonstrate the problematic nature of this criterion.

7.1.1 CASE OF UNDER-DESIGN

Suppose the cantilever beam whose parameters are listed in Table 2.1 of Chapter 2 is loaded by the random tip force described in Chapter 3, Section 3.5 and in Figure 3.5. A static load of 1 g is also applied to the same beam. The loads on the beam are therefore a static load of 1 g, perpendicular to the beam, and a random tip force whose RMS value is 1 kgf. Using the previous "equivalent static input design criterion," one has to statically compute the beam to the following loads:

$$
\begin{aligned}
\text{Equivalent static load} = {} & 1\,\text{g in the } y \text{ direction} \\
& + 3 \cdot \text{RMS of vertical tip force}
\end{aligned}
\tag{7.1}
$$

Solving each loading of the right-hand side of Eq. (7.1) results with the bending stress distribution shown in Figure 7.1, and designated "design criterion." The ANSYS® input file is **stat1.txt** (see Appendix) for both static cases.

The "true" dynamic solution for the random tip load case was given in Section 3.5. The three times RMS of the dynamic stress (3σ) covers 99.73% of the design history. This can be added to the static solution (which is part of the previous solution).

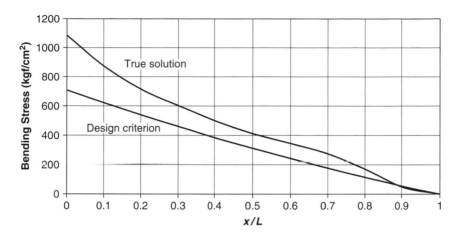

FIGURE 7.1 True solution and "design criterion" solution, static + dynamic loads, cantilever beam, a case of under-design!

Total bending stress = Stress due to $1\,\mathrm{g} + 3 \cdot$ RMS of dynamic stress (7.2)

The resulting bending stresses, marked "true solution," are also shown in Figure 7.1.

This is clearly a case of a dangerous under-design, where the structure is designed to withstand stresses that are lower than those that exist in the life cycle of the product. A structure designed with the "equivalent static input criterion" will fail during its life cycle.

7.1.2 CASE OF OVER-DESIGN

In this example, a concentrated mass whose weight is 1 kgf (and therefore its mass is $M_{\mathrm{tip}} = 1/980 = 0.001020408\,\mathrm{kgf} \cdot \mathrm{sec}^2/\mathrm{cm}$) is added to the tip of the cantilever beam. This structure is also subjected to a gravitational field of $1\,\mathrm{g}$ plus a random tip force excitation with the PSD given in Figure 3.5, whose RMS value is 1 kgf. Two kinds of computations are done using the ANSYS files listed in the Appendix. In the first computation (file **stat2.txt** of the Appendix), an equivalent static loading of $1\,\mathrm{g}$ over the whole structure plus a static tip force of 3 kgf (3σ) is applied. Results are shown in Figure 7.2, designated "design criterion." Next, the true response of the new structure is done, using file **stat3.txt** of the Appendix. Results for the RMS values are multiplied by 3, and added to the static results of $1\,\mathrm{g}$ loading. These are depicted in Figure 7.2 and designated "true solution."

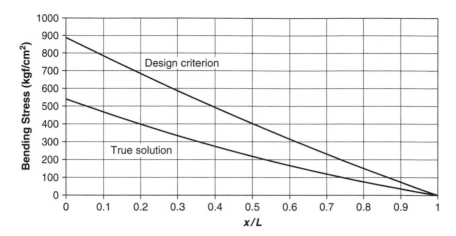

FIGURE 7.2 True solution and "design criterion" solution, static + dynamic loads, beam with tip mass, a case of over-design.

In the solution of the beam with the tip mass, new sets of resonance frequencies are obtained, which are compatible with the new structure. These new frequencies are determined during the modal analysis that is performed using file **stat3.txt**. This structure has the same rigidities as the cantilever beam, but has a different mass distribution.

This is clearly a case of over-design, where the structure is designed to withstand stresses that are higher than those that exist in the structure during a combined design loading. Although this design will not fail, an over-design is usually heavier and more expensive.

The Correct Criterion

Using the "equivalent static input design criterion," it was possible to present two completely different simple examples, which resulted in an over-design case and in (the much more dangerous case) an under-design case. Thus, the effect of using this criterion is not in one direction. The correct design criterion should never be based on the input, although inputs are usually the outcome of the product design specifications, and are part of the contract between the customer and the designers.

When a combination of static and vibration loads exists, the right design criterion to adopt is the one based on the outputs rather than the inputs.

$$\text{Combined output} = \text{Static output} + 3\sigma \text{ of the dynamic output} \qquad (7.3)$$

Of course, in order to use an output criterion, a model of the system should be built and solved (analytically or numerically). However, some kind of model is also required when the erroneous "equivalent static input design criterion" is used! The only difference is that use of the erroneous criterion implies a static analysis, whereas the use of the correct criterion demands performance of a dynamic analysis—something that many engineers prefer not to do.

Today, most of the structural analyses performed in the industry for a variety of structures are done using a finite element commercial code. Therefore, a model of the structure is built anyway for every case. It seems that many engineers fear to do dynamic analysis, even when the work is done in a finite element code. There are no reasons for such a "fear." The efforts required to build a model for a structural element or a structural system in order to perform a static analysis are almost identical whether only a static analysis is done, or both static and dynamic analyses are performed. For both cases, the main effort is to build the initial model, introduce the correct geometry, material properties, and boundary conditions, and decide on the failure criteria. Once this is done, running a dynamic analysis is a marginal effort!

A more realistic example, which may represent a practical design issue, is described here. The designed structure represents a payload carried externally by an aircraft. It is connected to the carrying vehicle by two hooks, and has three major structural elements: a nose structure, a main structure, and a tail structure, connected by two interconnections—one between the nose and the main structure, the other between the main structure and the tail. It is assumed that the load on the carried structure originates from two sources— the static maneuvering of the carrying aircraft, and vibrations transmitted to the external structure through the connecting hooks. Note that these are not the complete excitation in the real case. There is also the random vibrations major input with which the external aerodynamic flow excites the external payload. This input is not treated here, but the reader should remember to take it into account when solving the real structural design.

The external payload is described in Figure 7.3, and is treated as a beam. The stresses at the interfaces are the design criteria for the interconnections. The system is made of steel ($E = 2100000 \, \text{kgf/cm}^2$, $\rho = 7.959 \cdot 10^{-6} \, \text{kgf} \cdot \text{sec}^2/\text{cm}^4$, $\nu = 0.3$), with a square cross-section ($h = 10 \, \text{cm}$).

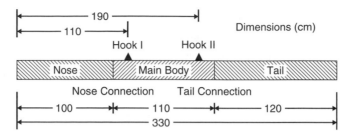

FIGURE 7.3 Geometry of the model for an external payload.

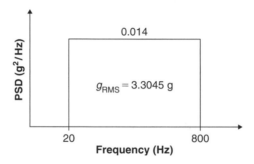

FIGURE 7.4 PSD of external excitation applied to the hooks.

The external loads applied on the structure are:

1. A static maneuver of 5 g.

2. A random acceleration input applied simultaneously to the two hooks. The PSD of the external load is shown in Figure 7.4.

The structure was calculated as a beam, using ANSYS (**env5.txt**, see Appendix) static and spectrum modules. Results for displacements (relative to the hooks) and bending stresses due to static loading are shown in Figure 7.5. Results for dynamic response are shown in Figure 7.6.

In Figure 7.6, RMS values of displacements and accelerations at the front and end tips of the structure are marked, as well as the RMS of the bending stresses at the two connections, which are required for the design of these connections.

From the outcome of these calculations, the design requirements for the two connections can be written as follows:

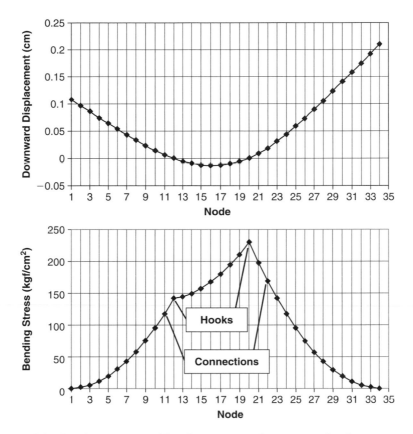

Figure 7.5 Displacements and bending stresses due to static loading.

For the nose-main structure connection:

Static stress of $\text{Stress}_s = 117\,\text{kgf/cm}^2$;

Dynamic narrow-band response with RMS value of $\text{Stress}_{rms} = 374\,\text{kgf/cm}^2$;

For three standard deviations, a total of $\text{Stress} = 117 + 3 \cdot 374 = 1239\,\text{kgf/cm}^2$.

For the tail-main structure connection:

Static stress of $\text{Stress}_s = 168\,\text{kgf/cm}^2$;

Dynamic narrow-band response with RMS value of $\text{Stress}_{rms} = 350\,\text{kgf/cm}^2$;

For three standard deviations, a total of $\text{Stress} = 168 + 3 \cdot 350 = 1218\,\text{kgf/cm}^2$.

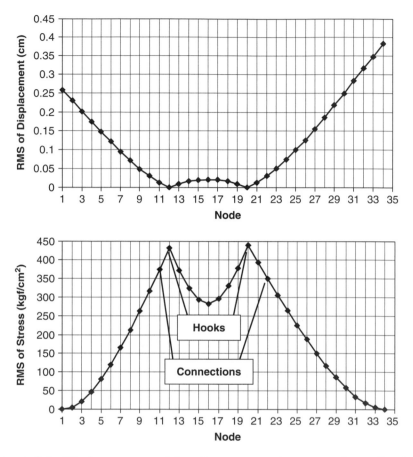

FIGURE 7.6 Displacements and bending stresses due to random vibration loading.

If the equivalent static input design criterion is used, the static computation should be done for a static load of g (static equivalent) $= 5\,g + 3 \cdot 3.045\,g = 14.91\,g$. Results for this computation, together with the results for the correct combination of static and dynamic loading are shown in Figure 7.7.

It can be seen that the use of the equivalent static input design criteria results in a significantly under-design, and using it in the design process will clearly result in a structural failure.

The importance of pre-determination of design criteria is an essential part of the design. A lot of thought and engineering efforts should be invested in the process of determination of these criteria.

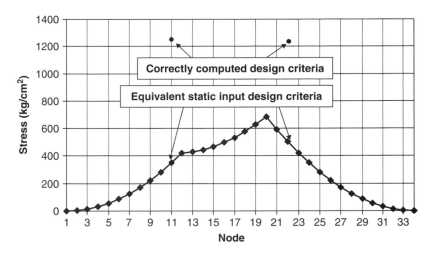

FIGURE 7.7 Equivalent static input design criteria and correct design criteria.

7.2 THE FACTOR OF SAFETY

7.2.1 FACTOR OF SAFETY

The factor of safety was introduced into the design process hundreds of years ago. It really was introduced in order to compensate the designers for unknown design parameters, or in other words, uncertainties. Thus, although structural probabilistic analysis started to gain access to design procedures only in the last 30–40 years, the early designer used *de facto* a "device" to compensate for uncertainties in their knowledge of the models and the structures' properties, especially the allowable stresses.

This book is not intended to describe all the aspects of the factor of safety. An extensive treatment of the subject, including history from ancient days, can be found in a recently published book [119]. Some of the reasons for the introduction of the safety factors are quoted here:

1. "To allow for accidental overloading … as well as for possible inaccuracies in the construction and possible unknown variables …" etc.

2. "A better term for this ratio is the factor of ignorance …" etc.

3. "Stresses are seldom uniform; materials lack the homogenous properties theoretically assigned to them; abnormal loads …" etc.

4. "Uncertainties in loading, the statistical variation in material strength, inaccuracies in geometry and theory and the grave consequences of failure of some structures ..." etc.

5. "Personal insurance for the design companies ..." etc.

Usually, factors of safety are part of formal design codes and of the customer's specifications. Although the probabilistic approach to safety factors has many benefits over the classical approach, not many of the present aerospace formal design codes and specifications have yet adopted this approach. As design establishments are formally tied to the formal requirements, the probabilistic approach is used today only in part of the design processes. Nevertheless, a combination of the classical safety factor approach with the structural probabilistic methods can be adopted in order to both comply with the formal requirements and enjoy the benefits of the probabilistic safety factor analyses.

The classical and the probabilistic safety factor procedures can best be explained using the "Stress-Strength" model used by many statisticians, and described in Chapter 5. Suppose a structure is under a load S, and has strength R. The load S is not necessarily the external load acting on the structure. S is understood as a required result of a structural parameter obtained in a structure of certain dimensions, material properties, and external loads. It can be the stress in a critical location, the displacement obtained in the structure, the stress intensity factor due to crack propagation, the acceleration in a critical mount location, etc. The strength is not necessarily the material allowable. It can be the yield stress, the ultimate stress, the maximum allowed displacement due to contact problems, the fracture toughness of a structure, the buckling load of compressed member, etc. One can define a failure function

$$g(R,S) = R - S \qquad (7.4)$$

and define a failure when

$$g(R,S) \leq 0 \qquad (7.5)$$

Both R and S can be functions of other structural parameters, thus the failure function, which describe a line in a 2-D space is, in general, a hyper-surface of n dimensions, which include all the variables on which R and S depend.

Using these notations, there are several possible definitions for the safety factor. The classical definition is

$$FS_{\text{Classical}} = \frac{R}{S} \qquad (7.6)$$

When this number is larger than 1, there is no failure. The design codes demand a minimum value for $FS_{Classical}$ to avoid failure when uncertainties exist, say $FS_{Classical} \geq 1.2$.

Another possible definition for the safety factor is for a worst-case design:

$$FS_{WorstCase} = \frac{R_{minimum}}{S_{maximum}} \qquad (7.7)$$

In this case, the minimum possible value of R (which may be a function of several variables, or a number with known dispersion) is computed, and divided into the maximum possible value of S. It is clear that $FS_{WorstCase}$ is smaller than $FS_{Classical}$. The advantage of $FS_{WorstCase}$ is that it takes into account known dispersions in the structure's parameters. Its disadvantage is that it assumes that all the worst-case parameters exist simultaneously (although this may be a rare case, with a very low probability), and therefore the resulted design is very conservative. Examples of the use of $FS_{Classical}$ and $FS_{WorstCase}$ were described in Chapter 5, which demonstrated that two different design procedures can lead to inconsistent consequences, and also discussed the modifications required for a certain design when the worst-case safety factor is used.

Another possible definition for the safety factor is for a mean value design, sometimes called a central design:

$$FS_{Central} = \frac{E(R)}{E(S)} \qquad (7.8)$$

$E(.)$ is the expected value (the mean) of a parameter $(.)$. In many cases, the mean values coincide with the nominal values, and the classical safety factor is identical to the central safety factor. For other cases where they do not coincide, the reader can explore the examples described in [119].

All the three safety factors defined (Eqs. (7.6–7.8)) are using deterministic values and the safety factor obtained is deterministic.

The stochastic safety factor was first introduced in [120, 121]. The same concept was adopted later by others and is sometimes named probabilistic sufficiency factor. The definition is

$$SF_{Stochastic} = \left(\frac{R}{S}\right) \qquad (7.9)$$

As both R and S may be random variables (and can depend on many structural parameters that are assumed random), the obtained stochastic safety factor is also a random number. Using the stochastic safety factor, one may answer the question, "What is the probability that a structure (with given uncertainties) has a factor of safety smaller than a given value?"

The preceding definitions are demonstrated by a simple "Stress-Strength" example. Suppose R has a normal distribution with mean $\mu_R = 1.2$ and standard deviation of $\sigma_R = 0.06$. S has a mean $\mu_S = 1.0$ and a standard deviation $\sigma_R = 0.05$. For both variables, the coefficient of variation is 5%. The classical safety factor is

$$FS_{\text{Classical}} = \frac{1.2}{1.0} = 1.2 \tag{7.10}$$

which is an acceptable safety factor. For the worst-case design, assume that $\pm 3\sigma$ are the upper/lower limits of the variables. The worst-case safety factor is

$$FS_{\text{Worst Case}} = \frac{1.2 - 3 \cdot 0.06}{1 + 3 \cdot 0.05} = 0.887 \tag{7.11}$$

which is unacceptable for the design.

The stochastic safety factor was computed using the ProFES® probabilistic program. Results for the CDF of this safety factor obtained using FORM are shown in Figure 7.8. The probability that the stochastic safety factor is equal or smaller than 1.2 is 0.5. The probability that it is equal or smaller than 1.0 is 0.0052226. It is interesting to check how the latter probability is changed if the distribution of the parameters is a truncated normal distribution, where truncation is done at $\pm 3\sigma$ values. In this case, the probability that the stochastic safety factor is smaller than 1 is 0.0048687, a relatively small decrease in the probability of failure. Truncated normal distribution for a normally distributed variable means that a screening process is performed, and all specimens outside the $\pm 3\sigma$ range are screened out. The designer can do a cost effectiveness analysis to determine whether the decrease obtained in the probability of failure is worth the much more expensive screening process.

When the coefficient of variation of both R and S is decreased, the effect of decreasing the dispersion in the random variables can be demonstrated. A decrease in the dispersion means tighter tolerances in the design and the production. In Figure 7.9, this effect demonstrates a significant decrease in

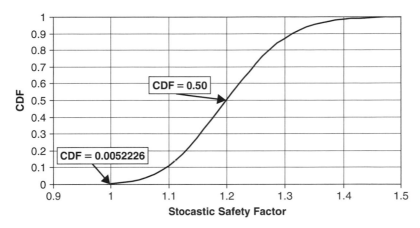

FIGURE 7.8   CDF of the stochastic safety factor.

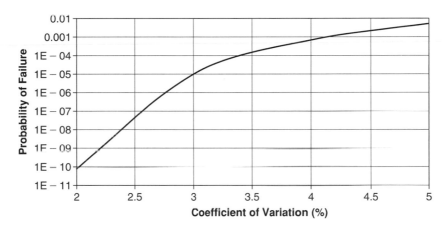

FIGURE 7.9   The effect of COV on the probability of failure.

the probability of failure (defined here as the probability that the stochastic safety factor is less than 1.0).

In Chapter 5, the probability of failure of a cantilever beam was demonstrated. The failure criterion of the problem was that the tip displacement of the beam does not exceed a given value. The random variables of the problem are the thickness of the beam, the Young's module of the beam's material, the applied external tip force, and the maximum allowable tip displacement. The length of the beam and its width were considered deterministic. Data for the variables are given in Table 5.2.

Here, the stochastic safety factor approach for the same problem is demonstrated. The tip deflection of a cantilever beam under a tip load P is ([26], from slender beam theory)

$$\delta_{tip} = \frac{4PL^3}{Ebh^3} \tag{7.12}$$

For the "stress-strength" model, R is the allowed tip deflection δ_0, and S is the actual tip deflection δ_{tip}, and the stochastic safety factor is therefore

$$FS_{Stochastic} = \frac{\delta_0}{\delta_{tip}} = \frac{\delta_0}{\left[\dfrac{4PL^3}{Ebh^3}\right]} \tag{7.13}$$

The data and the definitions of the stochastic safety factors were introduced into the ProFES program. The CDF of the stochastic safety factor is shown in Figure 7.10. The probability of failure, defined as the probability that the stochastic safety factor is equal or less than 1.0, is $0.00010868 = 0.010868\%$, which is similar to the result obtained in Chapter 5 (see Eqs. (5.6), (5.23), (5.26)). The probability that the stochastic safety factor is equal or less than 1.702 (the nominal value, see Chapter 5) is $0.5 = 50\%$.

In Figure 7.11, the importance factors are shown in a pie chart. It can be seen that the most important design parameter for this case is the thickness, and the second one is the allowed tip displacement. When changes of the

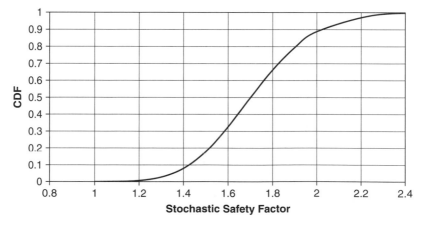

FIGURE 7.10 CDF of stochastic safety factor for the cantilever beam.

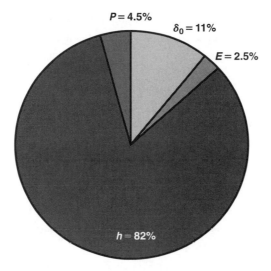

FIGURE 7.11 Importance factors of the random variables.

design are required, these two variables are those on which the designer should focus his efforts. The thickness is a direct design parameter, while the allowed displacement may be part of the customer specification, which might be negotiated when a design change is required. Changes in the data of the applied force (usually part of the specifications) and the Young's modulus (material replacement) will be much less effective.

Although the Stochastic Safety Factor is not included in most of the design specifications, it is good practice to compute it. It may build a bridge between the presently used factor of safety and the modern probabilistic approach. This bridge should be effective until probabilistic design criteria is introduced into the project's requirements.

7.3 RELIABILITY DEMONSTRATION OF STRUCTURAL SYSTEMS

A structure is designed as a part of a whole system. In aircrafts, the system includes the disciplines of structures, aerodynamics, propulsion, control, electronics, chemical engineering, avionics, human engineering, production and maintenance, as well as others. In missiles, explosives' technologies are also added to include the warhead design. During the design process of any system, a final reliability is required. This final reliability is obtained by a combination of the reliabilities of all the components, subassemblies and subsystems, as well as production processes and methods.

The progress of probabilistic analysis of structures achieved during the last three decades enables one to incorporate structural reliability in the system design process. The classical approach of the structural safety factor should be replaced with the reliability of the structure and its probability of failure. More on safety factors can be learned in the previous section of this chapter.

There are some unique features that characterize aerospace systems (as well as large civil engineering projects) from consumer goods that clearly have to be reliable. The main difference is that these large projects are of large scale and of multidiscipline efforts. Another major difference is that in many cases, the final design is manufactured in a very small number of products (or systems). There is a very small number of space shuttles. Only one Hubble space telescope was manufactured. Many satellites are "one of a kind" products. Only a small number of SR-71 intelligence aircrafts were manufactured. A relatively low number of ICBMs were produced. Only about seven dozen of Cargo C-5 (Galaxy) aircrafts were originally manufactured. The designers of such projects cannot rely on statistical results obtained by testing many specimens, contrary to the designers and manufacturers of consumer goods. For the latter, statistical data can be obtained in both the design and the manufacturing phase, and a tremendous amount of experience feedback is obtained from consumers. As the number of final products is small, the development costs highly increase the unit price of an aerospace product, preventing (economically) the possibility to perform large amounts of tests during this phase. Sometimes, the nature of the projects prevents testing in the final designed conditions; for instance, testing satellites in their space environment. In many cases, frequent modifications are introduced after the "end" of the development phase, and two manufactured specimens can differ. Performance envelopes are usually very large and varying for different carriers. Usually, the cost of failure is high, in both performances and costs—and sometimes, in lives.

The true problem of these projects is not the computation of their reliability by mathematical tools, but the verification of the reliability values, sometimes called "reliability demonstration."

The demands for end product reliability are traditionally expressed in a "required reliability" and a "required confidence level." It is common to find, in the customer's product specifications, a demand like "the required reliability is 90% with 90% confidence level." Although reliability engineers, statisticians, and mathematicians may understand such a sentence, it is not clear to design engineers and project managers. For instance, it is well known to statisticians and reliability engineers that when 22 successful tests of a system are performed, the 90% reliability with 90% confidence level is "demonstrated." It is less emphasized (and less commonly understood by project managers) that all these 22 tests should be performed using the same conditions. When the project performance envelope is wide, several extreme "working points" should be tested (each with 22 successful tests), and this emphasizes the limitation of the classical demonstration process for such projects.

Reliability of subsystems can be calculated ("predicted") today by many techniques and methodologies. Then, the total reliability can be evaluated and predicted by combining the individual contributions of these subsystems and sub-assemblies, declaring that the predicted "reliability of 90% with 90% confidence level" is reached. Nevertheless, the real problem is not the prediction or the computation of the reliability by mathematical tools, but the verification of the reliability values, sometimes called "reliability demonstration." There is no way to "prove" or demonstrate that this reliability is really obtained. It is also hardly possible to convince customers, project managers, and designers that the "confidence level" (a term they really do not understand) was also obtained.

It is well known that aerospace structures fail in service in spite of the extensive (and expensive) reliability predictions and analyses in which much engineering effort is spent. A recent example is the failure of the Columbia space shuttle in 2003. Regretfully, two space shuttles failed in a total number of 100 flights. These failures set the shuttle flight project back many years, with a tremendous financial penalty, on top of the life and morale costs. Analysis of failures in many aerospace projects usually reveals that more than 80% of "field failure" are the result of a "bad" design that could have been avoided. Thus, improving the design process may significantly cut the amount of product failures, increase the reliability, increase a product's safety, and tremendously decrease the costs. The main reasons for an erroneous design process are the use of inadequate design methodology and a wrong design of tests and experiments during the development (design) process.

A different approach to reliability demonstration of aerospace structures is required. Such an approach should be incorporated in the design process, by modeling the structural elements and structural systems, and by performing tests to validate the model, and not the product.

The reliability of the final product should be deeply incorporated in the design process; thus, the tremendously important role of the design engineers in the reliability demonstration process. The methodology of a highly improved design-to-reliability process must incorporate the expertise of the design engineers together with the expertise of the statisticians and reliability engineers.

The design to reliability methodology suggested is based on the following principles:

1. Incorporation of reliability demonstration (verification) into the design process.

2. Building of models for the structural behavior, and verifying them in tests.

3. Verification of the problem's parameters by experience, tests, and data collection.

4. Analysis of failure mechanism and failure modes.

5. Design of development tests in such a way that failure modes can be surfaced.

6. Design of development tests so that unpredicted failure modes can also be surfaced.

7. "Cleaning" failure mechanism and failure modes inside the required performance envelope.

8. Determination of required safety margins and the confidence in the models and the parameters.

9. Determination of the demonstrated reliability by "orders of magnitude" while applying engineering, and not merely mathematical considerations.

The first eight principles are self explanatory, while the last one may presently be controversial.

The "design-to-reliability" methodology suggested, based on these principles, is described and discussed below.

7.3.1 RELIABILITY DEMONSTRATION (VERIFICATION) IS INTEGRATED INTO THE DESIGN PROCESS

This principle implies a design team, which includes designers and reliability engineers, working together during the development phase. The reliability of the final product should be deeply incorporated in the design process; thus, there is a tremendously important role to the design engineers in the reliability demonstration process, a role that is usually neglected today. The methodology of a highly improved design-to-reliability process must incorporate the expertise of design engineers together with that of statisticians and reliability engineers. The approach of "we designers will design and you, reliability engineers, will compute the reliability" should be discouraged. A methodology for a design process, which includes the reliability prediction and verification during the development phase, is thus suggested.

7.3.2 ANALYSIS OF FAILURE MECHANISM AND FAILURE MODES

This is the most important phase in the design process, as it determines the design main features and failure criteria. The first analysis should be done during the conceptual design phase by the system engineers with the participation of the designers, and updated during the full development phase. It is highly recommended that an additional independent failure analysis will also be performed by experts who are not part of the project team, in order to use their experience. The project's system engineers and the design engineers are the professionals who can best contribute to the process of failure mechanism and failure modes analyses, and not the reliability engineers. The latter can contribute from their experience to the systematic process of failure mode analysis and direct the design engineers when performing this important design phase, but it is the responsibility of the designers to do the analysis. The failure modes analysis should direct the design process so that these failures will be outside the required performance envelope of the designed project.

Even when failure mode analysis is done by teams of experts, clearly there will still remain some unpredicted failure modes, mainly due to lack of knowledge ("we didn't think about it"). This implies that structural tests should be designed in configurations that would best simulate the mission profile of the system. This mission profile, whose preparation is also of major importance, should be prepared as early as possible in the project history, and should be based on the project's specification and experience with similar products and projects gained in the past.

7.3.3 MODELING THE STRUCTURAL BEHAVIOR, AND VERIFYING THE MODEL BY TESTS

During the development tests, special experiments for model's verification (rather than product's verification) are to be defined, designed, and performed. In many cases, it is relatively simple to prepare a structural model (analytical or numerical) with which the structural behavior is examined. The model is then corrected and updated by the results obtained in these tests. In addition, the parameters that influence the structural behavior should be defined and verified by tests, data collection, and the experience of both the design establishment and its designers. In cases where the structural model is not available, an empirical model can be built based on very carefully designed experiments that can check the influence of as many relevant parameters as possible. "Virtual tests" can then be performed using the updated model to check the structural behavior in many points of the required working envelope. Results of these "virtual tests" can be included in the information required to establish the structural reliability.

7.3.4 DESIGN OF STRUCTURAL DEVELOPMENT TESTS TO SURFACE FAILURE MODES

Structural tests must be designed to surface one or more possible failure modes. The design of structural tests is an integrated and important part of the structural design process. It is hardly possible to perform one structural test that can simulate all the real conditions of the "project envelope" in the laboratory. Static loadings can be separated from dynamic (vibration) loadings, if the experimental facilities available cannot perform coupled tests, as usually is the case. It seems advantageous to perform coupled tests in which many failure modes can be surfaced simultaneously, but usually these kinds of tests are more complex and more expensive, and results from one failure mode may obscure the outcomes of other failure modes. Some of the structural tests that cause failure modes to surface may be common to the tests for the model verification, in order to decrease costs of the structural specimens and decrease the structural testing schedule.

It is also recommended to establish experimentally a safety margin for the tested failure mode, as this margin can point out the extent of the structural reliability for the tested mode. Thus, structural tests should be continued until failure is obtained, unless a very high safety factor is demonstrated (when this happens, the design is not optimal).

This approach may look too expensive and time consuming to many project managers. In such cases, they should be encouraged to consider the price of failure at the more advanced stages of the development process, or after the product was already supplied to the customer.

7.3.5 DESIGN OF STRUCTURAL DEVELOPMENT TESTS TO SURFACE UNPREDICTED FAILURE MODES

Tests should be conducted in "as real as possible" conditions. Thus, load locations, experimental boundary conditions, and the tested structure should be designed as realistically as possible. The main difficulties may rise when vibration tests of subassemblies and a complete structure are performed. The need to introduce test fixtures contradicts the wish to perform realistic tests. Therefore, a new approach is called for the vibration tests methodology used presently and dictated by present specifications. This issue is a subject for quite a different discussion and will not be evaluated here, although some of the difficulties in performing realistic vibration tests was discussed in some detail in previous chapters.

7.3.6 "CLEANING" FAILURE MECHANISM AND FAILURE MODES

When failure in test occurs within the performance envelope (or outside it, but without the required safety factor), the design should be modified to "clean out" the relevant failure mode and the structure should be re-tested in order to verify the success of the "cleaning" process. The process of updating the design and its model must be repeated until no failure modes exist in the required performance envelope. This process may include "virtual tests" performed with the relevant model, verified by development tests. Such a process can assure, at the end of the development phase, that the reliability of the designed structure is very high, and qualitatively verified.

7.3.7 DETERMINATION OF THE REQUIRED SAFETY MARGINS, THE CONFIDENCE IN THE MODELS AND THE RELEVANT PARAMETERS

Safety margins or safety factors are defined at the beginning of a project, and depend on its characteristics, the formal specifications, and the past experience of the designers. It is recommended that in future projects, the approach of the stochastic safety factor [119–121], sometimes called probabilistic sufficiency factor, should be applied. Such an approach can bridge the gap between the classical safety factor used presently in most of the specifications, and the probabilistic approach that starts to gain recognition in the design community. The stochastic approach can provide a "translation method" between safety factors and reliability numbers. The confidence level is interpreted here as the confidence of the designer in the structural reliability obtained by the tests. This is not really a mathematical "statistical definition," but a design concept that has an engineering meaning, understood by customers, project managers, and designers.

7.3.8 DETERMINATION OF THE DEMONSTRATED RELIABILITY BY "ORDERS OF MAGNITUDE"

The demonstrated reliability should be determined by "orders of magnitude" and not by "exact" numbers, while applying engineering considerations. The concept, which may be controversial but starts to gain supporters in the design communities, is further discussed here.

In Figure 7.12, a flowchart of the design process is described using very general definitions. Of course, each of the blocks described in this chart can be further evaluated.

The demand for reliability demonstration defined by "order of magnitude" may be controversial, as it differs from the traditional "numerical demands" for reliability. Nevertheless, such a definition is much more realistic and

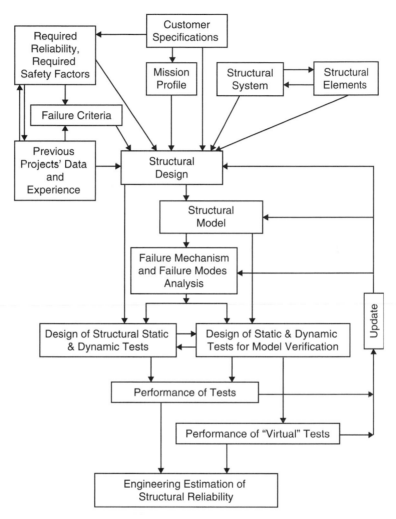

FIGURE 7.12 Flowchart of structural design to reliability process.

provides a much better engineering-oriented approach to the issue of reliability demonstration. Traditionally, customers define, in their requirements, a numerical value for the product's reliability and for the reliability confidence level (i.e., reliability of 90% with 90% confidence level). Reliability engineers also use the same definition. For the large aerospace projects discussed, there is no meaning for such requirements, as it is impossible to demonstrate, verify, or prove such values under the limiting circumstances of these projects. In addition, project managers and designers have some difficulties in translating the "confidence level" concept into practical engineering understanding.

Therefore, it is highly suggested to define reliability by "order of magnitude" like "very high (A)," "high (B)," "medium (C)," and "low (D)." The "confidence level" concept can be replaced by the "designer confidence in the estimated reliability" (engineering confidence), like "high confidence (a)," "medium confidence (b)," and "low confidence (c)." These definitions form a matrix, which is shown in Table 7.1.

The purpose of a high reliability oriented design is to "push" for the upper-left corner of the matrix. It can be argued that in order to "move" upward in the table, design modifications are required, while in order to "move" leftward in the table, more tests of the system and more evaluations of the model are required. There are no distinct sharp borders between the reliability levels and the confidence estimates in this matrix. In the last three rows of the table, numbers for traditional estimated reliability requirements are written. There is a difference in the demands from a subassembly, a subsystem (which comprise several subassemblies), and a system (several subsystems). The numbers are depicted in light gray to emphasize they are not supposed to be "exact," and are only a required estimate for the required reliability. There is really no difference between a system whose reliability was estimated to be 99.9% and a system with an estimated reliability of 99.8%. On the other hand, it is certain that when the reliability of two different optional systems is examined, the first one showing a reliability of 99% while the second one shows a reliability of 75%, then the first system is more reliable than the second. Exact numbers should therefore be used only as a qualitative comparison tool and not as an absolute quantitative tool.

There are no distinct borders between the reliability levels and the confidence estimates, as shown by the "undefined" lines in the table.

TABLE 7.1 Reliability and engineering confidence matrix.

Confidence Reliability	High (a)	Medium (b)	Low (c)	Sub- Assembly	Sub- System	System
Very High (A)	(A;a)	(A;b)	(A;c)	99.9	99	95
High (B)	(B;a)	(B;b)	(B;c)	99	95	90
Medium (C)	(C;a)	(C;b)	(C;c)	95	90	85
Low (D)	(D;a)	(D;b)	(D;c)	90	85	80

The described approach is much more realistic than the classical one, which cannot be verified ("proved") for the kind of projects described. It put much more emphasis on engineering considerations and concepts, therefore it is much easier for designers to understand and practice. The role of tests and experiments is major in the development process, and the importance of models and their verification becomes an important issue in the process.

The described (somehow controversial) approach was presented before an international audience in [122], and was well accepted by representatives of the industry in the audience. It was also applied successfully in the author's establishment.

Chapter 8 / Some Important Computer Programs for Structural Analysis

8.1 FINITE ELEMENTS PROGRAMS

Computer codes are the main tools of the modern engineering design process. There are thousands of computer programs developed during the last several decades for the use of engineers in the design establishments and the R&D institutes. Many programs were developed in the academic institutes, but these are generally undocumented and unsupported. On the other hand, commercially available codes for finite elements solutions, probabilistic analysis, and mathematical evaluations are well supported, debugged, and updated. Few of the most commonly used programs (most of them were used in the examples described in this book) are described here. The opinions expressed in this chapter are the author's, based on his experience with the described programs, and are not a recommendation to lease or purchase any of them. As these programs are updated periodically, the opinions expressed here are not updated, and are based on the situation that existed when this manuscript was prepared.

Most of the large computer codes used for the dynamic and probabilistic analyses have to be leased from the developer for an annual fee, which usually includes updates and technical support. Most of the mathematical solvers must be purchased with a specific license. Web site addresses of the codes providers are listed, and the interested reader should consult these sites for updated information.

The basic tools for structural dynamic analyses are the large finite elements codes. Two major programs are used today in more than 80% of the market—the ANSYS® and the NASTRAN®. Nevertheless, there are some other finite elements codes used in the industry, like ADINA™, ABAQUS®, and SESAM. The latter are not reviewed here. Usually, a specific design establishment leases only one finite element code, with the appropriate licensing agreement. The specific user must adjust other computational tools to comply with the finite element program used in his establishment.

Both ANSYS and NASTRAN can perform static, modal, buckling, transient, harmonic, and spectral analyses, as well as optimization algorithms. Both have a probabilistic module that can be used for structural probabilistic analysis using Monte Carlo simulation or a response surface analysis directly from the finite elements database file.

The NASTRAN is leased by the MSC Company (*http://www.mscsoftware.com*), which is also the provider of many other important programs. Consult the company web site for this and other programs.

The ANSYS is leased by the ANSYS Company (*http://www.ansys.com*), which is also the provider of many other important programs. An Educational Version, limited in modules, number of nodes, and elements is available for students and academic staff.

8.2 PROBABILISTIC ANALYSIS PROGRAMS

There are many structural probabilistic analysis codes. Many of them were developed for special-purpose researches in the academy, and are not available to the common user in the industry. One of the first available programs is the CalREL, developed in UC Berkeley by Professor Armen Der Kiureghian and his students in the Department of Civil & Environmental Engineering. CalREL is a general-purpose structural reliability analysis program. It is designed to work on its own or to operate as a shell program in conjunction with other structural analysis programs. Structural failure criteria are defined in terms of one or more limit-state functions. The specification is by the user in user-defined subroutines. CalREL is capable of computing the reliability of structural components and systems. Specific macro commands are available for the following types of analyses:

1. First-order component and system reliability analysis.
2. Second-order component reliability analysis by both curvature-fitting and point-fitting methods.

3. First-order reliability bounds for series systems.

4. First-order reliability sensitivity analysis with respect to distribution and limit-state function parameters.

5. Directional simulation for components and general systems, employing first or second-order fittings of the limit-state surfaces.

6. Importance sampling and Monte Carlo simulation for components and general systems.

CalREL has a large library of probability distributions for independent and dependent variables. Additional distributions may be included through a user-defined subroutine. CalREL is available for purchase from UC Berkeley in both object and source code. Details on the program's capabilities are available at *http://www.ce.berkeley.edu/~adk/*, where a contact address is given.

One of the first commercially available and technically supported probabilistic analysis programs is the PROBAN®, developed by DNV (Det Norske Veritas), a Norwegian nonprofit R&D organization. This is a general-purpose program for probabilistic, reliability, and sensitivity analysis, which was available for users many years before any U.S. developed programs were in the market. By complementing the hydrodynamic and structural analysis features, PROBAN forms a part of the powerful suite of SESAM (finite elements) programs for maritime and offshore engineering analysis. PROBAN contains state-of-the-art computational methods needed to perform sensitivity, reliability, and probability distribution analysis for components and systems. The methods include first and second order reliability methods (FORM/SORM), Monte Carlo simulation, Latin Hypercube sampling, and a number of other simulation methods. It contains an extensive statistical distribution library. PROBAN also contains the necessary features to perform Bayesian updating and parameter studies. PROBAN can be executed via an interactive graphical user interface. See *http://www.dnv.com/software/safeti/safetiqra/proban.asp*. This address may be changed frequently. In this case, the reader is encouraged to conduct a search on DNV.

A very famous and commonly used probabilistic computer code is the NESSUS®, developed under a contract from NASA Glenn Research Center at the Southwest Research Institute (SwRI) in San Antonio, TX. NESSUS is a modular computer software system for performing probabilistic analysis of structural/mechanical components and systems. NESSUS combines state-of-the-art probabilistic algorithms with general-purpose numerical analysis methods to compute the probabilistic response and reliability of engineered systems. Uncertainty in loading, material properties, geometry, boundary conditions, and initial conditions can be simulated. Many deterministic modeling tools can be used such as finite element, boundary element, hydro-codes, and user-defined Fortran subroutines. NESSUS offers a wide range of capabilities, a graphical user interface, and is verified using hundreds of test problems. NESSUS was initially developed by SwRI for NASA to perform probabilistic analysis of space shuttle main engine components. SwRI continues to develop and apply NESSUS to a diverse range of problems, including aerospace structures, automotive structures, biomechanics, gas turbine engines, geomechanics, nuclear waste packaging, offshore structures, pipelines, and rotor dynamics. To accomplish this, the codes have been interfaced with many well-known third-party and commercial deterministic analysis programs, like NASTRAN and ANSYS. A demonstration version of NESSUS is available for evaluation purposes at *http://www.nessus.swri.org/* for a period of three months.

Another commercially available probabilistic code is the ProFES®, developed and leased by ARA (Applied Research Associates Inc.), Southeast Division, located in Raleigh, NC. ProFES allows the user to quickly develop probabilistic models from his own model executables, analytical formulations, or finite element models. The user may use ProFES independently to perform probabilistic simulations using functions internal to ProFES, or functions he manually types in. He may use ProFES as an add-on to his own model executables so he can perform probabilistic studies from his deterministic models; he may also use ProFES as an add-on to the commercial finite element codes ANSYS and NASTRAN or to the commercial CAD package PATRAN; and he can assess the uncertainty inherent in any design or analysis situation and relate this uncertainty to product failure rates. The ProFES graphical user interface (GUI) is designed to make it easy to perform probabilistic analyses, especially for nonexperts. Random variables can be assigned to any of the variables that make up the analytical model or are input to the model executable. In the case of finite element analyses, random variables can be assigned to loads, material properties, element properties, and boundary conditions. In the case of CAD analyses, as is the case with PATRAN, random variables can also be given to geometry features (the features are obtained by importing non-graphic parts into PATRAN). Failure criteria (limit-states) are used to define failure as a function of the user's model results (e.g., output variables from user's own models, displacements, strains, stresses, etc.), programmed and compiled within the ProFES. ProFES also includes utilities for seamlessly post-processing the results (e.g., for probabilistic fatigue analysis). ProFES also provides a library of random variables and utilizes state-of-the-art probabilistic analysis methods. Probabilistic finite element analysis has advanced to the point that specialists can solve very complex problems. Computational methods for probabilistic mechanics are well developed and widely available, including FORM/SORM methods, static and adaptive response surface methods, simulation methods, and adaptive importance sampling. ProFES places these tools in the hands of practitioners and makes them usable without extensive re-training and software development. ProFES includes an innovative data-driven architecture that has the look and feel of commercial CAD and pre-processing packages. The ProFES approach supports transparent interfaces to commercially available finite element packages and allows the user to build his deterministic model in his preferred commercial pre-processing package. ProFES has been integrated with several major probabilistic computational methods along with an extensive library of random variable distributions. ProFES is designed to work with commercially available finite element software like the ANSYS and NASTRAN. Customer support is very efficient and prompt. See *http://www.profes.com* for more details.

8.3 CRACK PROPAGATION PROGRAMS

For computations of crack growth and other fracture mechanics problems, the most applied program is the NASGRO®. NASGRO 4.0 is a program developed by NASA Johnson Center in the 1980s. The first version, NASA/FLAGRO, was completed in 1986. The program was distributed by COSMIC since 1990, and was also included in the European Space Agency (ESA) crack propagation program ESACRACK. Currently, the program is maintained, updated, and distributed by the Southwest Research Institute (SwRI) in San Antonio, TX. NASGRO is an envelope program, which contains internal programs for fracture mechanics principles (NASFLA), critical crack size (NASCCS), stress intensity factors (NASSIF), glass-like material behavior (NASGLS), boundary elements for 2-D geometries (NASBEM), and material properties bank (NASMAT). An international consortium of users was founded, where each member contributes from its experience to the program. More technical details and leasing information are available at *http://www.nasgro.swri.org/*. A demonstration version of NASGRO 4.0 can be downloaded from this web site. The demonstration version is limited to two geometries and one material. Material properties for hundreds of materials and heat treatments in the program's data bank can be viewed in this demonstration version.

A very efficient computer code for the determination of 3-D crack propagation is under constant development in the Center for Aerospace Research and Education (CARE), in the Department of Mechanical and Aerospace Engineering in the University of California at Irvine, under the direction of Prof. S. N. Atluri. The program, named AGYLE, is based on a combination of finite elements (FE) and boundary elements (BE). The structure is modeled only once by finite elements, while the crack surface is modeled by a relatively small number of boundary elements. When loads are applied, 3-D crack propagation is computed, and for each load step (or cycle) only the boundary elements are modified automatically. This makes the AGYLE a very efficient program, with 3-D capabilities that cannot be found in any other crack propagation program. The AGYLE is not yet commercially available, and is still not supported for potential customers. It is believed that in the future this program (when development is completed) will replace many of the commercially available programs. Details can be obtained from Prof. Atluri, *satluri@uci.edu*.

8.4 MATHEMATICAL SOLVERS

When mathematical expressions are created, they can be solved using one of the many mathematical solvers that exist in the market. Most of these programs allow the user to solve for the numerical values of mathematical expressions, and to program the user's own programs inside the envelope of these codes. The mathematical numerical solutions performed in this book use two programs, MATLAB® and TK Solver™.

The well-known MATLAB program is a high-level technical computing language and interactive environment for algorithm development, data visualization, data analysis, and numerical computation. Using MATLAB, technical computing problems can be solved faster than with traditional programming languages, such as C, C++, and Fortran. MATLAB can be used in a wide range of applications, including signal and image processing, communications, control design, test and measurement, financial modeling and analysis, and computational biology. Add-on toolboxes (collections of special-purpose MATLAB functions, available separately) extend the MATLAB environment to solve particular classes of problems in these application areas. A toolkit for symbolic mathematics can also be included in the program, making it possible to evaluate analytically very complex mathematical expressions.

The MATLAB is a well-known mathematical tool, used by millions of users, whose experience can be shared through online forums. Details on MATLAB and its capabilities are shown in *http://www.mathworks.com/matlab/* and the various links provided in this site. There is a student edition of the MATLAB, (limited in capabilities but very inexpensive), which can later be upgraded to a full version. Users can join a Users' Forum, where many people contribute special-purposes files and where one can consult with many expert and other members of the MATLAB community.

TK Solver from Universal Technical Systems, Inc. (UTS) is one of the longest standing mathematical equation solvers on the market today. TK Solver readily solves simultaneous equations using iteration to significantly reduce computation and design hours. The user does not need to decide what variables will be inputs and which ones will be outputs when creating a mathematical model. This unique capability completely eliminates tedious "busy work" allowing the user to accomplish more in less time. The user can program his internal function within the TK Solver. The library of the TK Solver includes, among many others, computerized interactive version of Roark's Formulas for Stress and Strain [26], and a Dynamics and Vibration module that is a complete, fully computerized interactive version of Belvins' Formulas for Natural Frequencies and Mode Shapes. These modules can also be purchased as standalone software. Details can be found at *http://www.uts.com/software.asp*. There is an academic and student version, which is the full version (less the special libraries), under a special license agreement. This program is used quite extensively in some of the cases presented in this publication. A Users' Forum enables users to consult with experts and colleagues.

Chapter 9 / Conclusions—Do and Don't Do in Dynamic and Probabilistic Analyses

In this chapter, some practical recommendations, as learned from the previous chapters, are summarized. These recommendations, when applied, will result in better design methodologies and procedures—and hopefully, a better structural product.

It is our belief that the following two major recommendations are the main consequences of the material presented in this publication, and should be adopted by young engineers:

1. Do not be afraid of a structural dynamics analysis. Structural dynamics is a major cause in structural failures. Once understood physically, there are no reasons to avoid it in the design procedure of aerospace structures. In fact, performance of a structural dynamic analysis is no more complex than doing a classical static analysis.

2. Do not be afraid of a probabilistic analysis. Although much less frequent in a regular design procedure, probabilistic analysis of structures has many major benefits in understanding the practical behavior of a designed structure, the influence of design parameters on the structure behavior, and the estimation of the structural reliability. Such analysis should be an important building stone in a structural design methodology.

At the beginning of a structural design, when the structure's main features are determined, do a structural failure analysis. This is a major part of a correct design methodology, which determines the structural model and influences the design of structural computations and designed structural experiments during the development phase.

When you numerically analyze a structural system, be sure to invest the proper amount of effort in the preparation of the input data file. In fact, preparing the data input is the equivalence of preparing the structural model of the relevant design. At this stage, the modeled properties are determined, including geometry, material properties, boundary conditions, and loadings. It is emphasized again that the preparation of numerical data file for a structure is identical whether a static or any dynamic analysis is desired, and once such a file is prepared for static analysis, it can be used for structural dynamics analysis.

Do not be intimidated by high acceleration responses. A structure fails (statically or dynamically) due to high stresses and/or strains and not due to high accelerations. In many cases, structural solutions (and experimental results) show high accelerations in many locations. These high accelerations may be the result of the existence of high frequency resonances, and while these high accelerations exist in the structure (and may be responsible for a noisy behavior), the stresses (or strains) caused by them are very low and do not always endanger the structure—they do not cause failures. When a response to random excitations is required, be sure to compute the PSD of the stresses and/or strains, and not merely the PSD of the accelerations. High stresses and strains, either static or dynamic, are the reason for structural failure.

When computing a mean square value from a PSD curve, do not use trapezoidal areas as depicted in the PSD specification curves, which are usually described on a log-log paper. The straight lines in these curves are not straight, and the results for the mean square values will be erroneous.

Be consistent with the units used in an analysis. No matter which system of units is used, it is important to use the same unit throughout the whole analysis. Designers are warned to make a clear distinction between units of mass and units of force, and to be very clear in the definition of frequencies—circular frequencies given in cycles per second, or angular frequencies given in radians per second. Errors of a factor of 2π (and its powers) may result in an inconsistent use of these units. When analytical (or semi-analytical) solutions are performed, it is recommended to check in advance the units of the results in the mathematical (algebraic) equations to ensure the proper units of the required result are obtained. This may save the reader a lot of trouble when long algebraic evaluations are involved.

Be especially careful with the units used in crack growth problems. Because of the "strange" units of the Stress Intensity Factor (SIF) and the way data is given in many manuals and numerical codes, it is quite easy to "lose the way" and produce erroneous results.

Be aware of the behavior of a SDOF system, whether it is force or base excited. As the solution of an elastic structure may be performed by the use of numerous uncoupled SDOF equations, understanding the behavior of the SDOF system can give a tremendous insight into the behavior of the elastic system.

Most of the analyses described in this book are linear. Practical structures may include nonlinear effects, either in the material properties or in the geometry. Nevertheless, it is important to perform a linear analysis even in cases of a nonlinear system, as the linear behavior provides a tremendous insight into the behavior of the analyzed system.

Try to build a MDOF model of your structural element. The behavior of structural systems can often be described by a MDOF system. While such a description may sometimes be only approximate, it may yield quick approximate solutions in the early stages of the design, approximations that may give insight into the physical behavior of the structure, and into the influence of many structural parameters on the behavior of the true system.

It is important that you know the modal damping in order to get reliable results from a dynamic analysis, whether analytical or numerical solutions are performed. There is no reliable analytical method to compute realistic damping coefficients. Modal damping depends on the nature of the structure. They are usually small (order of magnitude of few percent) for "pure" metallic parts and higher for structures with riveted and bolted structures, and for composite materials and plastics. The best practical method to detect the damping coefficients of a structure is to measure them experimentally by one of the methods used for this purpose (e.g., ground vibration tests). It is imperative for a design establishment to collect and store results from such tests in order to use them in similar designs.

When designing a vibration test (on the laboratory shaker), it is important to take into account the fixture that connects the structural system to the shaker. There is no totally rigid fixture, and the elastic behavior of the fixture usually introduces "strange" behavior during the test. The fixture should be introduced also to the structural models!

When designing field tests (such as flight tests) to prototypes, insist on measuring accelerations (which is the "traditional way"), and strain measurements (with strain gages). Such measurements used to be difficult in the past, but are much simpler with today's equipment. Mount the strain gages at locations where failure is expected following the failure modes analysis.

You can solve static and dynamic contact problems with an existing finite element program only for a linear behavior. When you have to solve a dynamic contact problem, it is very important to measure experimentally the coefficient of resilience between the two contacting materials. This coefficient is also impossible to be analytically calculated, and has a major influence on the dynamic behavior of such structures. Caution: the finite elements codes do not take into account the coefficient of resilience, and assume perfectly elastic collisions. This means that results of finite elements runs are usually conservative.

When you analyze the dynamic behavior of any structure, compute the resonance frequencies and the normal modes. Usually, it is difficult to theoretically estimate the rigidities of clamps and interconnections between parts of a structural system. In these cases, the model should be updated using experimental results of ground vibration tests, usually by measuring the resonance frequencies and correcting the rigidities of the model. Stress modes and generalized masses should also be computed, as the knowledge of their relative magnitudes gives insight into the physical behavior of the structure. It also helps in the determination of the number of the modes that are to be included in the analysis of a given system, according to the nature of the system. One should know how the normal modes are normalized in an analytical solution or a numerical computation, and should be consistent in such definition throughout the design process. One should also bear in mind that static deformations and normal modes are not the same for a given structure. In order to understand the modal behavior of a structural element, use the animation provided by the finite elements code. This animation provides an excellent insight into the structural behavior.

If you set a design criterion for a combination of static plus dynamic loads, be sure to use the "output criterion" and not the "input" one. Therefore, the "equivalent static load" criterion described in Chapter 7 should be avoided.

Be aware that when a structure is subjected to base excitation or to a displacement excitation on one or more of its points, the resonance frequencies of the structure are obtained by assuming that the points of displacement excitations (whether on the base or any other location) are of a structure with constrained supports in these locations.

Design structural experiments according to the failure analysis, so that failure modes can be surfaced during the development phase. Try to persuade project managers to fail the structure in the structural development tests; otherwise, no factor of safety will be provided.

There is no chance that you will be able to simulate flight loads with the conventional equipment of an environmental test laboratory. This equipment is built traditionally to be controlled by acceleration or displacement excitation, while most of the flight excitation is originated from random forces or random pressure fluctuations. New environmental laboratory equipment may now enable force control of the tests. Sometimes, it is possible to simulate the real conditions only at certain locations and not on the entire structure. In such cases, design such an experiment so that only critical locations are tested. In order to cover all failure modes, several different tests may be required.

Update the structural models according to structural tests on subsystems, subassemblies, and prototypes. When a carefully performed experiment does not agree with the structural model, the latter is not correct, and should be updated.

In case the inspected structure has a small number of random variables, and you have no access to a probabilistic structural program, try to analyze the structure using the Lagrange multiplier or the Taylor expansion method.

When doing a probabilistic analysis in which Weibull and lognormal distributions are involved, be sure to check the distribution constants provided. There are different definitions in different books, and especially in computer codes. You may easily end with the wrong probability!

You can do a probabilistic analysis using the probabilistic module of either NASTRAN® or ANSYS®. You can also do a response surface analysis rather than a Monte Carlo computation, to save time for the project.

When experimental results agree in trend with the structural model, but dispersion in experimental results is observed, try to introduce an additional random variable or a stochastic process to the model. The distribution and the statistical moments of this variable should be determined from the dispersed data.

Try to reconstruct the examples presented in this book by using the files on the accompanying CD-ROM. If you do not have access to the ANSYS program, you can construct equivalent NASTRAN or another finite element program files by following the commands in the Notepad text files presented in the Appendix. After you do these examples, try to build a data file for your case by inserting your pre-processor file instead of the simple data file (and of course adjusting the solution and post-processing to your case).

Some small tips can make your finite element analysis easier:

1. Prefer to write a batch file as a **filename.txt**, and try to avoid using the graphical user's interface (GUI) when preparing the data file.

2. Use parametric language in such a file so changes in the structural parameters are easier, especially in the first phases of an analysis. In case there are some options to the desired computations, try to include them in your text file, and use the exclamation mark (!) to avoid performing the commands that do not belong to the analysis you do.

3. When performing a repeated analysis in cases where external force is involved, zero the previous load before applying another load. If this is not done, you may get a solution to a problem in which the previous load, and the present one, are included.

4. Insert as many comments as possible into the text file. This will help tremendously in an additional computation that is done later ("What did I do then?").

5. Use, if possible, a five-button mouse. You can program the two additional buttons for "copy" (from the text file) and "paste" into the finite element's command window.

The author will be glad to assist the reader in further understanding of the issues discussed in this book. Try the address *gioram@netvision.net.il*.

Appendix / Computer Files for the Demonstration Problems

A.1 Introduction

Computer files used in the numerous demonstration problems are included on the CD-ROM attached to this book. In this Appendix, methods to use the files, and a list of them, are described. This Appendix is also included on the CD-ROM, in the "Introduction" folder. The listings of the files may provide users who are using other finite elements programs rather than the ANSYS® to prepare compatible files for the other programs.

There are three types of files in this book and in the present Appendix:

1. Text files (in NOTEPAD) for ANSYS. These can be run by the ANSYS command /**input,filename,txt**. Each file has a /**eof** command at the end of the database. Also included, after the /**eof** command, the relevant commands for the solution and the post-processing phases of the relevant example. These can be copied from the NOTEPAD file and pasted into the reader's ANSYS working directory when required. Users of other finite elements codes may construct the suitable files by a proper "translation" of the listed files. Access to a licensed ANSYS program is required. The files should be copied to the working directory of the ANSYS, as defined by the user. All files can work with the ANSYS educational version.

2. MATLAB® m-files. The extension of these files is **filename.m**. These must be copied into the MATLAB working directory of the user. Access to a licensed MATLAB program (version 5.3 and higher) is required. In the MATLAB environment, a proper path to the directory

where the files are saved should be defined. Typing the m-file name in the MATLAB command window runs the files. Some information about the MATLAB program can be found in Chapter 8.

3. TK Solver™ files. The extension of these files is **filename.tkw**. These must be copied into the TK Solver working directory of the user. Access to a licensed TK Solver program is required. An inexpensive educational version of TK Solver is available to authorized persons in the academy—students and academic staff. Some information about the TK Solver program can be found in Chapter 8. The TKW files are not listed in this Appendix. When TK Solver is installed on the reader's computer, clicking on the relevant **filename.tkw** file will initiate the program and the relevant data.

In all the files, data of the last used example are listed. The user should correct these values to the parameters relevant to the required problem.

A.2 LIST OF FILES

FILES FOR CHAPTER 1

1. **duhamel1.tkw**, for analytical solution of the transient problem using Duhamel's integral.

2. **duhamel2.tkw**, an example of a numerical procedure for Duhamel's integral.

FILES FOR CHAPTER 2

1. **beam1.txt**, an ANSYS file for modal analysis of a cantilever beam.

2. **beamharm.txt**, an ANSYS file for the response of a cantilever beam to tip harmonic force.

3. **commass1.txt,** an ANSYS file for a beam with mounted mass.

4. **ssbeam.txt**, an ANSYS file for modal analysis of simply supported beam.

5. **ssplate.txt**, an ANSYS file for modal analysis and harmonic respone of simply supported plate.

6. **shell1.txt**, an ANSYS file for modal analysis of cylindrical shell.

FILES FOR CHAPTER 3

1. **beamrand_1.txt**, an ANSYS file for the response of a cantilever beam to tip random force.

2. **beamrand_2.txt**, an ANSYS file for the response of a cantilever beam to random base excitation.

3. **beamrand_3.txt**, an ANSYS file for the response of a cantilever beam to random tip displacement excitation.

4. **wing1.txt**, an ANSYS file for computation of a cantilevered skewed plate under random forces.

5. **bbplatc.txt**, an ANSYS file for the response of a beam-plate to acoustic excitation.

6. **frame1.txt**, an ANSYS file for the response of a frame to a random force excitation.

7. **commass1.txt**, an ANSYS file for a beam with mounted mass (random loading).

8. **ssplaterand.txt**, an ANSYS file for the computation of a simply supported rectangular plate.

FILES FOR CHAPTER 4

The following files were solved with ANSYS version 8.0. In the latest versions (V10 and V11), some of the contact elements of the programs no longer exist. The reader should check his version, and change the contact elements, where required.

1. **plate2.txt**, an ANSYS file for static contact problem of a cantilever beam.

2. **plate3.txt**, an ANSYS file for a dynamic contact problem, tip force excitation of a cantilever beam.

3. **plate4.txt**, an ANSYS file for a dynamic contact problem, base excitation of a cantilever beam.

4. **contact7.tkw**, a TK Solver file for an analytical solution of a SDOF problem with contacts.

5. **two2.txt**, an ANSYS file for a dynamic contact problem of two masses: collisions and gaps.

FILES FOR CHAPTER 5

1. **truss1.txt**, an ANSYS file for static solution of 3-members truss.

2. **probeam4.txt**, an ANSYS file for a probabilistic analysis using ANSYS.

3. **math.txt**, an ANSYS file that demonstrates solution of a "Stress-Strength" mathematical model, without using any finite elements.

FILES FOR CHAPTER 6

1. **corcrack1.m, corcrack2.m, corcrack3.m**, and **corcrack4.m**, MATLAB files for stochastic crack propagation (see Figure 6.9).

2. **virtest1.m, virtest2.m, virtest3.m**, and **virtest4.m**, MATLAB files for stochastic crack propagation (see Figure 6.9).

3. **cracknorm1.m**, data analysis for the stochastic crack propagation problem (see Figure 6.9).

4. **cracklognfit.m**, fitting a log-normal distribution to the stochastic crack growth problem.

5. **astar1.m**, computation of the random critical length of the crack.

6. **prop1.m**, a MATLAB file.

7. **weib1.tkw**, a TK Solver file for the distribution of local peaks in a stationary normal stochastic process.

8. **weib2.tkw**, a TK Solver file for the computation of the number of peaks above a given level.

9. **crack2.m**, solution of deterministic crack growth process for $0 < R < 1$.

10. **crack3.m**, solution of deterministic crack growth process for $R < 0$.

FILES FOR CHAPTER 7

1. **stat1.txt**, an ANSYS file for a static response of a cantilever beam to 1g static load.

2. **stat2.txt**, an ANSYS file for the dynamics of a cantilever beam to with tip mass.

3. **stat3.txt**, an ANSYS file for the random response of a cantilever beam with tip mass.

4. **env5.txt**, an ANSYS file for an aircraft carried beam.

A.3 Files Listing

CHAPTER 1

All files are of TKW+ program, and therefore are not listed.

CHAPTER 2

beam1.txt

!file beam1 for cantilever beam

/filnam,beam1

/title, Vibration of a cantilever beam

!units in kgf, cm, seconds

```
g = 980                    !gravity
L = 60                     !length of beam
b = 8                      !width of beam
h = 0.5                    !height of beam
E - 2.1e6                  !Young's modulus
ro = 7.959e-6              !mass density
W = L*b*h*ro*g             !total weight (reference only)
q = W/L                    !weight per unit length
A = b*h                    !cross section area
I = b*h**3/12              !area moment of inertia
€
/prep7
mp,ex,1,E                  !materials property
mp,dens,1,ro               !material property

et,1,beam3                 !type of element
r,1,A,I,h                  !real constants

!nodes
n,1,0,0
n,11,L,0
fill,1,11
```

```
!elements
en,1,1,2
engen,1,10,1,1,1,1

!boundary conditions
d,1,ux,0
d,1,uy,0
d,1,rotz,0
!d,11,uy,0

save
fini
/eof                        !end of input file

!modal analysis
/solu
antyp,modal
modop,subs,3                3! modes
mxpand,3,,,yes
solve
fini
/post1                      !general purpose post-processor
set,1,1                     !first mode
pldisp,2
set, 1,2                    !second mode
pldisp,2
set, 1,3                    !third mode
pldisp,2

fini
!animation of modes is possible using GUI
```

beamharm.txt

!file beamharm for cantilever beam

/filnam,beamharm

/title, Harmonic Respone of a cantilever beam

```
!units in kgf, cm, seconds

g = 980                          !gravity
L = 60                           !length of beam
b = 8                            !width of beam
h = 0.5                          !height of beam
E = 2.1e6                        !Young's modulus
ro = 7.959e-6                    !mass density
W = L*b*h*ro*g                   !total weight (reference)
q = W/L                          !weight per unit length
A = b*h                          !cross section area
I = b*h**3/12                    !area moment of inertia

/prep7

mp,ex,1,E                        !material property
mp,dens,1,ro                     !material property

et,1,beam3                       !type of element
r,1,A,I,h                        !real constant

!nodes

n,1,0,0
n,11,L,0
fill,1,11

!elements

en,1,1,2
engen,1,10,1,1,1,1

!boundary conditions
d,1,ux,0
d,1,uy,0
d,1,rotz,0
!d,11,uy,0

save
fini
/eof                             !end of input file
```

```
/solu                          !Harmonic Respone, full
antyp,harmic
hropt,full
hrout,off                      !Magnitude and Phase
outpr,all,1
harfrq,10,250                  !Frequency range
nsubs,240                      !Number of substeps
dmpr,0.02                      !Damoing coefficient
kbc,1                          !Stepped input each step
d,11,uy,1                      !1 kgf force at tip
save
solve
fini

/post1                         !General post-processor
set,list                       !see list of frequencies
fini

/post26                        !Frequency domain post processor
nsol,2,8,u,y,Y8                !displacement at tip
!nsol,3,1,u,y,Uclamp           !displacement at clamp
/grid,1
/axlab,x,Frequency Hz
/axlab,y,Displacement cm
plvar,2                        !plot displacement at tip
!esol,4,1,1,ls,3,Sbend         !Bending stress at clamp
!/axlab,y,Bending Stress kgf/cm²
!plvar,3                       !plot stress at clamped edge
fini
```

commass1.txt

```
!file commass1 for cantilever beam

/filnam,commass1

/title,Cantilever beam with 1 Mounted Mass

!DATA

L = 60                         !length, cm
b = 8                          !width, cm
```

```
h = 0.5                    !thickness, cm
E = 2.1e6                  !Young modulus
ro = 7.959e-6              !density
g = 980                    !value of g
W = L*b*h*ro*g             !total weight of the beam
q = W/L                    !beam weight per length
A = b*h                    !beam cross section area
I = b*h**3/12              !area moment of inertia
g = 980                    !g = 980 cm/sec 2
w1 = 1                     !mounted weight, node 14 (below node 4)
m1 = w1/g                  !mounted mass
k1y = 4                    !vertical spring, 4 to 14, case (a)
!k1y = 65                  !vertical spring, 4 to 14, case (b)

!a high rigidity spring in the x direction is added
!in order to avoid horizontal vibrations of the mass
k1x = 50000                !horiz. spring, 4 to 14 - very high stiffness

/prep7                     !pre-processor
mp,ex,1,E                  !material property, beam
mp,dens,1,ro               !material property, beam

et,1,beam3                 !beam element
et,2,mass21,,0,4           !first mass element
et,3,combin14,0,2          !spring y direction
et,4,combin14,0,1          !spring x direction

!real constants
r,1,A,I,h                  !beam
r,2,m1                     !mass
r,3,k1y                    !vertical spring
r,4,k1x                    !horizontal spring

!nodes for the beam
n,1,0,0
n,11,L,0
fill,1,11

!node for the mass
n,14,0.3*L,0               !node for mass
```

```
!elements
!for the beam
type,1
real,1
mat,1
en,1,1,2
engen,1,10,1,1,1,1

!first mass, element 11
type,2
real,2
en,11,14                          !at node 14

!vertical spring y, element 12
type,3
real,3
en,12,4,14                        !between 4–14

!horizontal spring x, element 13
type,4
real,4
en,13,4,14

!constrains (boundary conditions)
d,1,ux,0
d,1,uy,0
d,1,rotz,0

save
fini
/eof                              !end of input file

/solu                             !Modal Solution
antyp,modal
modop,subs,4    4!                modes, subspace method
mxpand,4,,,yes                    !expand 4 modes, calculate stresses
solve
fini
```

!RESPONSE TO HARMONIC EXCITATION-TIP FORCE
/solu
antyp,harmic
hropt,full
hrout,on
kbc,1

harfrq,2,250   !frequency range 2–250 Hz
nsubs,496   !every 0.5 Hz
f,11,fy,1   !unit vertical tip force
solve
fini

!frequency domain post-processor
/post26
numvar,200
/grid,1   !grid on
/axlab,x,Frequency, Hz
/axlab,y,Quantity
nsol,2,10,u,y,Y11   !disp. at tip
nsol,3,4,u,y,Y4   !disp. at node 4
nsol,4,14,u,y,Ymass   !disp at added mass

!RESPONSE TO RANDOM EXCITATION

/solu   !SPECTRAL ANALYSIS
antyp,spect   !analysis type - spectral
spopt,psd,3,yes   !psd with 3 modes, stresses also
                                                computed
psdunit,1,forc   !input in PSD of force (kgf^2/Hz)
psdfrq,1,,5,250   !input between 5 to 250 Hz
psdval,1,0.004081632,0.004081632   !values of PSD
kbc,1

psdres,disp,rel   !displacements relative to base
psdres,velo,rel   !velocities relative to base
psdres,acel,abs   !absolute accelerations

```
dmprat,0.02              !damping 2% for all modes
f,11,fy,1                !random input at tip
pfact,1,node             !nodal excitation
mcomb,psd                !modal combination for psd

solve
save
fini

/post1
set,list                 !see list of load steps (1,1;1,2 etc. for resonances)
set,3,1                  !for rms of displacements
                         !use this for stresses
                         !for beam elements an element table is required(!!!)
set,4,1                  !for rms of velocities
set,5,1                  !for rms of accelerations
                         !accelerations/980 = acel. in g's
fini

!frequency domain post processing
/post26
keep                     !keep post results for another glance
numvar,30                !prepare space for 30 variables
store,psd,10             !store frequencies, 10 to each side of resonance
/grid,1
/axlab,x,Frequency Hz
/axlab,y,PSD
nsol,2,11,u,y,Wtip       !disp. of tip
nsol,3,14,u,y,wmass      !disp. of mass
esol,4,1,1,ls,3,Sben     !bend. str. at element 1, node 1
                         !ls-3:bending, i side of element
rpsd,12,2,,1,2,Wtip      !PSD of rel.disp.-tip
rpsd,13,3,,1,2,wmass     !PSD of rel disp.-mass
rpsd,14,4,,1,2,bend      !PSD of Bend stress.

rpsd,22,2,,3,1,ACtip     !PSD of Acceleration, node 11 (in cm/sec²/Hz)
                         !for g²/Hz—divide by 980²
rpsd,23,3,,3,1,ACmass    !PSD of bending stress, node 1
```

ssbeam.txt

!file ssbeam for simply supported beam

/filnam,ssbeam

/title, Modes of a simply supported beam

!units in kgf, cm, seconds

```
g = 980                         !gravity
L = 60                          !length of beam
b = 8                           !width of beam
h = 0.5                         !height of beam
E = 2.1e6                       !Young's modulus
ro = 7.959e−6                   !mass density
W = L*b*h*ro*g                  total weight (reference only)
q = W/L                         !weight per unit length
A = b*h                         !cross section area
I = b*h**3/12                   !area moment of inertia

/prep7
mp,ex,1,E                       !material property
mp,dens,1,ro                    !material property

et,1,beam3                      !beam element
r,1,A,I,h                       !real constant

!nodes
n,1,0,0
n,11,L,0
fill,1,11

!elements
en,1,1,2
engen,1,10,1,1,1,1

!boundary conditions
d,1,ux,0                        !left end
d,1,uy,0
```

```
d,11,uy,0                          !right end
d,11,ux,0

save
fini
/eof                               !end of input file

/solu                              !modal analysis
antyp,modal
modop,subs,3                       !3 modes
mxpand,3,,,yes
solve
fini

/post1                             !look at mode shapes
set,1,1                            !first mode
pldisp,2
set,1,2                            !second mode
pldisp,2
set,1,3                            !third mode
pldisp,2
fini
```

!loading can be performed in static or dynamic
!solution phase using the specific input of the problem

ssplate.txt

!file ssplate for simply supported plate

/filnam,ssplate

/title, Simply Supported Plate

/config,nres,2000

!units-cm, kgf, sec.

```
g = 980                            !value of g
pi = 4*atan(1)                     !value of pi
a = 40                             !length (x)
b = 30                             !width (y)
```

```
h = 0.5                               !thickness
E1 = 2.1e6                            !Young modulus,x direction
E2 = 2.1e6                            !Young modulus,y direction
E3 = 2.1e6                            !Young modulus,z direction
nu = 0.3                             !Poisson ratio

ro = 7.959e-6                        !specific density
W = ro*g*a*b*h                       !total weight (for reference only)
/prep7                               !Prepare the model
et,1,shell63                         !3-D shell (plate) element
r,1,h                                !thickness

!material properties
mp,ex,1,E1                           !E in 3 directions
mp,ey,1,E2
mp,ez,1,E3
mp,dens,1,ro
mp,nuxy,1,nu
mp,nuyz,1,nu
mp,nuxz,1,nu

!Nodes
n,1,0,0
n,21,a,0
fill,1,21

ngen,17,100,1,21,1,0,b/16,0

!Elements
type,1
mat,1
real,1

en,1,1,2,102,101
engen,1,20,1,1,1,1
engen,100,16,100,1,20,1

!Boundary conditions
!edges are simply supported and don't move in plane
nsel,s,node,,1,21,1
```

```
d,all,uz,0
d,all,uy,0
nsel,all

nsel,s,node,,1601,1621,1
d,all,uz,0
d,all,uy,0
nsel,all

nsel,s,node,,1,1601,100
d,all,uz,0
d,all,ux,0
nsel,all

nsel,s,node,,21,1621,100
d,all,uz,0
d,all,ux,0
nsel,all

save
fini
/eof                          !end of input file

!Modal Solution for 6 first 6 modes
/solu
antyp,modal
modop,subs,6                  !6 modes, subspace method
mxpand,6,,,yes                !expand 6 modes, for stress calculations
solve
fini

!Response to vertical harmonic loads, 4 cases

/solu
antyp,harmic
hropt,full
hrout,on
kbc,1

harfrq,200,1250               !frequency range
nsubs,525                     !number of frequencies calculated
```

```
dmpr,0.02                        !damping ratio for all modes
fdel,all                         !required for consecutive runs

!for the required case, delete the relevant ! mark

!f,811,fz,1                      !case 1

f,416,fz,1                       !case 2

!f,416,fz,1                      !case 3
!f,1206,fz,1

!f,416,fz,1                      !case 4
!f,1206,fz,−1

solve
save
fini

!Frequency domain response
/post26
numvar,200
/grid,1
/axlab,x,Frequency, Hz
/axlab,y,Quantity

nsol,2,811,u,z,Z811
nsol,3,416,u,z,Z416
nsol,4,1206,u,z,Z1206

esol,5,315,416,s,x,SX           !sigma-x at element 315,node 416
esol,6,315,416,s,y,SY           !sigma-y at element 315,node 416
```

shell1.txt

```
!file shell1 for simply supported cylindrical shell
/filnam,shell1
/title,Simply Supported Cylindrical Shell
/config,nres,2000
!units - cm,kgf,sec.
```

```
g = 980                    !value of g
pi = 4*atan(1)             !value of pi

a = 10                     !radius
b = 60                     !height
h = 0.1                    !thickness

E1=2.1e6                   !Young modulus,x direction
E2 = 2.1e6                 !Young modulus,y direction
E3 = 2.1e6                 !Young modulus,z direction
nu = 0.3                   !Poisson ratio

ro = 7.959e-6              !specific density

/prep7                     !Prepare the model

et,1,shell63               !3-D shell(plate) element

r,1,h                      !thickness

!material properties
mp,ex,1,E1                 !E in 3 directions
mp,ey,1,E2
mp,ez,1,E3
mp,dens,1,ro
mp,nuxy,1,nu
mp,nuyz,1,nu
mp,nuxz,1,nu

csys,1                     !cylindrical coordinate system

!first layer of nodes
n,1,a,0,0
n,7,a,90,0
fill,1,7

n,13,a,180,0
fill,7,13

ngen,7,1,13,13,1,0,15,0
ngen,6,1,19,19,1,0,15,0
```

```
!next layers
ngen,17,100,1,24,1,0,0,b/16

!Elements
type,1
real,1
mat,1

!first layer,24 elements
en,1,1,2,102,101
engen,1,23,1,1,1,1
en,24,24,1,101,124

!next layers
engen,100,16,100,1,24,1

!boundary conditions
nsel,s,node,,1,24,1
d,all,ux,0
d,all,uy,0
d,all,uz,0
nsel,all
nsel,s,node,,1601,1624,1
d,all,ux,0
d,all,uy,0
!d,all,uz,0                      !held axially
nsel,all
save
fini
/eof                            !end of input file


!Modal Solution for 6 first 6 modes
/solu
antyp,modal
modop,subs,10                   !10 modes, subspace method
mxpand,10,,,yes                 !expand 10 modes, for stress calculations
solve
fini
```

CHAPTER 3

beamrand_1.txt

!file beamrand for cantilever beam

/filnam,beamrand_1

/title, Random Respone of a Cantilever Beam

!units in kgf, cm, seconds

```
g = 980                    !gravity
L = 60                     !length of beam
b = 8                      !width of beam
h = 0.5                    !height of beam
E = 2.1e6                  !Young's modulus
ro = 7.959e−6              !mass density
W=L*b*h*ro*g               !total weight (reference only)
q = W/L                    !weight per unit length
A = b*h                    !cross section area
I = b*h**3/12              !area moment of inertia

/prep7                     !same pre-processing as previous examples
mp,ex,1,E
mp,dens,1,ro
et,1,beam3
r,1,A,I,h

!nodes
n,1,0,0
n,11,L,0
fill,1,11

!elements
en,1,1,2
engen,1,10,1,1,1,1

!boundary conditions
d,1,ux,0
d,1,uy,0
d,1,rotz,0
```

```
save
fini
/eof                                    !end of input file

/solu                                   !Modal Solution
antyp,modal
modop,subs,3                            !3-D modes, subspace method
mxpand,3,,,yes !expand 3 modes, calculate stresses
solve
fini

/solu                                   !SPECTRAL ANALYSIS
antyp,spect                             !analysis type - spectral
spopt,psd,3,yes                         !psd with 3 modes,
                                         stresses also computed
psdunit,1,forc                          !input in PSD of force (kgf²/Hz)
psdfrq,1,,5,250                         !input between 5 to 250 Hz
psdval,1,0.004081632,0.004081632        !values of PSD

psdres,disp,rel                         !displacements relative to base
psdres,velo,rel                         !velocities relative to base
psdres,acel,abs                         !absolute accelerations

dmprat,0.02                             !damping 2% for all modes
f,11,fy,1                               !random input at tip
pfact,1,node                            !nodal excitation
mcomb,psd                               !modal combination for psd

solve
save
fini

/post1
set,list                                !see list of load steps (1,1;1,2 etc.
                                             for resonances)
set,3,1                                 !for rms of displacements
                                        !use this for stresses
                                        !for beam elements an element
                                             table is required!
```

set,4,1	!for rms of velocities
set,5,1	!for rms of accelerations
	!accelerations/980 = acel. in g's
fini	
/post26	
numvar,30	!prepare space for 30 variables
store,psd,10	!store frequencies, 10 to each side of resonance
/grid,1	
/axlab,x,Frequency Hz	
/axlab,y,PSD	
nsol,2,11,u,y,Wtip	!disp. node 11
esol,4,1,1,ls,3,Sben	!bend. str. at element 1, node 1
	!ls-3:bending, i side of element
rpsd,12,2,,1,2,Wtip	!PSD of rel. disp. node 11
rpsd,13,2,,3,1,ACtip	!PSD of Acceleration, node 11 (in cm/sec^2/Hz)
	!for g 2/Hz—divide by 980^2
rpsd,14,4,,1,2,BEND	!PSD of bending stress, node 1

!integrals of above variables give Mean Square values

int1,22,12,1,,MStip	!Mean square of tip displacement
int1,23,13,1,,MSaccel	!Mean square of tip accel.
int1,24,14,1,,MSbend	!Mean square of clamp bending stress.

beamrand_2.txt

!file beamrand_2 for cantilever beam

/filnam,beamrand_2

/title, Random Response of a Cantilever Beam to Base Excitation

!units in kgf, cm, seconds

g = 980	!gravity
L = 60	!length of beam
b = 8	!width of beam
h = 0.5	!height of beam
E = 2.1e6	!Young's modulus
ro = 7.959e−6	!mass density

```
W = L*b*h*ro*g                  !total weight (reference only)
q = W/L                         !weight per unit length
A = b*h                         !cross section area
I = b*h**3/12                   !area moment of inertia

/prep7                          !same preprocessing as in beamrand_1.txt
mp,ex,1,E
mp,dens,1,ro
et,1,beam3
r,1,A,I,h

n,1,0,0
n,11,L,0
fill,1,11

en,1,1,2
engen,1,10,1,1,1,1

d,1,ux,0
d,1,uy,0
d,1,rotz,0

save
fini
/eof                            !end of input file

/solu                           !Modal Solution
antyp,modal
modop,subs,3                    !3 modes, subspace method
mxpand,3,,,yes                  !expand 3 modes, calculate stresses
solve
fini


/solu                           !SPECTRAL ANALYSIS
antyp,spect                     !analysis type - spectral
spopt,psd,3,yes                 !psd with 3 modes,
                                stresses also computed

psdunit,1,disp                  !input in PSD of displacement (cm²/Hz)
psdfrq,1,,5,250                 !input between 5 to 250 Hz
psdval,1,0.000054128,0.000054128 !values of PSD
```

psdres,disp,rel	!displacements relative to base
psdres,velo,rel	!velocities relative to base
psdres,acel,abs	!absolute accelerations
dmprat,0.02	!damping 2% for all modes
d,1,uy,1	!random disp. input at clamp
pfact,1,base	!base excitation
mcomb,psd	!modal combination for psd
solve	
save	
fini	
/post1	
set,list	!see list of load steps (1,1;1,2 etc. for resonances)
set,3,1	!for rms of displacements. Use also for stresses
	!for beam elements an element table is required!
set,4,1	!for rms of velocities
set,5,1	!for rms of accelerations
	!accelerations/980 = acel. in g's
fini	
/post26	
numvar,30	!prepare space for 30 variables
store,psd,10	!store frequencies, 10 to each side of resonance
/grid,1	
/axlab,x,Frequency Hz	
/axlab,y,PSD	
nsol,2,11,u,y,Wtip	!disp. node 11
esol,4,1,1,ls,3,Sben	!bend. str. at element 1, node 1
	!ls-3:bending, i side of element
rpsd,12,2,,1,2,Wtip	!PSD of rel.disp. node 11
rpsd,13,2,,3,1,ACtip	!PSD of Acceleration, node 11 (in $cm/sec^2/Hz$)
	!for g^2/Hz—divide by 980^2
rpsd,14,4,,1,2,BEND	!PSD of bending stress, node 1
int1,22,12,1,,MStip	!Mean Square of tip displacement
int1,23,13,1,,Msaccel	!Mean Square of tip accel.
int1,24,14,1,,Msbend	!Mean Square of clamp bending stress.

beamrand_3.txt

!file beamrand_3 for cantilever beam

/filnam,beamrand_3

/title, Random Respone of a Cantilever Beam to Tip Disp. Exc.

!units in kgf, cm, seconds

```
g = 980                      !gravity
L = 60                       !length of beam
b = 8                        !width of beam
h = 0.5                      !height of beam
E = 2.1e6                    !Young's modulus
ro = 7.959e-6                !mass density
W = L*b*h*ro*g               !total weight (reference only)
q = W/L                      !weight per unit length
A = b*h                      !cross section area
I = b*h**3/12                !area moment of inertia

/prep7                       !same preprocessing as in beamrand_1
mp,ex,1,E
mp,dens,1,ro
et,1,beam3
r,1,A,I,h

n,1,0,0
n,11,L,0
fill,1,11

en,1,1,2
engen,1,10,1,1,1,1

d,1,ux,0
d,1,uy,0
d,1,rotz,0

save
fini
```

```
/solu                           !Modal Solution
antyp,modal
modop,subs,3                    !3 modes, subspace method
mxpand,3,,,yes                  !expand 3 modes, calculate stresses
d,11,uy,0
solve
fini

/solu                           !SPECTRAL ANALYSIS
antyp,spect                     !analysis type - spectral
spopt,psd,3,yes                 !psd with 3 modes, stresses also computed
psdunit,1,disp                  !input in PSD of displacement (cm²/Hz)
psdfrq,1,,5,250,250.001         !input between 5 to 250 Hz
psdval,1,0.00119152,0.00119152,0.0000001
kbc,1
psdres,disp,abs                 !displacements relative to base
psdres,velo,rel                 !velocities relative to base
psdres,acel,abs                 !absolute accelerations

dmprat,0.02                     !damping 2% for all modes
d,11,uy,1                       !random disp. input at tip
pfact,1,base                    !base excitation
mcomb,psd                       !modal combination for psd

solve
save
fini

/post1
set,list                        !see list of load steps (1,1;1,2 etc.
                                   for resonances)
set,3,1                         !for rms of displacements
                                !use this for stresses
                                !for beam elements an element
                                   table is required!
set,4,1                         !for rms of velocities
set,5,1                         !for rms of accelerations
                                !accelerations/980 = acel. in g's
fini
```

```
/post26
numvar,30                           !prepare space for 30 variables
store,psd,10                        !store frequencies, 10 to each side
                                        of resonance

/grid,1
/axlab,x,Frequency Hz
/axlab,y,PSD
nsol,2,11,u,y,Wtip                  !disp. node 11
esol,4,1,1,ls,3,Sben                !bend. str. at element 1, node 1
                                    !ls-3:bending, i side of element

rpsd,12,2,,1,1,Wtip                 !PSD of rel.disp. node 11
rpsd,13,2,,3,1,ACtip                !PSD of Acceleration,
                                        node 11 (in cm/sec²/Hz)
                                    !for g²/Hz—divide by 980²
rpsd,14,4,,1,1,BEND                 !PSD of bending stress, node 1

int1,22,12,1,,MStip                 !Mean Square of tip displacement
int1,23,13,1,,Msaccel               !Mean Square of tip accel.
int1,24,14,1,,MSbend                !Mean Square of clamp bending stress.
```

wing1.txt

```
!file wing1 for cantilevered wing

/filnam,wing1

/title,Cantilever Skewed Plate Under Random Forces

/config,nres,2000

!units-cm, kgf, sec.

g = 980                             !value of g
pi = 4*atan(1)                      !value of pi
a = 40                              !chord at clamp
b = 30                              !cantilever length
h = 0.5                             !thickness
E1 = 2.1e6                          !Young modulus,x direction
E2 = 2.1e6                          !Young modulus,y direction
E3 = 2.1e6                          !Young modulus,z direction
nu = 0.3                            !Poisson ratio
```

```
ro = 7.959e-6                          !specific density

W = ro*g*1.5*a*b*h                     !total weight (for information only)

/prep7                                 !Prepare the model
et,1,shell63                           !3-D shell (plate) element
r,1,h

!material properties
mp,ex,1,E1
mp,ey,1,E2
mp,ez,1,E3
mp,dens,1,ro
mp,nuxy,1,nu
mp,nuyz,1,nu
mp,nuxz,1,nu

!NODES
n,1,0,−a,0
ngen,9,10,1,1,1,0,a/8,0

n,9,b,−a/2,0
ngen,9,10,9,9,1,0,a/16,0

fill,1,9
fill,11,19
fill,21,29
fill,31,39
fill,41,49
fill,51,59
fill,61,69
fill,71,79
fill,81,89

!ELEMENTS
type,1
mat,1
real,1
```

```
en,1,1,2,12,11
engen,1,8,1,1,1,1
engen,10,8,10,1,8,1
```

!BOUNDARY CONDITIONS
```
nsel,s,node,,1,81,10
d,all,ux,0
d,all,uy,0
d,all,uz,0
d,all,roty,0
nsel,all
save
fini
/eof
```

!Modal Solution for 4 modes only
```
/solu
antyp,modal
modop,subs,4                    !4 modes, subspace method
mxpand,4,,,yes                  !expand 4 modes, for stress calculations
solve
fini
```

!response to random 2 forces
```
/solu                           !SPECTRAL ANALYSIS
antyp,spect                     !analysis type - spectral
spopt,psd,4,yes                 !psd with 4 modes,
                                stresses also computed
```

!select the relevant input and delete the ! marks and change as required
```
!psdunit,1,force                !input in PSD of force (kgf²/Hz)
!psdunit,2,force
!psdunit,1,accg,g
!psdunit,1,acel
psdunit,1,disp
```

!random force at node 89, 2 kgf rms
```
!psdfrq,1,,20.5,500
!psdval,1,4/479.5,4/479.5
!psdg,1
```

!random force at node 49, 1 kgf rms
!psdfrq,2,,20.5,500
!psdval,2,1/479.5,1/479.5
!psdg,2

!random base excitation for tip displacement
!psdfrq,1,,20.5,30,40,50,60,70,80
!psdfrq,1,,90,100,105,110,120,140,160
!psdfrq,1,,180,188.5,200,220,230,240,252.5
!psdfrq,1,,270,280,290.5,310,324,350,375
!psdfrq,1,,400,420,440,463,480,500

!psdval,1,0.000002469803,0.000001886208,0.000001298731,0.000000823921,
0.000000478475,0.000000251142,0.000000116346
!psdval,1,0.000000044353,0.000000015751,0.000000012934,0.000000015918,
0.000000035884,0.000000125390,0.000000353305
!psdval,1,0.000002283391,0.000011624921,0.000000799697,0.000000038448,
0.000000009721,0.000000001847,0.000000000155
!psdval,1,0.000000000904,0.000000001271,0.000000001346,0.000000000913,
0.000000000642,0.000000002499,0.000000012463
!psdval,1,0.000000046337,0.000000133019,0.000000498817,0.000003810632,
0.000000825332,0.000000169769
!psdg

psdfrq,1,,20.5,40,55,65,80,90,100
psdfrq,1,,105,111.5,130,140,150,158,165
psdfrq,1,,180,200,220,250,280,303,310
psdfrq,1,,320,344,370,400,430,439,450
psdfrq,1,,470,500

psdval,1,0.000030829827,0.000002206463,0.000000684833,0.000000327368,
0.000000136399,0.00000007709,0.000000042684
psdval,1,0.000000032634,0.000000027426,0.000000096991,0.000000208414,
0.00000031121,0.000000321516,0.000000278609
psdval,1,0.000000151223,0.000000056499,0.0000000226,0.000000006911,
0.000000002411,0.000000001070,0.000000000844
psdval,1,0.000000000568,0.000000000303,0.000000000812,0.000000006122,
0.000000124764,0.000000268451,0.000000114167
psdval,1,0.000000029178,0.000000009212
psdg

```
psdres,disp,abs          !displacements relative to base
psdres,velo,rel          !velocities relative to base
psdres,acel,abs          !absolute accelerations

dmprat,0.02              !damping 2% for all modes

!f,89,fz,1               !force at node 89
!pfact,1,node

!f,89,fz,0               !remove first force
!f,49,fz,1               !force at node 55
!pfact,2,node

nsel,s,node,,1,81,10
d,all,uz,1
nsel,all
pfact,1,base

psdcom,,4                !modal combination for 4 modes

solve
save
fini

/post1
set,list                 !see list of load steps (1,1;1,2 etc. for resonances)
set,3,1                  !for rms of displacements
                         !use this for stresses
set,4,1                  !for rms of velocities
set,5,1                  !for rms of accelerations
                         !rms of accelerations/980 = acel. in g's
save
fini

/post26
numvar,40                !prepare space for 30 variables
store,psd,10             !store frequencies, 10 to each side of resonance
/grid,1

/axlab,x,Frequency Hz
/axlab,y,PSD
```

```
nsol,2,89,u,z,Z89                !disp at node 89 (tip-rear)
nsol,3,49,u,z,Z49                !disp at node 49 (tip-midlle)
nsol,4,59,u,z,Z9                 !disp at node 9 (tip-forward)
!esol,5,5,5,s,x,Sbend1           !sigmax at mid clamp
esol,6,61,71,s,eqv,SEQV1         !sigmax at forward clamp
!nsol,7,51,u,z,Wy                !repeated...

rpsd,12,2,,1,1,Z89
rpsd,13,3,,1,2,Z49
rpsd,14,4,,1,2,Z9
rpsd,22,2,,3,1,A89
rpsd,23,3,,3,1,A49
rpsd,24,4,,3,2,A9
```

!for g^2/Hz - divide by 980^2
```
prod,32,22,,,g89,,,0.00000104123
prod,33,23,,,g49,,,0.00000104123
prod,34,24,,,g9,,,0.00000104123

!rpsd,15,5,,1,2,Sbend1
rpsd,16,6,,1,2,SEQV1
!rpsd,17,7,,3,1,Ay                !acceleration, node 51

/solu                            !harmonic response solution
antyp,harmic
hropt,full
hrout,on
kbc,1

harfrq,20,500                    !frequency range
nsubs,960                        !number of substeps-each 0.5 Hz
dmpr,0.02                        !damping ratio uniform for all modes

!f,89,fz,1                       !force excitation, node 89
!f,49,fz,1                       !force excitation, node 49
!base excitation
nsel,s,node,,1,81,10
d,all,uz,1
nsel,all
```

```
solve
save
fini

/post26
numvar,200
/grid,1
/axlab,x,Frequency, Hz
/axlab,y,Quantity
nsol,2,89,u,z,Z89
nsol,3,49,u,z,Z49
nsol,4,9,u,z,Z9
esol,5,61,71,s,eqv,SEQV

!results for response to a unit excitation provide
!transfer function between the input (force or displacement)
!and the response required.

!for printing or plotting the results of the transfer function,
!remember that there are Real and Imaginary parts.
!look at commands plcplx and prcplx in the help file.
```

bbplate.txt

```
/filnam,bbplate

!simply supported beam plate subjected to acoustic excitation

/title,Simply Supported Beam Plate,under acoustic excitation

!units in cm, kgf, seconds

/prep7

E = 2.1e6          !Young's modulus
ro = 7.959e-6      !mass density
nu = 0.3           !Poisson's ratio
xi = 0.01          !damping coefficient

L = 40             !length of beam
b = 5              !width of beam
h = 0.5            !thickness of beam
I = b*h*h*h/12     !cross section moment of inertia
```

prms = 0.03130   !rms of external pressure
!for its calculation, see text

q = prms*b   !force per unit length
!this parameter is required
!for the calculation of equivalent
!discrete forces calculation

!data on external uniform psd of excitation
!see text for the computations

psd=1e−6   !value of uniform psd
f1 = 20   !lower frequency of excitation
f2 = 1000   !upper frequency of excitation
delf = f2–f1   !delta frequencies
prms1 = sqrt(psd)   !rms value per 1Hz band
prms = sqrt(psd*delf)   !total rms of external pressure

!static deflection of beam under uniform load/cm
ymax = 5*q*L*L*L*L/(384*E*I)   !y at center (i.e. from Ref. [26])

ftot = prms*b*L   !total force

!the equvivalent rms force per node fn is obtained
!by equating the mid-beam deflection under
!uniform pressure and discrete equivalent forces
!these are computed using two static analysis shown at
!the end of the file, and searching (by trial and error)
!for the coefficient (1.0526 in this case) that
!match the two cases

fn = ftot/(38*1.0526)   !force at node (corrected)

psdf = fn*fn/delf   !uniform psd value for force excitation

et,1,shell63   !beam is described by shell element

!material properties
mp,ex,1,E
mp,dens,1,ro
mp,nuxy,1,nu

```
!real constant (uniform thickness)
r,1,h,h,h,h

!nodes
n,1,0,-b/2,0
n,101,0,b/2,0
ngen,21,1,1,1,1,L/20,0,0
ngen,21,1,101,101,1,L/20,0,0

!elements
type,1
real,1
mat,1
en,1,1,2,102,101
engen,1,20,1,1,1,1

eplot      !plot model

!boundary conditions:Simply Supported edges
!support at left edge
nsel,s,node,,1,101,100
d,all,uz,0
d,all,uy,0
d,all,ux,0
nsel,all

!support at right edge
nsel,s,node,,21,121,100
d,all,uz,0
d,all,uy,0
nsel,all

save
fini
/eof       !end of data file
           !the model can be loaded to ANSYS by /inp,bbplate,txt

!FIRST SOLUTION:RESPONSE TO PRESSURE LOADING
!for the second case, load AGAIN the input file and
!go to the SECOND SOLUTION
```

```
!modal analysis for pressure excitation
/solu
antype,modal
modopt,subs,4           !only 4 modes in the 20–1000 frequency range
mxpand,4,,,yes
sfe,all,1,pres,,prms1   !command required for pressure loading
solve
fini

!spectral analysis for pressure excitation
/solu
antyp,spectrum
spopt,psd,4,on          !4 modes superposition
psdunit,1,pres          !units of psd are pressure

psdp = 1                !a value of 1 is introduced
                        !scaling was prepared in the modal analysis
psdfrq                  !This command is required only
                        !for next computation. If introduced
                        !in the first computation, a warning
                        !is given. IGNORE IT!
psdfrq,1,,f1,f2
psdval,1,psdp,psdp
psdgraph,1

dmpr,xi                 !the same damping ratio for all resonances

lvscale,1               !scaling was performed already in modal anal

pfact,1,node            !participation factors computation

psdres,acel,abs         !absolute accelerations
psdres,velo,rel         !relative velocities
psdres,disp,rel         !relative displacements
solve

psdcom,,4               !4 modes combination
solve
fini

!from here move to the post-processor commands
```

!SECOND SOLUTION:RESPONSE TO DISCRETE FORCES LOADING
!modal analysis for force excitation

```
/solu
antype,modal
modopt,subs,4                    !4 modes in the 20–1000 range
mxpand,4,,,yes
solv
fini
```

!spectral analyses for force excitation

```
/solu
antyp,spectrum
spopt,psd,4,on
psdunit,1,force

psdfrq                           !This command is required only
                                 !for next computation. If introduced
                                 !in the first computation, a warning
                                 !is given. IGNORE IT!

psdfrq,1,,f1,f2
psdval,1,psdf,psdf
psdgraph

dmpr,xi                          !the same damping ratio for all resonances

nsel,s,node,,2,20,1
nsel,a,node,,102,120,1
f,all,fz,1
nsel,all

pfact,1,node                     !participation factors computation

psdres,acel,abs
psdres,velo,rel
psdres,disp,rel
solve

psdcom,,4
solve
fini
```

!POST PROCESSING (for both cases)
!general post-processor
/post1
set,3,1   !for displacements, stresses-rms values
set,4,1   !for velocities, stress rates-rms values
set,5,1   !accelerations rms values
                                !rms for accelerations in g's obtained
                                !by dividing with g value

!plot and/or list rms values of parameters using GUI

!FREQUENCY DOMAIN post-processor
/post26
numvar,40   !increase number of variables
store,psd,10

!variables definitions
nsol,2,11,u,z,CENTER   !disp. of mid-beam
nsol,3,6,u,z,QUART   !disp. of mid-beam
esol,4,10,11,s,x   !longitudinal stress at mid-beam

!psd computation
rpsd,12,2,,1,2,CENTER
rpsd,13,3,,1,2,QUART
rpsd,14,4,,1,1,SXCENTER

plvar,12
plvar,13
plvar,14

int1,22,12,1,,CENTER
int1,23,13,1,,QUART
int1,24,14,1,,SXCENTER

plvar,22
plvar,23
plvar,24

***!

!commands for the static load equivalent forces
!static load of discrete forces

```
!it is recommended to run these two sets of commands
!separately (each time with new input file)
!or delete the first case pressures before
!solving the second load, to avoid getting
!a superimposed case of the two cases together
!static load of prms
/solu
antyp,stat
sfe,all,1,pres,,prms
solv
fini

!discrete forces loading
/solu
antyp,stat
nsel,s,node,,2,20,1
nsel,a,node,,102,120,1
f,all,fz,fn
nsel,all
solve
fini

*****************************************************!
```

frame1.txt

```
!file frame1 for a frame

/filnam,frame1,txt

/title,Vibration of a frame

!units are in cm, kgf, seconds

!CASE DATA

!data for left member
L1 = 60                    !length
b1 = 8                     !width
h1 = 1                     !height
E1 = 2.1e6                 !Young's modulus
ro1 = 7.959e-6             !mass density
```

```
A1 = b1*h1                      !cross section area
I1 = b1*h1**3/12                !moment of inertia

!data for right member
L2 = 60
b2 = 8
h2 = 1
E2 = 2.1e6
ro2 = 7.959e-6
A2 = b2*h2
I2 = b2*h2**3/12

!data for top member
L3 = 60
b3 = 8
h3 = 1
E3 = 2.1e6
ro3 = 7.959e-6
A3 = b3*h3
I3 = b3*h3**3/12

/prep7

!elements types
et,1,beam3
et,2,beam3
et,3,beam3

!material properties for the 3 members
mp,ex,1,E1
mp,dens,1,ro1

mp,ex,2,E2
mp,dens,2,ro2

mp,ex,3,E3
mp,dens,3,ro3

!real constants for the 3 members
r,1,A1,I1,h1
```

```
r,2,A2,I2,h2
r,3,A3,I3,h3

!nodes
n,1,0,0
n,11,0,L1
fill,1,11

n,21,L3,L1
fill,11,21

n,31,L3,L1–L2
fill,21,31

!elements
type,1
mat,1
real,1

en,1,1,2
engen,1,10,1,1,1,1

type,3
mat,3
real,3

en,11,11,12
engen,1,10,1,11,11,1

type,2
mat,2
real,2

en,21,21,22
engen,1,10,1,21,21,1

!boundary conditions
!built in left support
d,1,ux,0
d,1,uy,0
d,1,rotz,0
```

```
!simply supported right support
d,31,ux,0
d,31,uy,0

save
fini
/eof                        !end of input file

!solution for modal analysis
!modal analysis
/solu
antyp,modal
modop,subs,6                !6 modes
mxpand,6,,,yes
solve
fini

!examination of normal modes
/post1                      !postprocessing, modal analysis
set,1,1                     !first mode
pldisp,2

set,1,2                     !second mode
pldisp,2

set,1,3                     !third mode
pldisp,2

set,1,4                     !fourth mode
pldisp,2

set,1,5                     !fifth mode
pldisp,2

set,1,6                     !sixth mode
pldisp,2

save
fini
```

```
!solution for random force excitation
/solu                   !Solution phase
antyp,spect             !analysis type - spectral
spopt,psd,4,yes         !psd with 4 modes, only 4 modes in the input range,
                        !stresses are also computed
psdunit,1,forc          !input in PSD of force (kgf 2/Hz)
psdfrq,1,,5,250         !input between 5 to 250 Hz
psdval,1,0.004081632,0.004081632
psdgraph                !plot input PSD

psdres,disp,rel         !displacements relative to base
psdres,velo,rel         !velocities relative to base
psdres,acel,abs         !absolute accelerations

!for uniform damping for all modes use the following command
!dmprat,0.02            !damping 2% for all modes

!for different damping coefficient for 4 modes
mdamp,1,0.03,0.01,0.005,0.01

f,11,fx,1               !random horizontal random force at upper left corner
pfact,1,node            !nodal excitation
psdcom,,4               !modal combination for psd
solve
save
fini

!postprocessing of random loading
!general postprocessing
/post1
set,list                !see list of load steps (1,1;1,2 etc. for resonances)
set,3,1                 !for rms of displacements
                        !use this for stresses
                        !for beam elements an element table is required!

!prepare element table for the bending stress in i&j nodes
etable,SBENDi,ls,3
etable,SBENDj,ls,6
pret                    !print element table
```

```
set,4,1                    !for rms of velocities
set,5,1                    !for rms of accelerations
                           !accelerations/980=acel. in g's
fini

!frequency domain postprocessing
/post26
numvar,30                  !prepare space for 30 variables
store,psd,10               !store frequencies, 10 to each side of resonance
/grid,1
/axlab,x,Frequency Hz
/axlab,y,PSD
nsol,2,11,u,x,Wtip         !disp. node 11
esol,4,1,1,ls,3,Sben       !bend. str. at element 1, node 1
                           !ls-3:bending, i side of element
esol,5,10,11,ls,3,Sben11   !bend. str. at element 10, node 11

rpsd,12,2,,1,2,Wtip        !PSD of rel. disp. node 11
rpsd,13,2,,3,1,ACtip       !PSD of Acceleration, node 11 (in cm/sec²/Hz)
                           !for g 2/Hz—divide by 980²
rpsd,14,4,,1,2,BEND1       !PSD of bending stress, node 1
rpsd,15,5,,1,2,BEND11      !PSD of bending stress, node 11

int1,22,12,1,,MStip        !Mean Square of tip displacement
int1,23,13,1,,Msaccel      !Mean Square of tip acceleration
int1,24,14,1,,MSbend1      !Mean Square of clamp bending stress.
int1,25,15,1,,MSbend11     !Mean Square of tip bending stress.
```

commass1.txt

!file commass1 for cantilever beam

/title,Cantilever beam with 1 Mounted Mass

!units in kgf, cm, seconds

```
g = 980                    !value of g
L = 60                     !length, cm
b = 8                      !width, cm
h = 0.5                    !thickness, cm
E = 2.1e6                  !Young modulus
```

```
ro = 7.959e−6                    !mass density
W = L*b*h*ro*g                   !beam total weight, kgf
q = W/L                          !beam weight per length
A = b*h                          !beam cross section area
I = b*h**3/12                    !beam area moment of inertia
g = 980                          !g = 980 cm/sec 2
w1 = 1                           !first mass, node 14 (below node 4)
k1y = 65                         !vertical spring, 4 to 14
k1x = 50000                      !horiz. spring, 4 to 14 - very high stiffness

m1 = w1/g                        !mounted mass

/prep7                           !pre-processor
mp,ex,1,E
mp,dens,1,ro

et,1,beam3
et,2,mass21,,0,4                 !first mass element
et,3,combin14,0,2                !spring y direction
et,4,combin14,0,1                !spring x direction

r,1,A,I,h                        !beam
r,2,m1                           !mass
r,3,k1y                          !ver. spring
r,4,k1x                          !hor. spring

!nodes for the beam
n,1,0,0
n,11,L,0
fill,1,11

!node for the mass
n,14,0.3*L,0                     !node for mass

!elements
!for the beam
type,1
real,1
mat,1
```

```
en,1,1,2
engen,1,10,1,1,1,1

!first mass, element 11
type,2
real,2
en,11,14                        !at node 14

!spring y, element 12
type,3
real,3
en,12,4,14                      !between 4–14

!spring x, element 13
type,4
real,4
en,13,4,14

!constrains
d,1,ux,0
d,1,uy,0
d,1,rotz,0

save
fini
/eof                            !end of input file

/solu                           !Modal Solution
antyp,modal
modop,subs,4                    !4 modes, subspace method
mxpand,4,,,yes                  !expand 4 modes, calculate stresses
solve
fini

/solu                           !SPECTRAL ANALYSIS
antyp,spect                     !analysis type - spectral
spopt,psd,3,yes                 !psd with 3 modes, stresses also computed
psdunit,1,forc                  !input in PSD of force (kgf²/Hz)
psdfrq,1,,5,250                 !input between 5 to 250 Hz
psdval,1,0.004081632,0.004081632
```

```
kbc,1
psdres,disp,rel              !displacements relative to base
psdres,velo,rel              !velocities relative to base
psdres,acel,abs              !absolute accelerations

dmprat,0.02                  !damping 2% for all modes
f,11,fy,1                    !random input at tip
pfact,1,node                 !nodal excitation
mcomb,psd                    !modal combination for psd

solve
save
fini

/post1
set,list                     !see list of load steps (1,1;1,2 etc. for resonances)
set,3,1                      !for rms of displacements
                             !use this for stresses
                             !for beam elements an element table is required
set,4,1                      !for rms of velocities
set,5,1                      !for rms of accelerations
                             !accelerations/980=acel. in g's
!fini

/post26
numvar,30                    !prepare space for 30 variables
store,psd,10                 !store frequencies, 10 to each side of resonance
/grid,1
/axlab,x,Frequency Hz
/axlab,y,PSD
nsol,2,11,u,y,Wtip           !disp. of tip
nsol,3,14,u,y,wmass          !disp. of mass
esol,4,1,1,ls,3,Sben         !bend. str. at element 1, node 1
                             !ls-3:bending, i side of element
rpsd,12,2,,1,2,Wtip          !PSD of rel.disp.-tip
rpsd,13,3,,1,2,wmass         !PSD of rel disp.-mass
rpsd,14,4,,1,2,bend          !PSD of Bend str.
```

rpsd,22,2,,3,1,ACtip   !PSD of Acceleration, node 11 (in cm/sec^2/Hz)
                            !for g^2/Hz—divide by 980^2
rpsd,23,3,,3,1,ACmass !PSD of bending stress, node 1

ssplaterand.txt

!file ssplaterand for simply supported plate

/filnam,ssplaterand

/title, Simply Supported Plate

/config,nres,2000

!units-cm, kgf, sec.

```
g = 980                 !value of g
pi = 4*atan(1)          !value of pi
a = 40                  !length x
b = 30                  !width y
h = 0.5                 !thickness
E1 = 2.1e6              !Young modulus,x direction
E2 = 2.1e6              !Young modulus,y direction
E3 = 2.1e6              !Young modulus,z direction
nu = 0.3                !Poisson ratio
ro = 7.959e-6           !specific density
W = ro*g*a*b*h          !total weight (for reference only)

/prep7                  !Prepare the model
et,1,shell63            !3-D shell (plate) element

r,1,h                   !thickness
!material properties
mp,ex,1,E1              !E in 3 directions
mp,ey,1,E2
mp,ez,1,E3
mp,dens,1,ro
mp,nuxy,1,nu
mp,nuyz,1,nu
mp,nuxz,1,nu

!Nodes
n,1,0,0
```

```
n,21,a,0
fill,1,21

ngen,17,100,1,21,1,0,b/16,0

!Elements
type,1
mat,1
real,1

en,1,1,2,102,101
engen,1,20,1,1,1,1
engen,100,16,100,1,20,1

!Boundary conditions
!edges are simply supported and don't move in plane
nsel,s,node,,1,21,1
d,all,uz,0
d,all,uy,0
nsel,all

nsel,s,node,,1601,1621,1
d,all,uz,0
d,all,uy,0
nsel,all

nsel,s,node,,1,1601,100
d,all,uz,0
d,all,ux,0
nsel,all

nsel,s,node,,21,1621,100
d,all,uz,0
d,all,ux,0
nsel,all

save
fini
/eof                            !end of input file
```

```
!Modal Solution for 6 first 6 modes
/solu
antyp,modal
modop,subs,6                     !6 modes, subspace method
mxpand,6,,,yes                   !expand 6 modes, for stress calculations
solve
fini

/solu                           !SPECTRAL ANALYSIS
antyp,spect                     !analysis type - spectral
spopt,psd,6,yes                 !psd with 6 modes, stresses also computed


psdunit,1,force                  !input in PSD of force (kgf²/Hz)


!psdfrq !for a second run, erase the previous frequencies!!!


psdfrq,1,,200,1250               !input between 200 to 1250 Hz
psdval,1,0.00095238,0.00095238   1! kgf rms value

!this is the place for another set of PSD if more nodes are excited:
!psdfrq,2,,200,1250
!psdval,2,0.00095238,0.00095238

psdres,disp,rel                  !displacements relative to base
psdres,velo,rel                  !velocities relative to base
psdres,acel,abs                  !absolute accelerations

dmprat,0.02                      !damping 2% for all modes

f,416,fz,1                       !random input at node 416
pfact,1,node                     !nodal excitation

!f,416,fz,0                      !zero the previous case
!f,1206,fz,1                     !add random force at node 1206
!pfact,2,node                    !nodal excitation (from table 2 above)
```

!REMEMBER TO ZERO IF A SECOND DIFFERENT RUN IS
REQUIRED
psdcom,,6   !modal combination for 3 modes
solve
save
fini

!DATA PROCESSING for RMS values
/post1
set,list   !see list of load steps (1,1;1,2 etc. for resonances)
set,3,1   !for rms of displacements
set,4,1   !for rms of velocities
set,5,1   !for rms of accelerations
                            !rms of accelerations/980 = acel. in g's
save
fini   !use this for stresses

!Data processing in the frequency domain
/post26
numvar,50   !prepare space for 50 variables
store,psd,10   !store frequencies, 10 to each side of resonance
/grid,1
/axlab,x,Frequency Hz
/axlab,y,Quantity

nsol,2,811,u,z,Z811   !displ. at mid-plate
nsol,3,416,u,z,Z416   !displ. under force
esol,4,315,416,s,x,SX   !X stress under force
esol,5,315,416,s,y,SY   !Y stress under force

rpsd,12,2,,1,2,Z811
rpsd,13,3,,1,2,Z416
rpsd,14,4,,1,2,SX416
rpsd,15,5,,1,2,SY416
rpsd,16,3,,3,1,AC416

!for g^2/Hz—multiply by $1/980^2$
prod,17,16,,,g2/Hz,,,0.00000104123

CHAPTER 4

contact7.tkw

A TKW+ file, not listed

plate2.txt

!file plate2 for cantilever beam with contact elements

/filnam,plate2

/title,Cantilever Beam by Plate Elements, with Contacts

!units kgf, cm, seconds

g = 980	!value of g
!beam data	
L = 60	!length
b = 8	!width
h = 0.5	!thickness
E1 = 2.1e6	!Young's modulus
ro1 = 7.959e-6	!mass density
nu = 0.3	!Poisson coefficient
W = L*b*h*ro1*g	!total weight
q = W/L	!weight/length
A = b*h	!cross section
I = b*h**3/12	!moment of inertia
P = 20	!total tip force
del = 3	!initial distance beam-anvil
!anvil data	
b1 = 1.25*b	!anvil's width
h1 = b1/5	!anvil's height
E2 = 2.1e10	!anvil's Young's modulus
ro2 = 7.959e-2	!anvil's density
!contact data	
k1 = 1000	!KN for contact
tn = 0.001	!penetration tolerance
ts = 5	!5% geometry change

```
/prep7
mp,ex,1,E1                              !material properties, beam
mp,dens,1,ro1
mp,nuxy,1,nu

mp,ex,2,E2                              !material properties, anvil
mp,dens,2,ro2
mp,nuxy,2,nu

mp,mu,3,0                               !material property, contact
r,3,k1,,tn,,ts                          !real constants, contact

et,1,solid45,0,,,1
et,2,solid45,0,,,1
et,3,contac48,0,1,0,,,,1                !contac48 elements

!nodes for beam
n,1,0,−h/2,b/2
n,11,L,−h/2,b/2

fill,1,11
ngen,5,20,1,11,1,0,0,-b/4
ngen,2,100,all,,,0,h,0

!nodes for anvil
n,301,L−b/4,del+h/2,b/2
ngen,5,1,301,301,1,0,0,−b/4
ngen,2,10,301,305,1,b/2,0,0
ngen,2,50,301,305,1,0,b/2,0
ngen,2,50,311,315,1,0,b/2,0

!elements for beam
type,1
real,1
mat,1

en,1,1,2,22,21,101,102,122,121
engen,1,10,1,1,1,1
engen,10,4,20,1,10,1
```

```
!elements for anvil
type,2
real,2
mat,2

en,101,301,311,312,302,351,361,362,352
engen,1,4,1,101,101,1

!contact elements
type,3
real,3
mat,3

en,201,311,301,111
en,202,312,302,131
en,203,313,303,151
en,204,314,304,171
en,205,315,305,191

!boundary conditions for beam
nsel,s,loc,x,−0.001,0.001
d,all,all,0
nsel,all

!boundary conditions for anvil
nsel,s,loc,y,(del+h/2+b/2−0.01),(del+h/2+b/2+0.01)
d,all,all,0
nsel,all

save
fini
/eof                    !end of input file

/solu                   !static analysis
antyp,stat
neqit,75                !max # of iterations
f,11,fy,P/5             !force at tip is P (total
f,31,fy,P/5             !divided to 5 nodes)
f,51,fy,P/5
f,71,fy,P/5
f,91,fy,P/5
```

```
nsubs,100                !number of sub steps
!autots,on
outres,all,all

solve
fini

!/post1
!this post-processor is used to check
!state of the solution in a given
!external force (load step). Remove ! marks as required

!/post26
!this post-processor is used to check
!behavior as a function of the external force.
!ANSYS interprets this force as "time"
!and proper adjustment is to be made
!so that the last "time" is at the last
!force.

!nsol,2,51,u,y,Y51 means that
                         !variable 2 is the y displacement
                         !at node 51, mid tip node()
!esol,3,20,51,s,y,SY51
                         !variable 3 is the Y stress at element 20, node 51
```

plate3.txt

```
!file plate3 for cantilever beam with contact elements

/filnam,plate3

/title,Cantilever Beam by Plate Elements, Dynamics:Tip Force Excitation

/config,nres,10000

!units kgf, cm, seconds

g = 980
!beam data
L = 60                   !length
b = 8                    !width
h = 0.5                  !thickness
E1 = 2.1e6               !Young's modulus of plate
```

```
ro1 = 7.959e-6                    !density
nu = 0.3                          !Poisson
W = L*b*h*ro1*g                   !total weight
q = W/L                           !weight/length
A = b*h                           !cross section
I = b*h**3/12                     !area moment of inertia
PP = 20                           !total tip force
pi = 4*atan(1)                    !value of pi
freq = 11.525                     !frequency of excitation
om = 2*pi*freq                    !angular frequency
del = 3                           !distance beam-anvil

!anvil data
E2 = 2.1e10                       !anvil's Young's modulus
ro2 = 7.959e-2                    !anvil's mass density

!contact data
k1 = 1000                         !KN for contact
tn = 0.001                        !penetration tolerance
ts = 5                            ! 5% geometry change

/prep7
mp,ex,1,E1                        !material properties, beam
mp,dens,1,ro1
mp,nuxy,1,nu

mp,ex,2,E2                        !material properties, anvil
mp,dens,2,ro2
mp,nuxy,2,nu

mp,mu,3,0                         !material property, contact
r,3,k1,,tn,,ts                    !real constants, contact

et,1,solid45,0,,,1
et,2,solid45,0,,,1
et,3,contac48,0,1,0,,,,1

!nodes for beam
n,1,0,-h/2,b/2
n,11,L,-h/2,b/2
```

```
fill,1,11
ngen,5,20,1,11,1,0,0,−b/4
ngen,2,100,all,,,0,h,0

!nodes for anvil
n,301,L−b/4,del+h/2,b/2
ngen,5,1,301,301,1,0,0,−b/4
ngen,2,10,301,305,1,b/2,0,0
ngen,2,50,301,305,1,0,b/2,0
ngen,2,50,311,315,1,0,b/2,0

!elements for beam
type,1
real,1
mat,1

en,1,1,2,22,21,101,102,122,121
engen,1,10,1,1,1,1
engen,10,4,20,1,10,1

!elements for anvil
type,2
real,2
mat,2

en,101,301,311,312,302,351,361,362,352
engen,1,4,1,101,101,1

!contact elements
type,3
real,3
mat,3
en,201,311,301,111
en,202,312,302,131
en,203,313,303,151
en,204,314,304,171
en,205,315,305,191

!boundary conditions for beam
nsel,s,loc,x,−0.001,0.001
```

```
d,all,all,0
nsel,all

!boundary conditions for anvil
nsel,s,loc,y,(del+h/2+b/2−0.01),(del+h/2+b/2+0.01)
d,all,all,0
nsel,all
save
fini
/eof                          !end of input file

/solu                         !transient excitation analysis
antyp,trans
trnop,full
t0 = 0.0001                   !initial time (very small !!!)
neqit,75                      !max number of iterations
time,t0
dt = 0.000125

PP = P*sin(om*t0)             !force at t0

f,11,fy,PP/5                  !force at tip is PP
f,31,fy,PP/5                  !divided to 5 nodes
f,51,fy,PP/5
f,71,fy,PP/5
f,91,fy,PP/5

nsubs,1                       !number of sub steps for initial load
!autots,on
outres,all,all
solve

!looping for "transient sine wave"
*do,n,1,200,1
      tc = t0+n*dt
      time,tc
      PP = P*sin(om*tc)
      f,11,fy,PP/5
      f,31,fy,PP/5
```

```
        f,51,fy,PP/5
        f,71,fy,PP/5
        f,91,fy,PP/5
        nsubs,1
        outres,all,all
        solve
*enddo

save
fini
/eof                                    !end of input file

!general purpose post-processor
/post1
set,list                                !for examination of load steps
set,X,1                                 !where X is the required step
!see disp, stresses etc. at a given time t.

!time domain post processor
/post26
numvar,30
!displacement at the mid-tip node
nsol,2,51,u,y,Y51

!Y stresses at node 51
esol,3,20,51,s,y,SY51

!Y stresses at node 151
esol,4,20,151,s,y,SY151

!X stress at node 141
esol,5,11,141,s,x,SX141
!etc...

/grid,1
plva,2
plva,3
plva,4
!etc....
```

plate4.txt

!file plate4 for cantilever beam with contact elements

/filnam,plate4

/title,Cantilever Beam by Plate Elements, Dynamics:Base Excitation

!units kgf, cm, seconds

g = 980	!value of g
!beam data	
L = 60	!length
b = 8	!width
h = 0.5	!thickness
E1 = 2.1e6	!Young's modulus
ro1 = 7.959e-6	!mass density
nu = 0.3	!Poisson coefficient
W = L*b*h*ro1*g	!total weight
q = W/L	!weight/length
A = b*h	!cross section
I = b*h**3/12	!area moment of inertia
del = 1	

!excitation data	
pi = 4*atan(1)	!value of pi
freq = 2*11.525	!frequency of excitation
om = 2*pi*freq	!angular frequency
ng = −3	!excitation with ng
a0 = ng*g	!amplitude of input acceleration
dd = −a0/om**2	!amplitude of input displacement
del = 1	!distance beam-anvil

!anvil data	
E2 = 2.1e10	!anvil's Young's modulus
ro2 = 7.959e-2	!anvil's mass density

!contact data	
k1 = 1000	!KN for contact
tn = 0.001	!penetration tolerance
ts = 5	!5% geometry change

```
/prep7
mp,ex,1,E1                    !material properties, beam
mp,dens,1,ro1
mp,nuxy,1,nu

mp,ex,2,E2                    !material properties, anvil
mp,dens,2,ro2
mp,nuxy,2,nu

mp,mu,3,0                     !material property, contact
r,3,k1,,tn,,ts                !real constants, contact

et,1,solid45,0,,,1
et,2,solid45,0,,,1
et,3,contac48,0,1,0,,,,1

!nodes for beam
n,1,0,−h/2,b/2
n,11,L,−h/2,b/2

fill,1,11
ngen,5,20,1,11,1,0,0,−b/4
ngen,2,100,all,,,0,h,0

!nodes for anvil
n,301,L−b/4,del+h/2,b/2
ngen,5,1,301,301,1,0,0,−b/4
ngen,2,10,301,305,1,b/2,0,0
ngen,2,50,301,305,1,0,b/2,0
ngen,2,50,311,315,1,0,b/2,0

!elements for beam
type,1
real,1
mat,1

en,1,1,2,22,21,101,102,122,121
engen,1,10,1,1,1,1
engen,10,4,20,1,10,1
```

```
!elements for anvil
type,2
real,2
mat,2

en,101,301,311,312,302,351,361,362,352
engen,1,4,1,101,101,1

!contact elements
type,3
real,3
mat,3

en,201,311,301,111
en,202,312,302,131
en,203,313,303,151
en,204,314,304,171
en,205,315,305,191

!boundary conditions for beam
nsel,s,loc,x,−0.001,0.001
d,all,all,0
nsel,all

!boundary conditions for anvil
nsel,s,loc,y,(del+h/2+b/2−0.01),(del+h/2+b/2+0.01)
d,all,all,0
nsel,all

save
fini
/eof                            !end of input file

/solu                           !transient analysis
antyp,trans
trnop,full
t0 = 0.0001                     !initial time (very small !!!)
neqit,75                        !max number of iterations
dt = 0.001
```

```
!initial time
time,t0
ddd = dd*sin(om*t0)                    !initial displacement

nsel,s,node,,301,305,1
nsel,a,node,,311,315,1
nsel,a,node,,351,355,1
nsel,a,node,,361,365,1
nsel,a,node,,1,81,20
nsel,a,node,,101,181,20

d,all,uy,ddd
nsel,all
nsubs,2
outres,all,all
solve

!additional times (transient)
*do,n,1,200,1
        tc = t0+n*dt
        time,tc
        nsel,s,node,,301,305,1
        nsel,a,node,,311,315,1
        nsel,a,node,,351,355,1
        nsel,a,node,,361,365,1
        nsel,a,node,,1,81,20
        nsel,a,node,,101,181,20
        ddd = dd*sin(om*tc)
        d,all,uy,ddd
        nsel,all
        nsubs,1
        outres,all,all
        solve
*enddo

fini

/post1                                 !general purpose post processor
set,list                               !for examination of load steps
```

set,X,1 !where X is the required step
!see disp, stresses etc at a given time

/post26
numvar,30
/grid,1
!displacement at the mid-tip node
nsol,2,51,u,y,Y51

!displacement at the mid-anvil
nsol,3,303,u,y,Y303

!displacement of the clamp
nsol,4,41,u,y,Y41

!relative displacement between tip and clamp
add,5,2,4,,dist,,,1,−1

!Y stresses at node 51
esol,6,20,51,s,y,SY51

!Y stresses at node 151
esol,7,20,151,s,y,SY151

!X stress at node 141
esol,8,11,141,s,x,SX141
!etc...

plva,2
plva,3
plva,4
!etc....

CHAPTER 5

probeam4.txt

!probeam4 for probabilistic analysis of a cantilever beam

!units in kgf, mm, seconds

!the looping file is the one between *create command and

! *end command, after the post processor!!!

/filnam,probeam4

*create,probeam4,pdan !begin probeam4.pdan, end with *end command

```
*set,b,10.0000          !width, mean value
*set,h,8.0000           !height, mean value
*set,L,400.0000         !length, mean value
*set,E,21000.0000       !Young's modulus, mean
*set,F,10.0000          !applied tip force, mean
*set,Yallow,27.0000     !allowed tip displacement, mean
*set,Sallow,40.0000     !allowed clamp stress

In = b*h*h*h/12            !area moment of inertia
Area = b*h                 !cross section area

/prep7
et,1,beam3                 !2-D beam element
mp,ex,1,E                  !material properties
r,1,area,In,h              !real constants
n,1,0,0                    !first node, clamp
n,11,L,0                   !last node, tip
fill,1,11
en,1,1,2 !elements
engen,1,10,1,1,1
d,1,ux,0                   !boundary conditions, clamped edge
d,1,uy,0
d,1,rotz,0
f,11,fy,F                  !applied force
fini

/solu                      !solution phase
antyp,stat
solve
fini

/post1                     !post-processing phase
etab,SBEN,nmisc,1          !prepare element table
```

```
*get,Smax,elem,1,etab,SBEN          !get bending stress at clamp
*get,Ymax,node,11,u,y              !get tip displacement

GF1 = Yallow-Ymax                   !first failure function (tip deflection)
GF2 = Sallow-Smax                   !second failure function (clamp stress)
SFSy = Yallow/Ymax                  !stochastic safety factor, displacement
SFSS = Sallow/Smax                  !stochastic safety factor, clamp stress
fini

*end                                !end of probeam4.pdan
/eof                                !end of file for /inp,probeam4,txt

/inp,probeam4,pdan      !define which is the loop file
/pds                    !enter probabilistic module (can be done in GUI)
/pdanl,probeam4,pdan    !use GUI!!!! Prob Design > Analysis File > assign

!definition of random variables (RV). Easily done with GUI
pdvar,b,gaus,10.0000,0.1            !define RV b
pdvar,h,gaus,8.0000,0.1            !define RV h
pdvar,L,gaus,400.0000,2.0000       !define RV L
pdvar,E,gaus,21000.0000,700.0000   !define RV E
pdvar,F,gaus,10.0000,0.3           !define RV F
pdvar,Yallow,gaus,27.0000,0.25     !define RV Yallow
pdvar,Sallow,gaus,40.0000,0.5000   !define RV Sallow
!RV's can now be plotted by pdplot command

!define response variables - random outputs in GUI
pdvar,Smax,resp                    !maximum stress at clamp
pdvar,Ymax,resp                    !maximum tip deflection
pdvar,GF1,resp                     !first failure function
pdvar,GF2,resp                     !second failure function
pdvar,SFSy,resp                    !stochastic safety factor, tip displ.
pdvar,SFSS,resp                    !stochastic safety factor, clamp stress.

!define probabilistic method. MC for Monte Carlo, lhs for latin hypercube
pdmeth,mcs,dir                     !probabilistic method - MC
pddmcs,1000,,all,,,,123457         !1000 simulations
pdexe                              ! execute simulations, solution takes time!!!
```

!Analysis of results. Done with GUI in Ansys Main Menu > Prob Design:
!Prob Results > Statistics > Cumulative DF   !CDF of results
!Prob Results > Statistics > Histograms   !Sampling Histograms
!Prob Results > Statistics > Sample History   !Sampling History

!prob Results > Statistics > Sample Probability !select the variable
                                                !in the NAME window
              !probability of Ymax>Yellow
                                                !is obtained, same for GF2.
!many other properties can be calculated.
!response surface method also available. See HELP file

math1.txt

!use of ANSYS Probabilistic Module for Mathematical Application
!Stress is given by $S = x1^2 + x2*x3$
!Failure function is GF1 = R-S, Failure if GF1 is
!smaller or equal zero

```
/filnam,math
*create,math,pdan        !begin creation of loop file
*set,x1,10.000           !mean of x1
*set,x2,5.000            !mean of x2
*set,x3,4.000            !mean of x3
*set,R,125.000          !mean of "strength"

/prep7
S = x1**2+x2*x3          !expression for "stress"
GF1 = R-S               !failure function

finish
*end

/eof
!run nominal case by /inp,math,txt
!then:

/inp,math,pdan
/pds
psanl,math,pdan
```

```
!definition of random variables
pdvar,x1,gaus,10.000,0.1
pdvar,x2,gaus,5.000,0.05
pdvar,x3,gaus,4.000,0.05
pdvar,R,gaus,125.000,1.4

!definition of response variables
pdvar,S,resp
pdvar,GF1,resp

!definition of analysis method
pdmeth,mcs,dir
pddmcs,1000,,all,,,,123457

!execution of computation
pdexe
```

!NO FINITE ELEMENT COMPUTATIONS ARE INVOLVED!!!

truss1.txt

```
!file truss1 for 3 member truss
/filnam,truss1
/title, Three Members Truss

!geometry parameters
pi = 4*atan(1)      !value of pi
L0 = 223            !length, L0
L1 = 100            !length, first member
L2 = 200            !length, second member

!computation of required terms, including L3
ca = (L2*L2+L0*L0−L1*L1)/(2*L2*L0)
cb = (L1*L1+L0*L0−L2*L2)/(2*L1*L0)
sa = sqrt(1−ca**2)
sb = sqrt(1−cb**2)
L3 = sqrt((0.5*L0)**2+L1**2−L1*L0*cb)

!diameter of members
d1 = 2.00
d2 = 0.2
```

```
d3 = 0.9
I1 = pi*d1**4/64          !area moment of inertia, member 1
a1 = pi*(d1**2)/4         !cross section areas
a2 = pi*(d2**2)/4
a3 = pi*(d3**2)/4

!Material properties (parameters)
E1 = 2.1e6                !Young's modulus
E2 = 2.1e6
E3 = 2.1e6

ro1 = 7.959e−6           !mass density
ro2 = 7.959e−6
ro3 = 7.959e−6

!nodes location (parameters)
x1 = 0
y1 = 0

x2 = L1*sb
y2 = L1*cb

x3 = 0
y3 = L0

x4 = 0
y4 = 0.5*L0

/prep7
et,1,link1                !elements types
et,2,link1
et,3,link1

!real constants
r,1,a1
r,2,a2
r,3,a3

!materials properties
mp,ex,1,E1
```

```
mp,ex,2,E2
mp,ex,3,E3

mp,dens,1,ro1
mp,dens,2,ro2
mp,dens,3,ro3

!nodes
n,1,x1,y1
n,2,x2,y2
n,3,x3,y3
n,4,x4,y4

!elements
type,1
mat,1
real,1
en,1,1,2

type,2
mat,2
real,2
en,2,3,2

type,3
mat,3
real,3
en,3,4,2

!three supports
d,1,all,0
d,3,all,0
d,4,all,0

save
fini
/eof            !end of input file

/solu           !static analysis
antype,stat
f,2,fy,-1200    !applied force at node 2
```

solve
fini

/post1
etab,sax,ls,1
*get,s1,elem,1,etab,sax
*get,s2,elem,2,etab,sax
*get,s3,elem,3,etab,sax

!Fcr = 0.9*pi**2*E1*I1/L1**2 (not required)
F1 = −s1*a1   !force in member 1 (absolute)
F2 = s2*a2   !force in member 2
F3 = s3*a3   !force in member 3
S1x = −s1   !stress in member 1
S2x = s2   !stress in member 2
S3x = S3   !stress in member 3
fini

CHAPTER 6

Because of text width difficulties, some program lines had to be broken. The reader should pay attention to these breaks in the listing lines. Please remember that a % sign marks a comment, and sometimes the second line of the comment is not preceded by the % sign. The reader should remember that in the attached CD-ROM, these lines are not broken.

astar.m

```
%astar1.m calculate a* from closed form equation.
%KIC is a normal random variable with mean=muk and
%sd=sigk. Suts is random variable with mean=mus and
%sd=sigs. acrit is the critical crack length. Its
%distribution is fitted by a lognormal distribution.
%mr is the mean and sr is the standard deviation
%of the data. xi and eps are the mean and standard
%deviation of log(data).

muk=4691;    %mean of KIC
sigk=155;    %SD of KIC
mus=1172;    %mean of UTS
sigs=40;     %SD of UTS
R=0.0;       %value of R
```

```
geom=sqrt(pi);      %expression for (geom)
K=20000;            %number of calculated points

for k=1:K
   acrit=((1-R)*normrnd(muk,sigk,1,1)/
         (geom*normrnd(mus,sigs,1,1)))^2;
   ac(k,:)=acrit;
end

data=ac;
[row,col] = size(data);
nbins = ceil(sqrt(row));
[n,xbin]=hist(data,nbins);

acm=mean(data);
acs=std(data);
sks=skewness(data);
mu=mean(data);            %mean of data
lnmu=log(mu);             %LN of mu
sig=std(data);            %standard deviation of data
lnsig=log(sig);           %LN of sig
dd=lnsig-lnmu;
dd2=2*dd;
ee=exp(dd2)+1;
epssq=log(ee);
eps=sqrt(epssq);          %epsilon in lognormal pdf
                            ("standard deviation")
xi=lnmu-0.5*epssq;        %xi in lognormal pdf("mean")
mr = mean(data);
sr = std(data);
x=(-3*sr+mr:0.1*sr:3*sr+mr)'; % Evenly spaced samples
                                    of the expected
                              %data range.
hh = bar(xbin,n,1); % Plots the histogram.
                         No gap between bars.
np = get(gca,'NextPlot');
set(gca,'NextPlot','add')
```

```
xd = get(hh,'Xdata'); % Gets the x-data of the bins.
rangex = max(xd(:)) - min(xd(:));% Finds the range of
                                % this data.
binwidth = rangex/nbins;% Finds the width of each bin.
y = lognpdf(x,xi,eps);

y = row*(y*binwidth); % Normalization necessary to
                        overplot the histogram.
hh1 = plot(x,y,'r-','LineWidth',2); % Plots density
                                    line over histogram
mr                      %mean of data
xi                      %"mean" of log-normal
sr                      %standard deviation of data
eps                     %"std" of log-normal
sks                     %skewness of data
set(gca,'NextPlot',np)
```

cocrack1.m

```
%corcrack1 is an ODE file for solution the correlated
%problem of normal stochastic process, with normally
%distributed KIC, where correlation between the
 ''load'' w
%and the ''strength'' astar is taken
%into account (through
%a common variable KIC). This file is called by
 virtest1.

function wd=corcrack1(t,w,flag,KIC,UTS)
switch flag
case"
end

m=2.7;          %m from literature
C=0.298e-11;    %C from literature
fs=150;         %circular frequency
q=0.25;
```

```
S=554.38;        %Smr-equivalent stress
geom=sqrt(pi); %geometry factor %parameter=B*sqrt(pi),
                 %B=1(infinite plate)
R=0.25;          %stress ratio

C1=C*fs*geom^m;
C2=(1-R)*KIC;
C3=geom;
mu=1;            %mean of stochastic normal %process
sigma=0.1;       %std of stochastic normal %process
wd(1)=((C1*(S ^m)*(w(1)^(m/2)))/(1-(C3*S*(w(1)^0.5))/
      ((1-R)*KIC))^q)*(normrnd(mu,sigma,1,1));
```

cocrack2.m

```
%corcrack2 is an ODE file for solution the correlated
%problem of normal stochastic process, with
 deterministic
%KIC and UTS where correlation between the ``load''
 w and
%the ``strength'' astar is taken into account
 (through a
%common variable KIC). This file is called by
%virtest1.

function wd=corcrack2(t,w,flag,KIC,UTS)
switch flag
case ''
end

m=2.7;           %m from literature
C=0.298e-11;     %C from literature
fs=150;          %circular frequency
q=0.25;
S=554.38;        %Smr-equivalent stress
geom=sqrt(pi);   %geometry parameter=B*sqrt(pi),
                 %B=1(infinite plate)
R=0.25;          %stress ratio
```

```
C1=C*fs*geom^m;
C2=(1-R)*KIC;
C3=geom;
mu=1;          %mean of stochastic normal %process
sigma=0.1;     %std of stochastic normal %process
wd(1)=((C1*(S^m)*(w(1)^(m/2)))/(1-(C3*S*(w(1)^0.5))/
      ((1-R)*KIC))^q)*(normrnd(mu,sigma,1,1));
```

cocrack3.m

```
%corcrack3 is an ODE file for solution the correlated
%problem of log-normal stochastic process, with
 normally
%distributed KIC, where correlation between the
 ''load'' w
%and the ''strength'' astar is taken into account
 (through
%a common variable KIC). This file is called by
 virtest3.

function wd=corcracktest4(t,w,flag,KIC,UTS)
switch flag
case "
end

m=2.7;              %m from literature
C=0.298e-11;        %C from literature
fs=150;             %circular frequency
q=0.25;
S=554.38;           %Smr-equivalent stress
geom=sqrt(pi);      %geometry parameter=B*sqrt(pi),
                    %B=1(infinite plate)
R=0.0;              %stress ratio

C1=C*fs*geom^m;
C2=(1-R)*KIC;
C3=geom;
```

```
mu=0;           %mean of stochastic log-normal %process
sigma=0.1;      %std of stochastic log-%normal %process
wd(1)=((C1*(S^m)*(w(1)^(m/2)))/(1-(C3*S*(w(1)^0.5))/
       ((1-R)*KIC))^q)*(lognrnd(mu,sigma,1,1));
```

cocrack4.m

```
%corcrack4 is an ODE file for solution the correlated
%problem of log-normal stochastic process, with
%deterministic KIC and UTS where correlation between
 the
%``load'' w and the ``strength'' astar is taken into
 account
%(through a common variable KIC). This file is called
 by
%virtest4.
function wd=corcrack4(t,w,flag,KIC,UTS)
switch flag
case "
end

m=2.7;              %m from literature
C=0.298e-11;        %C from literature
fs=150;             %circular frequency
q=0.25;
S=554.38;           %Smr-equivalent stress
geom=sqrt(pi);      %geometry parameter=B*sqrt(pi),
                    %B=1(infinite plate)
R=0.25;             %stress ratio

C1=C*fs*geom^m;
C2=(1-R)*KIC;
C3=geom;
mu=0;           %mean of stochastic log-normal %process
sigma=0.1;      %std of stochastic log-normal %process
wd(1)=((C1*(S^m)*(w(1)^(m/2)))/(1-(C3*S*(w(1)^0.5))/
       ((1-R)*KIC))^q)*(lognrnd(mu,sigma,1,1));
```

crack2.m

```
%crack2.m for deterministic crack propagation
  according to Forman Law
%modified according to ESACRACK, with R greater than
  0. m not equal to 2!
%C, m, q and KIC - material constants. w is crack
  length. initial crack length
%is a0, inserted in the run command as w0:
%[t,w]=ode23('crack2',tspan,w0);

function wd=crack2(t,w)
m=2.7;                    %m from literature
C=0.298e-11;             %C from literature
fs=150;                   %circular frequency
q=0.25;
S=554.38;                 %Smr-equivalent stress
geom=sqrt(pi);           %geometry parameter=B*sqrt(pi),
                           B=1(infinite plate)
%geom=2/sqrt(pi);         %for cylinder in tension,
                           center crack
KIC=4691;                 %fracture thoughness
R=0.0;                    %stress ratio
C1=C*fs*geom^m;
C2=(1-R)*KIC;
C3=geom;

[rows,cols]=size(w);wd=zeros(rows,cols);
wd(1)=(C1*(S^m)*(w(1)^(m/2)))*C2^q/
      ((C2-C3*S*(w(1)^0.5))^q);
```

cracklognfit.m

```
%cracklognfit.m fits a lognormal distribution to the
%crack half length at time = nt. Data is obtained
  after
%running cracknorm1.m
```

```
%the data of 500 virtual tests is put in nbins=number
 of
%bins. The distribution is fitted by a lognormal
%distribution. mr is the mean and sr is the standard
%deviation of the data at time nt.
%xi and eps are the mean and standard deviation of
 log(data).

nt=149;

ac=a(:,nt);

data=ac;
[row,col] = size(data);
nbins = ceil(sqrt(row));

[n,xbin]=hist(data,nbins);
acm=mean(data);
acs=std(data);
sks=skewness(data);
mu=mean(data);          %mean of data
lnmu=log(mu);           %LN of mu
sig=std(data);          %standard deviation of data
lnsig=log(sig);         %LN of sig
dd=lnsig-lnmu;
dd2=2*dd;
ee=exp(dd2)+1;
epssq=log(ee);
eps=sqrt(epssq);        %epsilon in lognormal pdf
                          ("standard deviation")
xi=lnmu-0.5*epssq;      %xi in lognormal pdf("mean")
mr = mean(data);
sr = std(data);
x=(-3*sr+mr:0.1*sr:3*sr+mr)'; %Evenly spaced samples of
                              %the expected data range
hh = bar(xbin,n,1);           %Plots the histogram.
                               No gap between bars.
```

```
np = get(gca, 'NextPlot');
set(gca,'NextPlot','add')

xd = get(hh,'Xdata');    % Gets the x-data of the bins.
rangex = max(xd(:)) - min(xd(:));    % Finds the range
                                       of this data.
binwidth = rangex/nbins; % Finds the width of each bin.
y = lognpdf(x,xi,eps);
y = row^(y*binwidth);    % Normalization necessary to
                           overplot the histogram.
hh1 = plot(x,y,'r-','LineWidth',2);  % Plots density
                                      %line over the
                                      %histogram
mr                       %mean of data
xi                       %"mean" of log-normal
sr                       %standard deviation of data

eps                      %"std" of log-normal
sks                      %skewness of data
set(gca,'NextPlot',np)
```

cracknorm1.m

```
%cracknorm1.m is a procedure that calculate, after
 crackstat2.m was run, the
%mean and standard deviations of the a vs. t lines
 at a given time t.
%It also compare the amean vs. t to the deterministic
 solution.

a=transpose(y);
amt=mean(a);             %row vector
am=amt';                 %column vector of MEANS
ast=std(a);              %row vector
as=ast';                 %column vector of STANDARD
                          DEVIATIONS
er=(yd-am)*100./yd;      %error between mean and
                          deterministic
```

```
plot(t,am,t,yd),grid,xlabel('Time, sec'),
ylabel('a_mean and a_deter.')
figure
plot(t,as),grid,xlabel('Time,sec'),
ylabel('standard deviation')
aupr=am+3.*as;
alwr=am-3.*as;
figure
plot(t,alwr,t,aupr,t,am),grid,xlabel('Time,sec'),
ylabel('-3sigma, +3sigma, mean')
figure
plot(t,y),grid,xlabel('Time, sec'),
ylabel('Differential Equation Solution')
%The command
%histfit(a(:,177))
%shows an histogram of a's at time # 177, and also
 shows a normal distribution
%with the same mean and standard deviation.
```

prop1.m

```
%prop1 is the crack propagation file for example in
 crack chapter

clear %CAUTION: Clears the previous %workspace

[tspan]=[0:50:1800]; %time from 0 to 1800 sec, every
 50 seconds

%[tspan]=[0:1:131]; %time from 0 to 131 sec, every
 11 seconds (tension)

w0=2;

[t,w]=ode45('crack2',tspan,w0); %solve deterministic
  case

yd=w; %deterministic solution
plot(t,yd,'k'),grid,xlabel('Time, sec'),
ylabel('Half Crack Length, mm (DE Solution)')
```

virtest1.m

```
%virtest1.m is a program that calculates the stochastic
%crack propagation using modified Forman equation and
%stochastic normal process (corcrack1.m). the solution
%{yd} to the deterministic crack propagation ODE is
%fitted with a k-th order polynom,and a column matrix
 of
%the polynom's coefficients is created. t is time,
 tspan
%is the solution range in time, {w} is half crack
%length, [y] is a matrix of N solutions, {YD} is the
%approximated deterministic polynomial solution(a
%column). The deterministic problem is done with
 crack2.m
%file next {y}'s are for stochastic problem, done with
%corcrack1.m file
```

```
clear                   %CAUTION: Clears the %Previous
                         Workspace
k=9;                    %order of polynom
M=1;                    %insert number of stochastic
                         solutions
[tspan]=[0:0.5:82];     %from 0 to 88 sec, every
                        %half second
                        %these numbers should be modified
                         for other computations
                        %by changing the numbers in the
                         command.

N=length(tspan);        %number of time points
w0=0.1;                 %initial crack length, mm.
R=0.25
options=odeset('AbsTol',1e-6);
geom=sqrt(pi);
%solution of the deterministic case with crack2.m for R
```

```
%between zero(included) and 1
[t,w]=ode45('crack2',tspan,w0);   %solve deterministic
                                    case
yd=w;                             %deterministic
                                   solution
pd=polyfit(t,yd,k);               %deterministic row of
                                  %coefficients
YD=polyval(pd,t);                 %deterministic
                                   polynomial
                                  %approximation
 %repeated solutions for normal stochastic process,
 from m=1 to M

 for m=1:M
   muk=4691;                 %mean of KIC
   sigk=155;                   %standard deviation of KIC
   KIC=normrnd(muk,sigk,1,1); %fracture thoughness
   mus=1172;                  %mean of UTS
   sigs=40;                     %standard deviation of UTS
   UTS=normrnd(mus,sigs,1,1);    %UTS
[t,w]=ode45('corcrack1',tspan,w0,options,KIC,UTS);
acrit=((1-R)*KIC/(geom*UTS))^2; %critical length of
      crack
y(:,m)=w;
astar(:,m)=acrit;
end
%plot the crack length vs.time for all samples + the
%deterministic solution (black)

plot(t,y,'k',t,yd,'k'),grid,xlabel('Time, sec'),
ylabel('Half Crack Length, mm (DE Solution)')

%the following command plots the above information
 + the
%dispersion in astar.
%plot(t,y,t,yd,t(150),astar(150,:),'k+'),grid,xlabel
('Time, sec'),ylabel('Crack Length, mm (DE Solution)')
```

virtest2.m

```
%virtest2.m is a program that calculates the stochastic
%crack propagation using modified Forman equation and
%stochastic normal process (corcrack2.m). The solution
%{yd} to the deterministic crack propagation ODE is
%fitted with a k-th order polynom,and a column matrix
 of
%the polynom's coefficients is created. t is time,
 tspan
%is the solution range in time, {w} is half crack
 length,
%[y] is a matrix of N solutions, {YD} is the
 approximated
%deterministic polynomial solution(a column).The
%deterministic problem is done with crack2.m file.next
%{y}'s are for stochastic problem, done with
 corcrack2.m %file

clear           %CAUTION: Clears the Previous Workspace
k=9;            %order of polynom
M=1;            %INSERT the number of
                %stochastic solutions
[tspan]=[0:0.5:82]; %from 0 to 88 sec, every half
                    second
%these numbers should be modified for other
 computations
%by changing the numbers in the command.
N=length(tspan);           %number of time and
                           %cracklength points
w0=0.1;                    %initial crack length, mm.
R=0.25
options=odeset('AbsTol',1e-6);
geom=sqrt(pi);
%solution of the deterministic case with crack2.m for R
```

```
%between zero(included) and 1
[t,w]=ode45('crack2',tspan,w0);   %solve deterministic
                                    case
yd=w;                             %deterministic
                                   solution
pd=polyfit(t,yd,k);               %deterministic row of
                                  %coefficients
YD=polyval(pd,t);                 %deterministic
                                   polynomial
                                  %approximation

%repeated solutions for normal stochastic process,
from m=1 to M

for m=1:M
   muk=4691;        %mean of KIC
   sigk=155;        %standard deviation of KIC %(unused)
   KIC=muk;         %fracture thoughness
   mus=1172;        %mean of UTS
   sigs=40;         %standard deviation of UTS %(unused)
   UTS=mus;         %UTS
   [t,w]=ode45('corcrack2',tspan,w0,options,KIC,UTS);
   acrit=((1-R)*KIC/(geom*UTS))^2; %critical crack
          length
   y(:,m)=w;
   astar(:,m)=acrit;
end
%plot the crack length vs. time for all samples + the
%deterministic solution (black)

plot(t,y,'k',t,yd,'k'),grid,xlabel('Time, sec'),
ylabel('Half Crack Length, mm (DE Solution)')

%the following command plots the above information
 + the
%dispersion in astar.
%plot(t,y,t,yd,t(150),astar(150,:),'k+'),grid,xlabel
('Time, sec'),ylabel('Crack Length, mm (DE Solution)')
```

virtest3.m

```
%virtest3.m is a program that calculates the stochastic
%crack propagation using modified Forman equation and
%stochastic log-normal process (corcrack3.m).the
 solution
%{yd} to the deterministic crack propagation ODE is
%fitted with a k-th order polynom,and a column matrix
 of the
%polynom's coefficients is created. t is time, tspan
%is the solution ramge in time, {w} is half crack
 length,
%[y] is a matrix of N solutions, {YD} is the
 approximated
%deterministic polynomial solution (a column).The
%deterministic problem is done with crack2.m file.next
%{y}'s are for stochastic problem, done with
 corcrack3.m file


clear       %CAUTION: Clears the %Previous Workspace
k=9;        %order of polynom
M=500;      %INSERT the number stochastic solutions
[tspan]=[0:0.5:86];   %from 0 to 88 sec, every %half
                      second
%these numbers should be modified for other
 computations
%by changing
%the numbers in the command.
N=length(tspan);  %number of time and crack-%length
                  points
w0=0.1;           %initial crack length, mm.
R=0.0
options=odeset('AbsTol',1e-6);
geom=sqrt(pi);
%solution of the deterministic case with crack2.m for R
```

```
%between zero (included) and 1
[t,w]=ode45('crack2',tspan,w0);   %solve deterministic
                                    case
yd=w;                             %deterministic
                                   solution
pd=polyfit(t,yd,k);               %deterministic row of
                                  %coefficients
YD=polyval(pd,t);                 %deterministic
                                   polynomial
                                  %approximation

%repeated solutions for normal stochastic process,
from m=1 to M

for m=1:M
   muk=4691;       %mean of KIC
   sigk=155;       %standard deviation of KIC
   KIC=normrnd(muk,sigk,1,1);  %fracture thoughness
   mus=1172;       %mean of UTS
   sigs=40;        %standard deviation of UTS
   UTS=normrnd(mus,sigs,1,1);  %UTS
   [t,w]=ode45('corcrack3',tspan,w0,options,KIC,UTS);
   acrit=((1-R)*KIC/(geom*UTS))^2; %critical crack
          length
   y(:,m)=w;
   astar(:,m)=acrit;
end
%plot the crack length vs. time for all samples + the
%deterministic solution (black)

plot(t,y,'k',t,yd,'k'),grid,xlabel('Time, sec'),
ylabel('Half Crack Length, mm (DE Solution)')

%the following command plots the above information
 + the
%dispersion in astar.
%plot(t,y,t,yd,t(150),astar(150,:),'k+'),grid,xlabel
 ('Time, sec'),ylabel('Crack Length, mm (DE Solution)')
```

virtest4.m

```
%virtest4.m is a program that calculates the stochastic
%crack propagation using modified Forman equation and
%stochastic normal process (corcrack4.m).the solution
%{yd} to the deterministic crack propagation ODE is
%fitted with a k-th order polynom,and a column
%matrix of the polynom's coefficients is created. t is
 time, tspan
%is the solution range in time, {w} is half crack
 length,
%[y] is a matrix of N solutions, {YD} is the
 approximated
%deterministic polynomial solution(a column).The
%deterministic problem is done with crack2.m file.next
%{y}'s are for stochastic problem, done with
 %corcrack4.m file

clear       %CAUTION: Clears the %Previous Workspace
k=9;        %order of polynom
M=500;      %INSERT the number of %stochastic solutions
[tspan]=[0:0.5:82];   %from 0 to 88 sec, every %half
                      second
%these numbers should be modified for other
 computations
%by changing the numbers in the Command.
N=length(tspan);    %number of time and crack-length
                     points
w0=0.1;             %initial crack length, mm.
R=0.25
options=odeset('AbsTol',1e-6);
geom=sqrt(pi);
%solution of the deterministic case with crack2.m
 for R
%between zero (included) and 1
```

```
[t,w]=ode45('crack2',tspan,w0);    %solve deterministic
                                     case
yd=w;                              %deterministic
                                    solution
pd=polyfit(t,yd,k);                %deterministic row of
                                   %coefficients
YD=polyval(pd,t);                  %deterministic
                                    polynomial
                                   %approximation


%repeated solutions for normal stochastic process,
 from m=1 to M

for m=1:M
   muk=4691;    %mean of KIC
   sigk=155;    %standard deviation of KIC (Unused)
   KIC=muk;     %fracture thoughness
   mus=1172;    %mean of UTS
   sigs=40;     %standard deviation of UTS %(Unused)
   UTS=mus;     %UTS
   [t,w]=ode45('corcrack4',tspan,w0,options,KIC,UTS);
   acrit=((1-R)*KIC/(geom*UTS))^2; %critical crack
          length
    y(:,m)=w;
   astar(:,m)=acrit;
end
%plot the crack length vs. time for all samples + the
%deterministic solution (black)

plot(t,y,'k',t,yd,'k'),grid,xlabel('Time, sec'),
ylabel('Half Crack Length, mm (DE Solution)')

%the following command plots the above information
 + the
%dispersion in astar.
%plot(t,y,t,yd,t(150),astar(150,:),'k+'),grid,xlabel
('Time, sec'),ylabel('Crack Length, mm (DE Solution)')
```

weib1.m

```
%weib1 generates Weibull distribution numbers for TTCI
in Yang's paper

clear %clear workspace

alfaF=4.9174;    %alfa from Yang paper for WPF(33)
betaF=15936;     %beta from Yang paper for WPF(33)
alfaFB=5.499;    %alfa from Yang paper for WPFB(37)
                     -XWPF(37) in the paper
betaFB=11193;    %beta from Yang paper for WPFB(37)
                     -XWPF(37) in the paper

aF=(1/betaF)^alfaF;       %a for standard MATLAB
                             computations, WPF(33)
bF=alfaF;                 %b for standard MATLAB
                             computations, WPF(33)
aFB=(1/betaFB)^alfaFB;    %a for standard MATLAB
                             computations, WPFB(37)
bFB=alfaFB;               %b for standard MATLAB
                             computations, WPF(37)

NN=5000;    %number of samples

%Computations for WPF(33)
for n=1:NN
   RF(n)=weibrnd(aF,bF,1,1);   %Weibull random number
                                  generator, WPF(33)
end
RWF=RF';      %transpose of line matrix to column
YF=log(RWF);  %LN(t)
weibplot(RWF) %Weibull paper plot of random numbers
figure
hist(RWF,20),xlabel('Time to Crack Initiation, hours,
    WPF(33)')%Histogram of random numbers
figure
```

```
hist(YF,20),xlabel('Y=LN(RW), WPF(33)')
    %Histogram of resulting Y
figure

%Computations for WPFB(37)
for n=1:NN
  RFB(n)=weibrnd(aFB,bFB,1,1); %Weibull random number
                                    generator, WPFB(37)
end

RWFB=RFB';     %transpose of line matrix to column
YFB=log(RWFB); %LN(t)
weibplot(RWFB) %Weibull paper plot of random numbers
figure
hist(RWFB,20),xlabel('Time to Crack Initiation, hours,
    WPFB(37)')%Histogram of random numbers
figure
hist(YFB,20),xlabel('Y=LN(RW), WPFB(37)')
    %Histogram of resulting Y
figure

%Computing CDF for Weibull distribution
t=[0:500:25000];
MM=length(t);

for m=1:MM
   cdfF(m)=1-exp(-aF*t(m)^bF);
   cdfFB(m)=1-exp(-aFB*t(m)^bFB);
   pdfF(m)=aF*bF*t(m)^(bF-1)*exp(-aF*t(m)^bF);
   pdfFB(m)=aFB*bFB*t(m)^(bFB-1)*exp(-aFB*t(m)^bFB);
end

plot(t,cdfF,'ko-',t,cdfFB,'bx-'),grid,xlabel
('Time to Crack Initiation, hours'),ylabel('CDF')
legend('WPF','XWPF',0)
figure
plot(t,pdfF,'ko-',t,pdfFB,'bx-'),grid,xlabel
('Time to Crack Initiation, hours'),ylabel('PDF')
legend('WPF','XWPF',0)
```

weib2.m

```
%weib2 computes EIFS distribution from Yang's results
%for WPF(33) and XWPF(37)

%Data from Table 3, Yang's paper.
%for WPF(33), Y is added. For XWPF(37) X is added.

clear              %clear workspace

a0=0.03;           %inch. Crack initiation length

bY=0.9703;
bX=0.9620;

cY=bY-1;
cX=bX-1;

QY=0.0002381;
QX=0.000309;

alfaY=4.9174;
alfaX=5.499;

betaY=15936;
betaX=11193;

epsY=0;
epsX=0;

TTCIY=[6563;9312;10629;10892;
   11357;11637;12053;13379;13544;13762;13773;13939;
   14098;14123;14149;14262;14350;14400;14436;15499;
   15600;15798;16128;16141;17109;17134;17185;17507;
   17820;17839;18068;18357;19154];
NN=length(TTCIY);

TTCIX=[6106;6147;7457;7545;7721;
   8425;8538;8636;8908;8968;9078;9085;9316;9973;
   10253;10457;10908;11045;11051;11071;11370;11493;
```

```
   11564;11571;11698;11708;11920;12051;12062;12118;
   12302;12389;12505;12629;12655;13339;16008];
MM=length(TTCIX);

%calculations for WPF(33)
for n=1:NN
   EIFSYL(n)=a0/(1+a0^cY*cY*QY*TTCIY(n))^(1/cY);
end
eifsY=EIFSYL';     %EIFS for WPF(33) data

%xYmax=1.2*max(eifsY);      %range for computation is
                            20% higher than max value
                            of eifs
xYmax=6e-3;
xuY=(a0^(-cY))^(-1/cY);    %upper limit of crack
                           (crack initiation length xu)
xYL=[0:(xYmax/500):xYmax]; %division to 100 values of
                           x, line matrix
xY=xYL';                   %division to 100 values of
                           x, column matrix
LL=length(xY);

for l=1:LL
   FaYL(l)=exp(-((xY(l)^(-cY)-a0^(-cY))/
            (cY*QY*betaY))^alfaY); %probability,
            line matrix
end
FaY=FaYL';   %probability (eifs=< xY)

%Figure 1 :CDF for EIFS, WPF(33),regular paper
plot(xY,FaY,'ko-'),grid,xlabel('EIFS WPF(33), in.'),
ylabel('CDF')
title('1:CDF for EIFS, WPF(33),regular paper')
figure

%Figure 2 :CDF for EIFS, WPF(33),semilog paper
semilogx(xY,FaY,'ko-'),grid,xlabel('EIFS WPF(33),
in.'), ylabel('CDF')
title('2:CDF for EIFS, WPF(33),semilog paper')
```

```
figure

%Computing PDF function by numerical differentiation
dxY=xYmax/500;
dFaY(1)=0;
pdfY(1)=0;
for l=2:LL
   pdfY(l)=(FaY(l)-FaY(l-1))/dxY;
end

%Figure 3 :PDF for EIFS, WPF(33),regular paper
plot(xY,pdfY,'ko-'),grid,xlabel('EIFS WPF(33), in.'),
ylabel('PDF')
title('3:PDF for EIFS, WPF(33),regular paper')
figure
%Figure 4 :PDF for EIFS, WPF(33),semilog paper
semilogx(xY,pdfY,'ko-'),grid,xlabel('EIFS WPF(33),
in.'), ylabel('PDF')
title('4:PDF for EIFS, WPF(33),semilog paper')
figure

%Computing the mean and std of EIFS for WPF(33) by
 integration (integral x*pdf)
for l=1:(LL-1)
   dmY(l)=(pdfY(l)+pdfY(l+1))*(xY(l)+xY(l+1))*dxY/4;
end
mY=sum(dmY)

for l=1:(LL-1)
   dvY(l)=(pdfY(l)+pdfY(l+1))*(((xY(l)+xY(l+1))/
          2-mY)^2)*dxY/2;
end
vY=sum(dvY);
stdY=sqrt(vY)

%calculations for XWPF(37)
for m=1:MM
   EIFSXL(m)=a0/(1+a0^cX*cX*QX*TTCIX(m))^(1/cX);   %
end
eifsX=EIFSXL';       %EIFS for XWPF(37) data
```

```
%xXmax=1.4*max(eifsX);        %range for computation is
                               30% higher than max value
                               of eifs
xXmax=6e-3;
xuX=(a0^(-cX))^(-1/cX);       %upper limit of crack
                               (crack initiation length
                               xu)
xXL=[0:(xXmax/500):xXmax];    %division to 100 values of
                               x, line matrix
xX=xXL';                      %division to 100 values of
                               x, column matrix
KK=length(xX);

for k=1:KK
   FaXL(k)=exp(-((xX(k)^(-cX)-a0^(-cX))/
      (cX*QX*betaX))^alfaX); %probability, line matrix
end
FaX=FaXL';        %probability (eifs=< xX)

%Figure 5 : CDF for EIFS, XWPF(37), regular paper
plot(xX,FaX,'ko-'),grid,xlabel('EIFS XWPF(37), in.'),
ylabel('CDF')
title('5:CDF for EIFS, XWPF(37), regular paper')
figure

%Figure 6 : CDF for EIFS, XWPF(37), semilog paper
semilogx(xX,FaX,'ko-'),grid,xlabel('EIFS XWPF(37),
in.'),ylabel('CDF')
title('6:CDF for EIFS, XWPF(37), semilog paper')
figure

%Computing PDF function by numerical differentiation
dxX=xXmax/500;
dFaX(1)=0;
pdfX(1)=0;
for k=2:KK
   pdfX(k)=(FaX(k)-FaX(k-1))/dxX;
end
```

```
%Figure 7 : PDF for EIFS, XWPF(37), regular paper
plot(xX,pdfX,'ko-'),grid,xlabel('EIFS XWPF(33), in.'),
ylabel('PDF')
title('7:PDF for EIFS, XWPF(37), regular paper')
figure

%Figure 8 : PDF for EIFS, XWPF(37), semilog paper
semilogx(xX,pdfX,'ko-'),grid,xlabel('EIFS XWPF(33),
in.'),ylabel('PDF')
title('8:PDF for EIFS, XWPF(37), semilog paper')
figure

%Computing the mean and std of EIFS for XWPF(37) by
 integration (integral x*pdf)
for k=1:(KK-1)
   dmX(k)=(pdfX(k)+pdfX(k+1))*(xX(k)+xX(k+1))*dxX/4;
end
mX=sum(dmX)

for l=1:(KK-1)
   dvX(k)=(pdfX(k)+pdfX(k+1))*(((xX(k)+xX(k+1))/
   2-mX)^2)*dxX/2;
end
vX=sum(dvX);
stdX=sqrt(vX)

xYt=4Y(2:501);     %from 1 to 500
FaYt=FaY(2:501);

alfaeY=0.918;
%alfaeY=0.8952274;
betaeY=7.3205820401e-4;
epse=0;
for m=1:500
   FWY(m)=1-exp(-(xYt(m)/betaeY)^alfaeY);
end

plot(xYt,FaYL,'k-',xYt,FWY,'r-'),grid,xlabel('EIFS,
in.'),ylabel('CDF')
```

```
title('9:CDF for EIFS, WPF(black), Weibull Appr.(red)')
figure

semilogx(xYt,FaYt,'k-',xYt,FWY,'r-'),grid,xlabel
('EIFS, in.'),ylabel('CDF')
title('10:CDF for EIFS, WPF(black), Weibull Appr.(red),
semilog paper')
figure
xXt=xX(2:501); %from 1 to 500
FaXt=FaX(2:501);

alfaeX=1.;
betaeX=8.38062e-4;
epse=0;
for m=1:500
    FWX(m)=1-exp(-(xXt(m)/betaeX)^alfaeX);
end
plot(xXt,FaXt,'k-',xXt,FWX,'r-'),grid,xlabel('EIFS,
in.'),ylabel('CDF')
title('11:CDF for EIFS, XWPF(black),
Weibull Appr.(red)')
%figure

%plot(xXt,FaXt,'k-',xXt,FWX,'r-'),grid
%figure
%semilogx(xXt,FaXt,'k-',xXt,FWX,'r-'),grid
%figure

%plot(xY,pdfY,'k-',xX,pdfX,'r-'),grid,xlabel ('EIFS,
in.'),ylabel('PDF')
%title('12:PDF of Both Cases, WPF(black), XWPF(red)')
%figure

%mu=-7.57188784207213;
%mu=-2.55;

%sigma=0.8;
%sigma=0.90114406762865;
%ss=1/(sigma*sqrt(2*pi))
```

```
%for n=1:500
%pdflog(n)=(ss/xYt(n))*exp(-(0.5/sigma^2)
            *(log10(xYt(n))-mu)^2);
%end
%plot(xYt,pdfYt,'kx-',xYt,pdflog,'ko-'),grid
```

CHAPTER 7

env5.txt

/filnam,env5

/title,Beam Under an Aircraft

/prep7

!Beam data, dimensions in cm, kgf, sec

L = 330	!length
ro = 7.959e-6	!mass density
E = 2.1e6	!Young's modulus
nu = 0.3	!Poisson coefficient
b = 10	!width
h = 10	!height
A = b*h	!area
I1 = b*h*h*h/12	
g = 980	!value of g

et,1,beam3	!beam element

r,1,A,I1,h	!real constants
mp,ex,1,E	!material properties
mp,dens,1,ro	
mp,nuxy,1,nu	

!nodes
n,1,0,0
n,34,L,0
fill,1,34

```
!elements
type,1
mat,1
real,1
en,1,1,2
engen,1,33,1,1,1,1

!boundary conditions (hooks)
d,12,ux,0
d,12,uy,0
d,20,ux,0
d,20,uy,0

save
fini
/eof                            !end of input file

!Modal analysis (6 modes between 20 to 800 Hz)
/solu
antyp,modal
modopt,subs,6
mxpand,6,,,yes
solve
fini

!Static Analysis, 5g vertical loading
/solu
antype,stat
!acel,,(5+3*3.3045)*g         !(for equivalent loading)
acel,,5*g
solve
fini

!static analysis data processing
/post1

etab,sdir,ls,4
etab,sbup,ls,5
etab,sbdn,ls,6
```

```
pret
etab,eras

!spectral analysis, input-accelerations in the hooks (both equal)
/solu
antype,spectrum
spopt,psd,6,on

psdg1 = 0.0014
psdg2 = 0.014
psdunit,1,accg,980
psdunit,2,accg,980

psdfrq,1,,19.999,20,800,800.001
psdval,1,0.00001,psdg1,psdg2,0.00001

psdfrq,2,,19.999,20,800,800.001
psdval,2,0.00001,psdg1,psdg2,0.00001


dmprat,0.015

d,12,uy,1
pfact,1,base

d,12,uy,0                !zero previous d beforc inserting second d
d,20,uy,1
pfact,2,base

psdres,acel,abs          !absolute accelerations
psdres,velo,rel          !relative velocities
psdres,disp,rel          !relative displacements
solve
psdcom,,6                !combine 6 modes
solve
fini

!excitation of force in both ends, of the beam, for 4 modes
/solu
antype,spectrum
spopt,psd,4,on
```

```
psdf1=142
psdf2=144
psdunit,1,force
psdunit,2,force
!psdfrq                                    !required only for additional run
psdfrq,1,,19.999,20,400,400.001
psdval,1,0.00001,psdf1,psdf1,0.00001

psdfrq,2,,19.999,20,400,400.001
psdval,2,0.00001,psdf2,psdf2,0.00001

dmprat,0.015
!f,34,fy,0                                 !zero f2 before running AGAIN f1
f,1,fy,1
pfact,1,node

f,1,fy,0                                   !zero f1 before applying f2
f,34,fy,1
pfact,2,node

psdres,acel,abs
psdres,velo,rel
psdres,disp,rel
solve

psdcom,,4                                  !response combined, 4 modes
solv
fini

!view spectral analysis results for rms
/post1
set,3,1                                    !for displacements, stresses
!set,4,1                                   !for velocities, rate of stresses
!set,5,1                                   !for accelerations, rate of rate...
etab,bnup,ls,5
ctab,bndn,ls,6
pret
etab,eras                                  !erase e table

!Analysis in the frequency domain
/post26
```

numvar,51
store,psd,5
nsol,2,1,u,y,n1
nsol,3,11,u,y,n11
nsol,4,22,u,y,n22
nsol,5,34,u,y,n34
esol,6,10,11,ls,5,bn11
esol,7,21,22,ls,5,bn22
nsol,8,12,u,y,nhook
rpsd,12,2,,1,2,d1
rpsd,13,3,,1,2,d11
rpsd,14,4,,1,2,d22
rpsd,15,5,,1,2,d34
rpsd,16,6,,1,2,st11
rpsd,17,7,,1,2,st22

rpsd,22,2,,3,1,a1
rpsd,23,3,,3,1,a11
rpsd,24,4,,3,1,a22
rpsd,25,5,,3,1,aa34
rpsd,28,8,,3,1,aa12

quot,32,22,,,g1,,,1/980,980
quot,33,23,,,g11,,,1/980,980
quot,34,24,,,g22,,,1/980,980
quot,35,25,,,g34,,,1/980,980
quot,38,28,,,ghook,,,1/980,980

stat1.txt

!file stat1 for cantilever beam

/filnam,stat1

/title,cantilever beam, static load 1g

!also for tip,load of 3 kgf

!units in cm, kgf, seconds

g = 980   !value of gravity
L = 60   !length of beam

```
b = 8                          !width of beam
h = 0.5                        !height of beam
E = 2.1e6                      !Young's modulus
ro = 7.959e-6                  !mass dendity
W = L*b*h*ro*g                 !weight of beam
q = W/L                        !weight per unit length
A = b*h                        !cross section area
I = b*h**3/12                  !area moment of inertia

/prep7
mp,ex,1,E
mp,dens,1,ro
et,1,beam3
r,1,A,I,h

n,1,0,0
n,11,L,0
fill,1,11

en,1,1,2
engen,1,10,1,1,1,1

d,1,ux,0                       !boundary conditions
d,1,uy,0
d,1,rotz,0
!d,11,uy,0

save
fini

/solu                          !solution for inertia loading
antyp,stat
acel,,-g
solve
fini

!/solu                         !solution for tip force loading
!antyp,stat
!f,11,fy,3
!solve
!fini
```

stat2.txt

!file stat2 for cantilever beam with tip mass

/filnam,stat2

/title,cantilever beam+tip load static load 1g

!units in cm, kgf, seconds

g = 980	!value of gravity
L = 60	!length of beam
b = 8	!width of beam
h = 0.5	!height of beam
E = 2.1e6	!Young's modulus
ro = 7.959e-6	!mass dendity
W = L*b*h*ro*g	!weight of beam
q = W/L	!weight per unit length
A = b*h	!cross section area
I = b*h**3/12	!area moment of inertia

/prep7
mp,ex,1,E
mp,dens,1,ro
et,1,beam3
et,2,mass21,,0,4

r,1,A,I,h
r,2,mtip

n,1,0,0
n,11,L,0
fill,1,11

type,1
real,1
mat,1
en,1,1,2
engen,1,10,1,1,1,1 !cantilever beam

type,2
real,2

```
mat,2
en,11,11              !tip mass

d,1,ux,0
d,1,uy,0
d,1,rotz,0
!d,11,uy,0

save
fini

/!solu                !inertia loading
!antyp,stat
!acel,,−g
!solve
!fini

/solu                 !tip force loading
antyp,stat
f,11,fy,3
solve
fini
```

stat3.txt

```
!file stat3 for cantilever beam

/filnam,stat3,txt

/title,cantilever beam+tip mass, random response

!units in cm, kgf, seconds

g = 980               !value of gravity
L = 60                !length of beam
b = 8                 !width of beam
h = 0.5               !height of beam
E = 2.1e6             !Young's modulus
ro = 7.959e-6         !mass dendity
W = L*b*h*ro*g        !weight of beam
q = W/L               !weight per unit length
```

```
A=b*h                    !cross section area
I=b*h**3/12              !area moment of inertia

/prep7
mp,ex,1,E
mp,dens,1,ro
et,1,beam3
et,2,mass21,,0,4

r,1,A,I,h
r,2,mtip

n,1,0,0
n,11,L,0
fill,1,11

type,1
real,1
mat,1
en,1,1,2
engen,1,10,1,1,1,1       !cantilever beam

type,2
real,2
mat,2
en,11,11                 !tip mass

d,1,ux,0
d,1,uy,0
d,1,rotz,0

save
fini

/solu                    !Modal Solution
antyp,modal
modop,subs,3             !3 modes, subspace method
mxpand,3,,,yes           !expand 3 modes, calculate stresses
solve
fini
```

```
/solu                   !SPECTRAL ANALYSIS
antyp,spect             !analysis type - spectral
spopt,psd,3,yes         !psd with 3 modes, stresses also computed
psdunit,1,forc          !input in PSD of force (kgf²/Hz)
psdfrq,1,,5,250         !input between 5 to 250 Hz
psdval,1,0.004081632,0.004081632

psdres,disp,rel         !displacements relative to base
psdres,velo,rel         !velocities relative to base
psdres,acel,abs         !absolute accelerations

dmprat,0.02             !damping 2% for all modes
f,11,fy,1               !random input at tip
pfact,1,node            !nodal excitation
mcomb,psd               !modal combination for psd

solve
save
fini

/post1
set,list                !see list of load steps (1,1;1,2 etc. for resonances)
set,3,1                 !for rms of displacements
                        !use this for stresses
                        !for beam elements an element table is required!
set,4,1                 !for rms of velocities
set,5,1                 !for rms of accelerations
                        !accelerations/980 = acel. in g's
fini

/post26
numvar,30               !prepare space for 30 variables
store,psd,10            !store frequencies, 10 to each side of resonance
/grid,1
/axlab,x,Frequency Hz
/axlab,y,PSD
nsol,2,11,u,y,Wtip      !disp. node 11
esol,4,1,1,ls,3,Sben    !bend. str. at element 1, node 1
                        !ls-3:bending, i side of element
```

```
rpsd,12,2,,1,2,Wtip      !PSD of rel.disp. node 11
rpsd,13,2,,3,1,ACtip     !PSD of Acceleration, node 11 (in cm/sec²/Hz)
                         !for g²/Hz—divide by 980²
rpsd,14,4,,1,2,BEND      !PSD of bending stress, node 1

int1,22,12,1,,MStip      !Mean square of tip displacement
int1,23,13,1,,Msaccel    !Mean square of tip accel.
int1,24,14,1,,MSbend     !mean square of clamp bending stress.
```

References

[1] Thomson, W. T. and D. Dillon. *Theory of Vibrations with Applications*, 5th ed. Upper Saddle River, NJ: Prentice Hall, 1997.

[2] Den Hartog, J. P. *Mechanical Vibrations*, New York: McGraw-Hill, 1956, 1985.

[3] Biggs, J. M. *Introduction to Structural Dynamics*, New York: McGraw-Hill, 1964.

[4] Timoshenko, S. and D. H. Young. *Vibration Problems in Engineering*, 3rd ed. Princeton, NJ: D. Van Nostrand Company, Inc., 1956.

[5] Ewins, D., S. S. Rao, and S. G. Braun. *Encyclopedia of Vibrations*, Boston, MA: Academic Press, 2001.

[6] Rao, S. S. *Mechanical Vibrations*, Boston: Addison-Wesley, 1995.

[7] Inman, D. J. *Engineering Vibrations*, Upper Saddle River, NJ: Prentice Hall, 1995.

[8] Meirovitch, L. *Fundamentals of Vibrations*, New York: McGraw-Hill, 2000.

[9] Ginsberg, J. H. *Mechanical and Structural Vibrations*, New York: John Wiley & Sons, 2001.

[10] Hurty, W. C. and M. F. Rubinstein. *Dynamics of Structures*, Upper Saddle River, NJ: Prentice Hall, 1964.

[11] Graham, K. S. *Fundamentals of Mechanical Vibrations*, New York: McGraw-Hill, 1992.

[12] Vierck, R. K. *Vibration Analysis*, New York: Harper & Row, 1979.

[13] De Vries, G. "Emploi de la Methode Vectorielle d'Analyse dans les Essais de Vibration," *La Recherche Aeronautique*, 74: 41–47 (1960).

[14] Beatrix, C. "Les Procedes Experimentaux de l'Essai Global de Vibration d'Une Structure," *La Recherche Aeronautique*, 109: 57–64 (1965).

[15] Elishakoff, I. *Probabilistic Methods in the Theory of Structures*, New York: John Wiley & Sons, 1983.

[16] Maymon, G. "Some Engineering Applications in Random Vibrations and Random Structures," *AIAA Inc., Progress in Astronautics and Aeronautics Series*, 178 (1998).

[17] Zienkiewicz, O. C. *Finite Elements Methods in Engineering Science*, New York: McGraw-Hill Education, 2nd rev. ed. 1972.

[18] Zienkievicz, O. C. and R. L. Taylor. *Finite Element Method for Solid and Structural Mechanics*, 6th ed. Boston: Butterworth-Heinemann, 2005.

[19] Timoshenko, S. *Mechanics of Materials*, 3rd ed. Melbourne, FL: Krieger Publishing Company. 1976.

[20] Mott, R. L. *Applied Strength of Materials*, Upper Saddle River, NJ: Prentice Hall, 2002.

[21] Patnaik, S. and D. Hopkins. *Strength of Materials, A New Unified Theory for the 21st Century*, Boston, MA: Elsevier, 2003.

[22] Case, J., L. Chilver, and C. Ross. *Strength of Materials and Structures*, Boston: Elsevier, 1999.

[23] Shames, R. H. *Mechanics of Deformable Solids*, Upper Saddle River, NJ: Prentice Hall, 1994.

[24] Center of Beam Theory and Dynamical Systems, Michigan State University, http://www.bt.pa.msu.edu/index.htm.

[25] Han, S. M., H. Benaroya, and T. Wei. "Dynamics of Transversally Vibrating Beams Using Four Engineering Theories," *Journal of Sound and Vibrations* 225(5), 935–988 (1999).

[26] Young, W. C. *Roark's Formulas for Stress and Strain*, 7th ed. New York: McGraw-Hill Professional, 2001.

[27] Maymon, G. "Response of Geometrically Nonlinear Plate-like Structures to Random Acoustic Excitation, Part I: Theoretical Evaluation of the Method," Research Report LG82RR0001, Lockheed Georgia Company, March 1982.

[28] Maymon, G. "Response of Aeronautical Structures to Random Acoustic Excitation—A Stress Mode Approach" in *Random Vibration—Status and Recent Development* (Elishakoff, I., Lyon, R. H., editors), The Stephen Harry Crandall Festschrift, Studies in Applied Mechanics, vol. 14. Boston: Elsevier, 1986.

[29] Timoshenko, S. P. and S. Woinowsky-Kreiger. *Theory of Plates and Shells*, 2nd rev. ed. New York: McGraw-Hill Publications, International, 1964.

[30] Kraus, H. *Thin Elastic Shells*, New York: John Wiley & Sons, Inc., 1967.

[31] MIL-STD-810F, Department of Defense Test Method Standard for Environmental Engineering Considerations and Laboratory Tests, January 2000.

[32] Lutes, L., and S. Sarkani. *Random Vibrations*, Boston: Elsevier, 2003.

[33] Wirsching, P. H., T. L. Daez, and K. Ortez. *Random Vibrations, Theory and Practice*, Mineola, NY: Dover Publications, 2000.

[34] Nigam, N. C. and S. Narayam. *Applications of Random Vibrations*, New York: Springer-Verlag, 1994.

[35] Prazen, E. *Modern Probability Theory and Its Application*, New York: John Wiley & Sons, 1960.

[36] NASA HDBK 7004B, "Force Limited Vibration Testing," NASA Technical Handbook, January 2003.

[37] Scharton, T. D. "Force Limited Vibration Testing Monograph," NASA Reference Publication RP 1403, JPL, 1997.

[38] Scharton, T. D. "Vibration Test Force Limits Derived From Frequency Shift Method," *AIAA Journal of Spacecrafts and Rockets*, 2 (2): 312–316 (March 1995).

[39] Crocker, M. J. "The Response of Supersonic Transport Fuselage to Boundary Layer and Reverberant Noise," *Journal of Sound and Vibration*, 9 (1): 6–20 (1969).

[40] Maestrello, L. "Measurement and Analysis of the Response Field of Turbulent Boundary Layer Excited Panels," *Journal of Sound and Vibration*, 2 (3): 270–292 (1965).

[41] http://www.uts.com/software.asp, website of TKW+ provider.

[42] Kolsky, H. *Stress Waves in Solids*, Mineola, NY: Dover Publications, 2003.

[43] Lin, Y. K. *Probabilistic Theory of Structural Dynamics*, New York: McGraw-Hill, 1967.

[44] Madsen, H. O., S. Krenks, and N. C. Lind. *Methods of Structural Safety*, Mineola, NY: Dover Publications, 2006.

[45] Melcher, R. E. *Structural Reliability Analysis and Predictions*, New York: John Wiley & Sons, 1999.

[46] Thoft-Christensen, P. and M. Baker. *Structural Reliability and Its Applications*, New York: Springer-Verlag, 1982.

[47] Frangopol, D. H., R. B. Corotis, and R. Rackwitz (editors). "Reliability and Optimization of Structural Systems," Working Conference, Boulder Co. 1996. Boston: Elsevier, 1997.

[48] http://www.nessus.swri/, The Nessus Structural Probabilistic Analysis Code homepage.

[49] http://www.profes.com/, The ProFES Structural Probabilistic Analysis Code homepage.

[50] http://www.dnv.com/software/safeti/safetiqra/proban.asp, The PROBAN Structural Probabilistic Analysis Code homepage.

[51] Hasofer, A. M. and N. C. Lind. "Exact and Invariant Second Moment Code for Materials," *Journal of Engineering Mechanics*, vol. 100(1), 1974, pp. 111–121.

[52] Rackwitz, R. and B. Fiessler. "Structural Reliability Under Combined Random Load Sequences," *Computers and Structures*, vol. 9(5), 1978, pp. 489–494.

[53] Hoenbichler, M. and R. Rackwitz. "Non Normal Dependant Vectors in Structural Safety," *Journal of Engineering Mechanics*, vol. 107(6), 1981, pp. 1127–1138.

[54] Wu, Y. T. "Efficient Methods for Mechanical and Structural Reliability Estimates," Ph.D Dissertation, Dept. of Mechanical Engineering, University of Arizona, Tucson, AZ, April 1984.

[55] Maymon, G. "Probability of Failure of Structures without a Closed-Form Failure Function," *Computers and Structures*, vol. 49(2), 1993, pp. 301–313.

[56] Montgomery, D. C. *Introduction to Linear Regression Analysis*, New York: John Wiley & Sons, 1982.

[57] Maymon, G. "Direct Computation of the Design Point of a Stochastic Structure Using a Finite Element Code," *Structural Safety*, vol. 14, 1994, pp. 185–202.

[58] Madsen, H. O. and L. Tvedt. "Methods for Time-Dependant Reliability and Sensitivity Analysis," *Journal of Engineering Mechanics*, vol. 116(10), 1990, pp. 2118–2135.

[59] Wen, Y. K. and H. C. Chen. "On Fast Integration for Time Variant Structural Reliability," Probabilistic Engineering Mechanics, vol. 2(3), 1987, pp. 156–162.

[60] Timoshenko, S. and J. Gere. *Theory of Elastic Stability*, New York: McGraw-Hill, 1961.

[61] Popov, E. P. *Engineering Mechanics of Solids*, Prentice-Hall, 1990.

[62] Paris, P. C. and E. Erdogan. "A Critical Analysis of Crack Propagation Laws," *Journal of Basic Engineering*, vol. 85(4), 1963, pp. 528–534.

[63] FLAGRO version 3.0 Demonstration Help File, 2003. Demonstration version can be obtained from http://www.nasgro.swri.org/.

[64] Vasudevan, A. K., K. Sadananda, and G. Glinka. "Critical Parameters for Fatigue Damage," *International Journal of Fatigue*, vol. 23(s1), 2001, pp. s39–s53.

[65] Sadananda, K. and A. K. Vasudevan. "Multiple Mechanism Controlling Fatigue Crack Growth," Fatigue & Fracture of Engineering Materials & Structures, vol. 26(9), 2003, pp. 835–845.

[66] Virkler, D. A. et al. "The Statistical Modeling Nature of a Fatigue Crack Propagation," *Journal of Engineering Materials and Technology*, vol. 10(4), 1979, pp. 143–153.

[67] Ghenom, H. and S. Dore. "Experimental Study of Constant Probability Crack Growth under Constant Amplitude Loading," *Engineering Fracture Mechanics*, vol. 27(1), 1987, pp. 1–66.

[68] Lawrence, M., W. K. Liu, G. Besterfield, and T. Belytchko. "Fatigue Crack Growth Reliability," *Journal of Engineering Mechanics*, vol. 116(3), 1990, pp. 698–708.

[69] Lin, Y. K. and J. N. Yang. "A Stochastic Theory of Fatigue Crack Propagation," *AIAA Journal*, vol. 23(1) 1984, pp. 174–124.

[70] Yang, J. N. and R. C. Donath. "Statistical Crack Propagation in Fastener Holes Under Spectrum Loading," *Journal of Aircrafts*, vol. 20(12), 1983, pp. 1028–1032.

[71] Tanaka, H. and A. Tsurui. "Reliability Degradation of Structural Components in the Process of Fatigue Crack Propagation Under Stationary Random Loading," *Engineering Fracture Mechanics*, vol. 27(5), 1987, pp. 501–516.

[72] Tsurui, A., J. Nienstedt, G. I. Schueller, and H. Tanaka. "Time Variant Structural Reliability Analysis Using Diffusive Crack Growth Models," *Engineering Fracture Mechanics*, vol. 34(1), 1989, pp. 153–167.

[73] Maymon, G. "Virtual Crack Propagation Tests with Application to Probability of Failure Estimation," AIAA 44th SDM Conference, Norfolk VA, 2003.

[74] MIL-HDBK-5H, Metallic Materials and Elements for Aerospace Vehicle Structures, DOD Handbook, prepared by the USAF AFRL/MLSC, Wright Patterson AFB, OH, 1998.

[75] Bogdanoff, J. "A New Cumulative Damage Model," Parts 1–3, *Journal of Applied Mechanics*, vol. 46, 1979, pp. 245–257, 733–739.

[76] Bogdanoff, J. and A. Kozin. "A New Cumulative Damage Model, Part 4," *Journal of Applied Mechanics*, vol. 47, 1980, pp. 40–44.

[77] Sobcyzk, K. and B. F. Spencer. *Random Fatigue, From Data to Theory*, Boston, MA: Academic Press, 1992.

[78] Spencer, B. F. and J. Yang. "Markov Model for Fatigue Crack Growth," Journal of Engineering Mechanics, vol. 114(12), 1988, pp. 2134–2157.

[79] Yang, J., G. C. Salivar, and C.G. Annis. "Statistical Modeling of Fatigue Crack Growth in Nickel Based Superalloy," Engineering Fracture Mechanics, vol. 18(2), 1983, pp. 257–270.

[80] Noronha, P. J. et al. "Fastener Hole Quality, I and II," U.S. Air Force Flight Dynamics Lab. Technical Report. AFFDL TR-78-206, Wright Patterson AFB, OH, 1979.

[81] Lin, Y. K., J. N. Wu, and J. N. Yang. "Stochastic Modeling of Fatigue Crack Propagation," in Proceedings of the IUTAM Symposium On Probabilistic Methods in Mechanics of Solids and Structures, Stockholm, Sweden, 1984, pp. 103–110.

[82] Sobcyzk, K. "Stochastic Modeling of Fatigue Crack Growth," in Proceedings of the IUTAM Symposium On Probabilistic Methods in Mechanics of Solids and Structures, Stockholm, Sweden, 1984, pp. 111–119.

[83] Sobcyzk, K. "Modeling of Random Fatigue Crack Growth," *Engineering Fracture Mechanics*, vol. 24(4), 1986, pp. 609–623.

[84] Murukami, Y. et al. *Stress Intensity Factors Handbook*, Vol. 1 and Vol. 2, Oxford: Pergamon Press, 1990.

[85] Bellinger, N. C., J. P. Komorowski, and T. J. Benak. "Residual Life Predictions of Corroded Fuselage Lap Joints," *International Journal of Fatigue*, vol. 23(S1), 2001, pp. 349–356.

[86] Barter, S. A., P. K. Sharp, G. Holden, and G. Clark. "Initiation and Early Growth of Fatigue Cracks in an Aerospace Aluminum Alloy," *Fatigue & Fracture of Engineering Materials & Structures*, vol. 25(2), February 2002, pp. 111–115.

[87] Tokaji, K. and T. Ogawa. "The Growth Behavior of Microstructurally Small Fatigue Cracks in Metals," in *Short Fatigue Cracks*, ESIS 13, Edited by K. J. Miller and E. R. Delos Rios, Mechanical Engineering Publications, London, pp. 85–99, 1997.

[88] Yang, J. N. and S. D. Manning. "Distribution of Equivalent Initial Flaw Size," presented at the Annual Reliability and Maintainability Symposium (RAMS), January 1980, San Francisco, CA, pp. 112–120.

[89] Yang J. N. et al. "Stochastic Crack Propagation in Fasteners Holes," *Journal of Aircraft*, vol. 22(9), 1985, pp. 810–817.

[90] DeBartolo, E. A. and B. M. Hillbery. A Model of Initial Flaw Sizes in Aluminum Alloys, *International Journal of Fatigue*, vol. 23(1), 2001, pp. 79–86.

[91] Liao, M. and Y. Xiong. "Risk Analysis of Fuselage Splices Containing Multi-Site Damage and Corrosion," *Journal of Aircraft*, vol. 38(1), 2001, pp. 181–187.

[92] Maymon, G. "A 'Unified' and a $(\Delta K^+ \cdot K_{max})^{1/2}$ Crack Growth Models for Aluminum 2024-T351," *International Journal of Fatigue*, vol. 27(6), 2005, pp. 629–638.

[93] Maymon, G. "Probabilistic Crack Growth Behavior of Aluminum 2024-T351 Alloy Using the 'Unified' Approach," *International Journal of Fatigue*, vol. 27(7), 2005, pp. 828–834.

[94] Elber W. "Fatigue Crack Closure Under Cyclic Tension." *Engineering Fracture Mechanics*, vol. 2(1) (1970), pp. 37–44.

[95] Elber W. "The Significance of Fatigue Crack Closure," in *Damage Tolerance in Aircraft Structures*, ASTM-STP 486, ASTM, PA, 1971, pp. 230–242.

[96] Newman, J. C., Jr. "A Crack Closure Model for Predicting Fatigue Crack Growth Under Aircraft Spectrum Loading," in *Methods and Models for Predicting Fatigue Crack Growth Under Random Loading*, ASTM-STP 748, ASTM PA, (1981), pp. 53–84.

[97] Newman, J. C., Jr. "FASTRAN II: A Fatigue Crack Growth Structural Analysis Program," NASA TM-104159, 1992.

[98] Harter, J. A. *AFGROW Users Guide and Technical Manual*, AFRL-VA-Wright-Patterson TR-1999-3016, 1999.

[99] Phillips, E. P. "Results of the Round Robin on Opening Load Measurement Conducted by ASTM Task Group E24.04.04 on Crack Closure Measurements and Analysis," NASA TM-101601, 1989.

[100] Newman, J. C., Jr. and W. Elber. "Mechanics of Fatigue Crack Closure," in Proceedings of the International Symposium on Fatigue Crack Closure, ASTM-STP 982, American Society of Testing Materials, PA, 1988.

[101] Clerivet, A. and C. Bathias. "Study of Crack Tip Opening Under Cyclic Loading Taking Into Account the Environment and *R* Ratio." *Engineering Fracture Mechanics*, vol. 12(4), 1979, pp. 599–611.

[102] Shih, T. T. and R. P. Wei. "A Study of Crack Closure in Fatigue," *Engineering Fracture Mechanics*, vol. 7(1), 1974, pp. 19–32.

[103] Sadananda, K. and A. K. Vasudevan. "Short Crack Growth Behavior," in *Fatigue and Fracture Mechanics*, vol. 27, ASTM-STP 1296, ASTM PA, 1997, pp. 301–316.

[104] Sadananda, K. and A. K. Vasudevan. "Analysis of Small Crack Growth Behavior Using Unified Approach," in *Small Fatigue Cracks: Mechanics, Mechanisms and Application*, R. O. Ritchie and Y. Murakami (eds.). Boston: Elsevier, 1999:73–83.

[105] Vasudevan, A. K., K. Sadananda, and K. Rajan. "Role of Microstructure on the Growth of Long Fatigue Cracks." *International Journal of Fatigue*, vol. 19(93), 1997, pp. 151–159.

[106] Sadananda, K., A. K. Vasudevan, R. L. Holtz, and E. U. Lee. "Analysis of Overload Effects and Related Phenomena," *International Journal of Fatigue*, vol. 21(s1), 1992, pp. S233–S246.

[107] Sadananda, K. and A. K. Vasudevan. "Multiple Mechanism Controlling Fatigue Crack Growth," *Fatigue Fracture Engineering Materials and Structures*, vol. 26(9), 2003, pp. 835–845.

[108] Sadananda, K. and A. K. Vasudevan. "Short Crack Growth and Internal Stresses," *International Journal of Fatigue*, vol. 19(93), 1997, pp. S99–S108.

[109] Kujawski, D. "A New $\left(\Delta K^+ \cdot K_{max}\right)^{0.5}$ Driving Force Parameter for Crack Growth in Aluminum Alloys," *International Journal of Fatigue*, vol. 23(8), 2001, 23(8), pp. 733–740.

[110] Kujawski, D. "Correlation of Long—and Physically Short—Crack Growth in Aluminum Alloys," *Engineering Fracture Mechanics*, vol. 68(12), 2001, pp. 1357–1369.

[111] Kujawski, D. "A Fatigue Crack Driving Force Parameter with Load Ratio Effects," *International Journal of Fatigue*, vol. 23(s1), 2001, pp. S239–S246.

[112] Pang, C. M. and J. H. Song. "Crack Growth and Closure Behavior of Short Fatigue Cracks," *Engineering Fracture Mechanics*, vol. 47(3), 1994, pp. 327–343.

[113] Lee, S. Y. and J. H. Song. "Crack Growth and Closure Behavior of Physically Short Crack Under Random Loading," *Engineering Fracture Mechanics*, vol. 66(3), 2000, pp. 321–346.

[114] Schra, L. et al. "Engineering Property Comparison for 2324-T39 and 2024-T351 Aluminum Alloy Plates," NLR TR 84021U, National Aerospace Laboratory, Amsterdam, the Netherlands, 1984.

[115] Wanhill, R. J. H. "Low Stress Intensity Fatigue Crack Growth in 2024-T3 and T351," *Engineering Fracture Mechanics*, vol. 30(2), 1988, pp. 233–260.

[116] Sinclair, G. B. and R. V. Pieri. "On Obtaining Fatigue Crack Growth Parameters from the Literature," *International Journal of Fatigue*, vol. 12(1), 1990, pp. 57–62.

[117] Kanninen, M. F. and C. H. Popelar. *Advanced Fracture Mechanics*, Oxford University Press, 1985 (Chapter 11).

[118] Maymon, G. "Crack Propagation Due to Combined Static and Dynamic Random Loading," in the Proceedings of the 43rd Israel Annual Conference on Aerospace Sciences, Tel-Aviv, Haifa, February 2003.

[119] Elishakoff, I. *Safety Factors and Reliability: Friends or Foes?*, Norwell, MA: Kluwer Academic Publishers, 2004.

[120] Maymon, G. "The Stochastic Safety Factor—A Bridge Between Deterministic and Probabilistic Structural Analysis," the 5th International Conference on Probabilistic Safety Assessment and Management (PSAM5), Osaka, Japan, November 2000.

[121] Maymon, G. "The Stochastic Factor of Safety—A Different Approach to Structural Design," the 42nd Israel Annual Conference on Aerospace Sciences, Tel-Aviv, Haifa, Israel, February 2002.

[122] Maymon, G. "Reliability Demonstration of Aerospace Structures—A Different Approach," in Proceedings of ICOSSAR05 Conference, G. Augusti, G. I. Schueller, M. Ciampoli (eds.), Rome, Italy, June 2005, pp. 3239–3244.

Index